POST HARVEST TECHNOLOGY

POST HARVEST TECHNOLOGY

Editor

Sonja Boehler

scitus academics

Post Harvest Technology

Edited by **Sonja Boehler**

This Book is Printed on Acid-Free Paper. ♾

ISBN: 9781681179391
Library of Congress Control Number: 2017949078

SCITUS Academics LLC,
616, Corporate Way, Suite 2, 4766,
Valley Cottage, NY 10989

www.scitusacademics.com

Notice

Printed in United States of America.

Preface

Horticultural crops not only provide nutritional and healthy foods to human beings, but also generate a considerable cash income for growers. However, horticultural crops typically have high moisture content, tender texture and high perishability. If not handled properly, a high value nutritious product can deteriorate and rot in a matter of days or hours. Losses occur after harvesting is known as post-harvest losses. It starts first from the field, after harvest, in grading and packing areas, in storage, during transportation and in the wholesale and retail markets. Several losses occur because of poor facilities, lack of know-how, poor management, market dysfunction or simply the carelessness of farmers. The most important goals of post-harvest handling are keeping the product cool, to avoid moisture loss and slow down undesirable chemical changes, and avoiding physical damage such as bruising, to delay spoilage. Sanitation is also an important factor, to reduce the possibility of pathogens that could be carried by fresh produce, for example, as residue from contaminated washing water. Healthy seeds and grains are the demanding enterprise of the recent era for the production of high yield in the next season. The seeds must be stored for the maintenance of high-yielding crop. During storage, major losses of seeds are caused by various biological and nonbiological factors. There is a need to examine reasonable factors of these crop losses, which ultimately affect the market value and quality of the seed. The quality of seeds can be maintained by using careful postharvest handling techniques. There is need to establish the well-suited methods to assess the losses during the process and to use the best technique to minimize the loss and to ensure the quality and safety of the crop. Importance of Postharvest Technology lies in the fact that it has capability to meet food requirement of growing world population by eliminating avoidable losses making more nutritive food items with higher values by proper processing, storage, packaging, transport and marketing.

This book 'Post Harvest Technology' covers postharvest food preservation and processing methods, with an emphasis on grains. It emphasizes on the aspects and postharvest techniques that are used to maintain seed quality. A comprehensive review of the better, economical, convenient, and productive methods is provided, focused on the needs of developing countries but also with relevance in more industrialized countries.

Table of Contents

CHAPTER 1

Post-Harvest Physiology and Technology in Orchids

De LC[1*], Vij SP[2] and Medhi RP[1]

[1]NRC for Orchids, Pakyong, Sikkim, India

[2]Scientist Emeritus, Department of Botany, Punjab University, Chandigarh, India

ABSTRACT

Orchids account for a large share of global floriculture trade and are estimated around 10% of international fresh cut flower trade. They have taken a significant position in cut flower industry due to its attractiveness, diversity in forms, shape and color, high productivity, right season of bloom, and easy packing and transportation. Postharvest life of orchid cut flowers is influenced by pre-harvest factors like varietal or species differences, light intensity, sugar level of flowers, temperature and water loss. It is also affected by harvest factors such as time and stage of harvest and postharvest factors viz. ethylene production, precooling, pulsing, use of preservatives, packaging and storage.

The hybrids of Dendrobium, Vanda and Mokara remain perfect from 7 days to 30 days. The flowers of Cattleya and Phalaenopsis remain fresh for 1 to 4 weeks whereas Aranda lasts for 18 to 28 days. Higher sugar levels of flowers improve longevity of cut flowers. The optimum harvesting stage of commercial orchids is fully open and mature flowers. In Cymbidium hyb. 'PCMV', harvest at two buds opened stage had maximum vase life (66.8 days). Ethylene is the main factor responsible for early senescence. In Cymbidium hybrid 'Red Princess' pulsing with 5% sucrose increases vase life upto 56 days. Pulsing with 4 mM STS for 10 minutes in Aranda and 0.5 mM STS for 24 hours in Phalaenopsis blocks the deleterious effect of ethylene.

In tropical orchids like Dendrobium and Oncidium, AgNO3 (10-30 ppm) and HQS (50-100 ppm) extends vase life and bud opening of cut flowers. In Cymbidium, 1-MCP and AVG are superior to STS in prolonging the vase life of cut flowers. In Cymbidium 'PCMV', highest per cent of fully opened buds (75%) and maximum vase life (45 days) were recorded with the chemical combination of sugar 4% + salicylic acid 200 ppm.

In orchids, cut spikes are inserted in tube containing water or water with preservatives and bunch of 5 or more or individual spikes are placed inside the CFB box in alternate fashion. Cool growing orchids are stored at lower temperature even at 5°C in cold chambers whereas tropical and subtropical orchids are stored at 7-10°C and 90-95% relative humidity.

Keywords: Orchids; Ethylene; Vase life; Pre-harvest; Post-harvest factors

INTRODUCTION

Orchids comprise the largest family of flowering plants with 25,000 to 35000 species belonging to 600-800 genera [1]. They are prized for their incredible diversity in the size, shape and color and attractiveness of their flowers and high keeping qualities even upto 10 weeks. Most of the orchids have originated from tropical humid forests of Central and South America, India, Sri Lanka, Burma, South China, Thailand, Malayasia, Philippines, New Guinea and Australia. Brazilian Cattleya, Mexican Laelia and Indian Cymbidium, Vanda and Dendrobium have played a major role in developing present day beautiful hybrid orchids which numbers more than 200000. In the international trade, among top ten cut flowers, orchids rank the sixth position and among orchids Cymbidium ranks the first position and in floricultural crops it accounts for 3% of the total cut flower production [2].

In India, orchid comprises 158 genera and 1331 species which grow upto an elevation of 5000 m [3]. Indian terrestrials are located in humus rich moist earth under tree shades in North Western India. Western Ghats harbour the small flowered orchids. Epiphytic orchids are common in north eastern India which grows up to an elevation of 2000 m from sea level. Indian orchid species having high ornamental values are Aerides multiflorum, Aerides odoratum, Arundina graminifolia, Arachnis, Bulbophyllum, Calanthe masuca, Coelogyne elata, Coelogyne flavida, C. corymbosa; Cymbidium aloifolium, C. lowianum, C. devonianum, C. hookerianum, C. lancifolium; Dendrobium aphyllum, D. nobile, D. chrysanthum, D. farmeri, D. chrysanthum, D. densiflorum, D. moschatum, D. fimbriatum, D. jenkinsii; Paphiopedilum venustum, P. spicerianum, P. hirsutissimum, P. insigne, Phaius wallichii, Pleione praecox, Renanthera imschootiana, Rhyncostylis retusa, Thunia alba, Vanda cristata, Vanda coerulea and Vanda coerulescens [4].

In India, some of native genera like Cymbidium, Paphiopedilum, Vanda, Arachnis and Dendrobium are cultivated on a large scale for cut flower production. The Cymbidium is mainly grown in NEH Region, Sikkim, Darjeeling hills and Arunachal Pradesh. Tropical orchids are cultivated in Kerala and some parts of Tamil Nadu. We should prefer those species which flower during winter and spring months to export flowers to temperate regions from December to May. The orchids have taken a significant position in cut flower industry due to its attractiveness, long shelf life, high productivity, right season of bloom, and easy packing and transportation but lack a complete information module starting from pre-harvest to harvest and subsequently postharvest techniques including cooling, pulsing, impregnation, bud opening, holding solutions, storage, packaging and transportation of different commercial genera of orchids. In the present investigation, the works on post-harvest physiology and technologies of different orchid genera done by various relevant research workers were reviewed.

Hybrids and Varieties for Cut Flower

Among the cut flower crops, Cymbidium, Dendrobium, Phalaenopsis, Odontoglossum, Oncidium, Cattleya, Paphiopedilum, Vanda, Aeridovanda, Aranda, Mokara, Arachnis, Vascostylis, Renanthera, Rhyncicentrum, Rhyncovanda etc. are important.

A good quality cut flower of an orchid should have the following characteristics [5]:

- Minimum eight standard blooms per stem
- Flowers must be cleaned, evenly colored and free from physiological disorders
- Stem must have flowers evenly arranged and around the stem.
- Two third of the stem should be covered with the flowers.
- Flowers must have a firm texture and a luminescent sheen
- Stems must be firm when held up
- The minimum base diameter of the stem should be of 10mm

Important varieties and hybrids under different genera of orchids for cut flower purposes are listed in Table 1 [6].

Table 1: Important varieties and hybrids under different genera of orchids.

Sl No.	Genera	Hybrids/Varieties
1	*Aeridovanda*	Doctor Poyck, Vieng Ping, 'Bensiri', 'Noreen', 'Early Bird', 'Shiv Sidhu', 'New Dawn', Harrison Luke Somsri Sunlight'
2	*Aranda*	'City of Singapore', 'Hilda Galistan', 'Urmila Nandey', 'Christine', 'Thailand Sunspot', 'Millenium Dawn', 'Broga Giant', 'Salaya Red', 'Propine White', 'Propin Spot', 'Lueng Cholburi', 'Ishbel Manisaki', 'Baytown', 'Chao Praya Blue', Chao Praya Dot Com', 'Chao Praya Beauty', ' Ethan Pride', 'Taksari Chandrabir'
3	*Arachnis*	Ishabel, Maggie Oei
4.	*Ascocenda*	Apinantat Red Berry, Pralor Tuyen, Pak-Kred, Bangkok, Surin, Kamada, Crownfox, Sundancer, Laksi 'Red Ruby',Guo Chia Long 'Spotty', Fuchs Angel frost, 'Carol Belk', 'Renuka Angle', 'Joyce Bevins', 'Adisak Blue', 'Renu Gold', 'Tipi Blue Boy', 'Bobs Fortune', 'Rubychai', Shah Rukh Khan', Yang Sophia Firuz', 'Abdul Ghani Othman', 'Chunika', 'Fuch's Star'
5.	*Cattleya and allied*	Lovely Bangkok, Admiration, Bob Belts, General Patton, Joyce Hannington, Little angel, Margaret Stewart, Nillie Roberts, Paerl Harbour, Primma Donna, Queen Sirkhit 'Diamond Crown', Secret Love, Ladda Belle 'Pink Pearl', Maikai, Pastoral, Robert, Prism Palette 'Tricolour Magic', 'Ahmad Seikhi', 'Hsinging Catherine'
6.	*Cymbidium*	Levis Duke Bella Vista, Madrid Forest King, Sparkle Late Green, Angelica December Gold, Sleeping Nymph, Pine Clash Moon Venus, Soul Hunt, Dr. H. C. Aurora, Susan Highes, Tia Gaig Suther Land, Miss Sanders, Amesbury, Kenny Wine, Red Star, Red Princess, Show Girl , Jungfrau 'Snow Queen', Jungfrau 'Dos Pueblos', Lilian Stewart 'Coronation', Lilian Stewart 'Party Dress', Orkney 'Pink Heather', Ensikhan 'Alpha Orient', 'Fire Storm Ruby'
7.	*Dendrobium*	Emma White, Thongchai Gold, July, Erika, Sonia-17, Sonia-28, Kasem White, Madam Pompadour, Bangkok Blue, Ann, Gold Twist, Candy Stripe Pink, Genting Blue, Bengal Beauty, Sakura Pink, Candy Stripe, Burana Charming, Blue Fairy, Channel, Nette White, 'Dang Saard'
8.	*Mokara*	Walter Oumae 'Seksan', Thailand, Sayan, Walter Oumae 'Royal Sapphire', Susan 'Orange', Walter Oumae 'Calypso', Eng Ling, Madame Panne, Mak Chin On, Bangok Gold, Bibi, Chao Praya Gold, Chark Kuan Orange, Chark Kuan Pink, Chark Kuan Rose, Chark Kuan Super, Dinah Shore, Kelvin Red, Kelvin Orange, Luenberger Gold, Margaret Thatcher, Pink Star, Sayan, Sayang Pink, Walter Oumae, WTO, Jiti, Happy Beauty, Salaya Gold
9.	*Odontoglossum*	Carroll, Ismene, Palnina, 'Italian Job', 'Joyce Stewart', 'Pepe Gerald', 'Purple Rain', 'Katherine Jenkins', 'Roy Wittwer', 'Laguna Blanca', 'Precocious', 'Snow Fall', 'Pesky', 'Queen of Mars', 'Stam Point', 'Holiday Yellow', 'Ronald Norman', 'Bridget Ring Lawless', 'John Hanry Hanson', 'Point Pesky'
10.	*Oncidium*	Aloha Iwanga Dogasima, Goldiana, Gower Ramsey, Golden Shower, Sum Lai Who Jungle Queen, Taka H & R, Sharry Baby Sweet Fragrance AM/AOS, Golden Glow, Popki Red, Irine Gleason Red, Vision Brownish Red, Catherine Wilson x New Calidonia Brownish Red, Robson Orchid Glad
11.	*Paphiopedilum*	Niveum, Concolor, *P. rothschildianum* (3 to 5 flowers), *P. sanderianum* (3 to 5 flowers), Prince Edward of York, Michel Koopwitz, Saint Swithin, Mount Toro, Sorcerer's Apprentice, Grande, Don Wimber, Elizabeth March, Hanne Popow, Jason Fischer, Living Fire
12.	*Phalaenopsis*	Taisuco Crane, Taisuco Kochdian, Cygnus, Yukimai, Sogo Musadian, White Dream, Florida Snow, Nobby's Pink Lady, Minho Valentine, Minho King Beauty, New Cinderella, Taisuco Firebird, Sogo Smith, Carol Campbell, Emil Giles, Brother Lawrence, Taipei Gold, Golden Bells, Sogo Managers, Brother Passat, Be Glad, Cassandra, Vilind, Carmelas Pixie, Zuma's Pixie, Timothy Christopher, Be Tris, Quevedo, Strawberry
13.	*Renanthera*	Brookie Chandler, Manila T-Orchids, Kilauea, Mok Yark-Seng, Poipu, Tom Thumb, 'Red Leopard', 'Scarlet Belle', 'Chanachae', 'Serdang', 'Brady Crocker', '20th WOC Singapore-2011', 'Bart Motes'
14.	*Rhynchovanda*	'Wilton Hill', 'Jammie Harper', 'Apichart', 'Noo Noi', 'Peter Draper', ' Brighton's Albino', 'Prairie Lady'
15.	*Vanda*	Annette Jones, Antonio Real, Golamcos Blue Magic, Fuch's Charmer, Jimmy Millers RF Orchids, Keree Delight, Memoria Lyle Swanson, Motes Indigo x Merrillii, Motes Honeybun, Motes Primerose, Miss Joaquim.V. Rothschidiana, VTMA –Red, Pink, White, Vasco, Johnny Miller,Veerawan, Roberts Delight, Rasriprai, Pat Delight, Pakchong Blue, Mimi Plammer, Manuvade, Lumpini Red, Kultana Gold x Thongchai Gold, Fuchs Delight, Charles Goodfellow, Pine River,Adisak, Doctor Anek, John Club, Bill Sutton, Ellen Noa, Emily Notley, Evening Glow, Honomu, Honolulu, Hilo Blue
16.	*Vascostylis*	Paragon Joy x Kasems Delight, Precious, Veeraphool, Crown Fox 'Red Yen', Aroon Fairy, Viboon Velvet

Table 2: Vase life of orchids due to species and varietal differences [6].

Name of Species/hybrids	Vase life (days)
Aerides odoratum, Aerides multiflorum, Cymbidium iridioides, Dendrobium nobile, Renanthera imschootiana	28-56 days
Paphiopedilum hirtussimum, P. wardianum	56 days
Phaius tankervillae	28-42 days
Vanda coerulea, Vanda teres, Zygopetalum intermedium	14-21 days
Cymbidium hybrids	20-55 days
Dendrobium hybrids	14-21 days
Vanda, Mokara hybrids	14-30 days
Cattleya hybrids	10- 20 days
Phalaenopsis hybrids	25-30 days
Aranda	18-28 days

Physiology of cut flowers

Vase life or longevity of a cut flower was determined on the basis of attributes like diameter and length of florets, opening of flowers, changes in fresh weight, diameter or length of stem or pedicel, senescence pattern, color of petals, total longevity and foliage burning. In general, cut flowers completed their life cycle in two distinct phases, (i) bud swelling to bud opening and (ii) maturation, senescence and wilting. Flower bud development to swelling involved growth or change in orientation of petals or subtending tissues and may also require abscission of protective structure [7]. When an inflorescence was cut from the plant, a number of physiological processes were affected which included the supply of water, depletion of stored substrates and production of ethylene. The most common symptom of flower senescence was wilting i.e. loss of turgor pressure of the cells due to failure of water uptake [8]. The failure of water uptake as a result of stem blockage might be due to air blockage, microbial growth or physiological plugging. Otherwise senescence was accompanied by a dramatic increase in the leakage of several molecules such as aminoacids, sugars, inorganic ions, anthocyanins and activity of petal ACC synthetase and disintegration of tonoplast and mitochondria [9]. Two major metabolic and biochemical changes occurring in senescing petals were increase in respiration and hydrolysis of cell components. Vase life of cut flowers depends upon the rate of transpiration through open stomata of leaves and solutes present in vase water [10]. Experimental evidences have shown that cuticular transpiration played an important role in the water loss of orchid flowers. The transpiration rate of tropical orchid flowers ranged from 0.15 to 0.17 mg water $cm^{-2}h^{-1}$ or 0.4 to 1.9 g of water per inflorescence per day depending upon the total floral surface area. Halevy [11] classified flowers into climacteric or non-climacteric based on presence or absence of an increased rate of ethylene production associated with petal senescence. In most vegetative tissues, the overall synthesis of ethylene was the conversion of SAM (S-adenosylmethionine) to ACC (1-aminocyclopropane-1- carboxylic acid). Water loss in orchid flowers was considerably lower than that reported for roses and carnation due to the absence of supporting leaves in orchid sprays. It has been found that carbohydrate levels in mature flowers were lower than the levels in the tight buds. Moreover, the level of carbohydrates in the flower decreased markedly with time after harvest as reflected in the decreasing rate of respiration. This problem could be overcome by the exogenous supply of sucrose.

In orchid flowers, ethylene production was an autocatalytic process and ethylene level of 1 ppm caused premature fading of flowers [12,13]. The premature fading of petals might be induced by pollination and by removing pollinia [14]. In Cymbidium, fading of flowers due to pollination was characterized by the formation of anthocyanin in both column and lip, swelling of column, stigmatic closure and wilting of sepals and petals. The anthocyanin level began to decrease with age [15,16]. Developmental processes associated with post-pollination events included senescence of perianth, pigmentation changes, ovary maturation, ovule differentiation and female gametophyte development [17]. Flowers of Dendrobium 'Pompadour' developed premature petal and sepal senescence following pollination. Pollination induced an ethylene climacteric accompanied by a small respiratory climacteric, epinasty and increased flower and inflorescence fresh weight and water uptake [18]. The orchid flowers harvested in the tight bud stage had a lower rate of respiration than open flowers. The respiration rates continuously declined during the post-harvest period until the flower faded [19].

Da Silva, JA Teixeira [20] reviewed the some aspects of changes and programmed cell death occurred in petal senescence.

CELLULAR STRUCTURAL CHANGES

- Membrane rupturing and increased in cytoplasm debris and loss of permeability and fluidity due to oxidation.
- Invagination of tonoplast and endocytosis of cytoplasmic contents and disappearance of cortical microtubules.
- Reduction in cytoplasm volume, cessation of cytoplasmic streaming and change in proton flux across plasma membrane.
- Degeneration and collapse of organelles and increase in number of peroxisomes.

Biochemical and Structural Molecular Changes

- Increase in proteinases and nucleases; upregulation of phospholipases, acyl hydrolases and lipoxygenase neutral lipids; sterol/phospholipid ratio; lipid peroxidation; reactive oxygen species; water leakage and cell wall cross linking.
- Decrease in phospholipids, chlorophylls, proteins, thiol groups, nucleic acids and RNA.

Post-harvest life of orchid cut flowers is influenced by pre-harvest factor like varietal differences, light intensity, sugar level of flowers, temperature, nutrition and water loss; harvest factors like time, method and stage of harvest and post-harvest factors including ethylene production, pre-cooling, water quality, pulsing, use of preservatives, bud opening, packaging and storage.

Pre-harvest factors

It was estimated that a third of the post-harvest life of flowers is programmed by pre-harvest handling.

Varietal differences: Varietal differences in cut flowers had been reported due to variations in water uptake, fresh weight, flower diameter, stems lignification, vase life and senescence behaviour. Among different species, the vase life ranges from Lycaste spp. (9 days), Phaius tankervillae (24 days), Zygopetalum intermedium (18 days), Aerides multiflorum, A. odoratum, Cymbidium irridioides, Dendrobium nobile and Renanthera imschootiana (28-56 days), Paphiopedilum hirtussimum, P. wardianum(56 days) and Vanda coerulea, Vanda teres (28-42 days) (Table 2). Out of nine hybrids of Cymbidium, 'PCMV', 'Red Princess', 'White Beauty', 'H.C. Aurora', 'Sun Gold', 'Ensikhan, Florance', 'Valley Legend' and 'Platinum Gold', evaluated at NRCO, Pakyong, Sikkim during 2008-2009, 'Florance' had the highest vase life of 54 days followed by 'White Beauty' (53 days) and lowest in 'Platinum Gold' (22 days) (Figure 1) [21]. The present day orchid hybrids of Dendrobium, Vanda and Mokara remained perfect for 7 to 30 days. The flowers of Cattleya and Phalaenopsis remained fresh for 1 to 4 weeks whereas Aranda lasted for 18 to 28 days.

Cymbidium, 'Fire Storm Ruby' for cut flower

Dendrobium, 'Dang Saard' for cut flower

Vanda, 'Roberts Delight Blue' for cut flower

Oncidium, 'Sharry Baby Sweet Fragrance' for cut flower

Blc, 'Hsinging Catherine' for cut flower

Phalaenopsis, 'Strawberry' for cut flower

Figure 1: Varietal differences in cut flowers.

Light intensity: Light was found to determine the carbohydrate levels before harvest which in turn influence the keeping quality. Flowers containing relatively higher amounts of carbohydrates especially mobile sugars lasted longer in the vase. Plants having few leaves, or leather like leaves (like most cattleyas and oncidiums), required a high-light environment. If the leaves are soft and limp (like some phalaenopsis and most paphiopedilum), the plants were probably very light-sensitive, and should not be placed in a sunny south-facing window. Most orchid's prefered indirect or filtered light and 50% shading.

- Low light orchids (1200-2000 f.c.): Phalaenopsis, Calanthe
- Medium light orchids (2000-3000 f.c.): Cattleya, Laelia, Brassovola
- High light orchids (3000 f. c. or more): Cymbidium, Vandaceous groups.

Temperature: Generally, higher temperature resulted in higher level of respiration. Cooling was essential to reduce other metabolic changes such as enzymatic activity and to slow the maturation of flowers. Cooling prior to packaging and transport reduced ethylene production and improved longevity. Based on temperature requirements, orchids were classified into three groups:

- Warm orchids (Aerides, Vanda, Rhyncostylis and Dendrobium): 32.2°C day temperature and 15.5°C night temperature
- Intermediate orchids (Cattleya, Laelia, Oncidium, Miltonia): 26.6°C day temperature and 12.8°C night temperature
- Cool orchids (Cymbidium, Odontoglossum, Cypripedium): 24°C day temperature and 10°C night temperature

Humidity: As a thumb rule, orchids required 80-85% humidity for satisfactory growth. Monopodial orchids required higher humidity than sympodial ones. Many sympodial orchids like Cattleya, Oncidium and Dendrobium form pseudo bulbs, which were swollen shoots that store water and nutrients to help the plant survive periods during prolonged drought. Insufficient humidity during summer might lead to shriveling of pseudo bulb. Excessive humidity during winter might lead to spotting of flowers usually caused by Botrytis. Most orchids preferred water of pH 5.0-6.5. Watering with lower or higher pH or with high levels of dissolved minerals could hamper nutrient uptake. Rain water was the best. Regular watering was essential under high sunlight and high temperature conditions. Sprinkling or misting might be practiced during hot summer. Watering should be reduced in late summer and keep the plants barely moist during winter. Watering the plants with thick leaved orchids having CAM activities such as Aranda and Dendrobium in the late afternoon prior to harvesting season improved the keeping quality.

Nutrition: Orchids were light feeders and they required nitrogen from beginning to two-third of their life cycle. During rest period, they did not need any fertilizers. During flower initiation and inflorescence development plant were fed with less nitrogen, more phosphorus and potassium. In orchids, foliar feeding was found to be ideal. Frequent application of fertilizers in low concentrations was the best way of feeding orchids. A concentration of 0.2 to 0.3 % of 30:10:10 (N: P: K) at vegetative stage and 10:20:20 (N: P: K) at blooming stage were applied for quality flower production. Sometimes, fresh coconut water, diluted cow urine and fish meal emulsions were also useful as foliar spray.

Harvest factors

Time of harvest: Flowers should be harvested in mild temperature because high temperature causes rapid respiration rates and excessive water loss. Flowers should be harvested in the early morning or in the evening. In the early morning, flowers remained turgid due to transpiration at night and higher sugar levels. Similarly, flowering stems retain a higher amount of stored carbohydrates if cut in the afternoon and retained more vase life.

Method of harvest: Sharp tools or secateurs were always used to detach the stem of flowers from the mother plant. The angle of the cut was given in slanting position and the stem was not crushed during harvesting, especially hard wood stems. The spikes were dipped in a bucket containing water immediately after harvest.

Stage of harvest: The optimum harvesting stage of the commercial orchids was fully open and mature flowers. The stage of harvest, spike length and number of flowers of some commercial orchids are given in Table 3.

Table 3: Stage of harvest, spike length and no of flowers of some commercial orchids.

Orchid hybrid	Commercial stage of harvest	Spike length (cm)	No of Flowers
Aranda	50% bloom	45-60cm	8-10
Cattleya	2-4 days before bud open	25-40 cm	1 or more
Cymbidium	75% bloom or two buds open stage	60-90 cm	10-15
Dendrobium	All flowers except top bud	40-60cm	8-12
Oncidium	80 % bloom	60 cm	Many
Paphiopedilum	3 to 4 days after opening of flowers	25-40 cm	1-5
Phalaenopsis	Fully open flowers	40-60 cm	8-10
Vanda	Fully open flowers	50-75 cm	8-15

Out of three Cymbidium hybrids namely 'Pine Clash Moon Venus', 'Valley Legend Steff' and 'Pure Inca Gold', flower spikes were harvested at four stages like fully open, 75% open, 50% open and 25% open to standardize the stage of harvesting, and enhanced vase life. Vase life was noticed highest in 75 % open stage. Maximum vase life of 59 days was recorded in 'Pine Clash Moon Venus' and followed by 48.83 days in 'Valley Legend Steff' and 53 days in 'Pure Inca Gold'.

Post-harvest factors

Temperature: Opening of flower buds and rate of senescence accelerated at higher temperatures. At lower temperature, the respiration came down and the flowers produced a lesser amount of ethylene. Temperature played an important role for flowers harvested at the immature stage for full expansion of buds and the flower buds were kept at temperatures as low as 0.5 to 4.0°C in Cymbidium and Paphiopedilum, 5-7°C in Dendrobium and 7-10°C in Cattleya.

Light: Light was essential for long distance transport or prolonged storage of cut flowers. Similarly, high light intensity was essential for opening of tight bud cut flowers. Florists had to maintain a light intensity of 2000-3000 lux for 12-24 hours in their shops for illuminations for most of cut flowers.

Humidity: Cut flowers were kept at 90-95% relative humidity for maintaining turgidity. Flowers started showing wilting symptoms when they had lost 10-15% of their fresh weight. The rate of transpiration from leaves was reduced with the increase of high relative humidity.

Water quality: Water quality is defined as pH and EC value, hardness contents of phytotoxic elements and microorganisms causing vascular occlusions affecting longevity of cut flowers. Saline water decreased the vase life of cut flowers. In case of cut gladiolus, the longevity of flowers decreased when the concentration of salts in the water reaches 700 ppm, whereas for cut roses, chrysanthemum and carnations, 200 ppm is harmful. The increase in concentration of salts over 200 ppm shortens vase life by half a day for each 100g per litres increase in salinity. Basic ions like Ca^{++}, Mg^{++} present in hard water were less harmful to flowers than soft water containing sodium ions. Fluorine is very toxic to most of the cut flowers and causes injury to freesias, gladiolus and gerberas at 1 ppm, and chrysanthemums, roses, poinsettias and snapdragons at 5 ppm. Flowers like lilacs, cymbidiums and daffodils are resistant to fluorine ions. Vase life increased in tap or well water passed through a de-ionizer. The importance of low pH of the holding solutions is well known for improving vase life. A holding solution of pH 3.0-5.0 was optimum for increasing vase life of cut flowers.

Ethylene: Ethylene played an important role in the regulation and co-ordination of senescence in climacteric flowers. Production of this hormone was less and stable in floral buds and young flowers. A sharp increase in ethylene evolution was found during flower maturation, opening and senescence. Afterwards, ethylene production decreased and remained static. Basically, ethylene was first produced in the pistil and the evolved ethylene acted on the petals and induces expression of genes for ACC synthase, ACC oxidase and cysteine proteinases, resulting in the auto-catalytic ethylene production from the petals, in-rolling of petals and wilting of flowers [22]. The gynoecium has been shown to produce a significant amount of ethylene before its production in the petals possibly induced by the factors like ABA or IAA suggesting its importance in controlling ethylene production in the flower during natural and pollination induced senescence, with emasculation hastening the release of ethylene [23]. A wide range of flowers was affected by ethylene with showing of some typical symptoms such as sleepiness of carnation and kalanchoe petals, fading and in-rolling of the corolla of Ipomoea, fading and wilting of sepal tips in orchids and induction of anthocyanin formation in female reproductive parts and abscission of flowers and petals. Orchid flowers were highly sensitive to ethylene. High level of ethylene production was due to herbivore damage, mechanical injuries and pollination. Decapped and emasculated flowers produced more ethylene than untreated ones. Sometimes, forced unfolding of flower buds in orchids reduced vase life.

Diseases and insect pests: Fungi, bacteria and insects affected the quality of cut flowers by producing higher amounts of ethylene. Microbes accelerated flower senescence by the plugging of xylem vessels with pectin degraded products, by producing ethylene and toxic compounds. Among bacteria genera, Alcaligenes, Pseudomonas, Enterobacter, Erwinia, Bacillus, Corynaebacteria, Aeromonas, Acetinobacter and Flavobacterium were commonly found in vase water. Some fungal species Botryris cinerea, Fusarium oxysporum, Mucor, Penicillium spp, Rhizopus, Aspergillus spp, Alternaria alternata and Acremonium strictum were responsible for early senescence of flowers and wilting and decaying of potted plants [24,25].

Treatments for Improving Longevity of Cut Orchid Flowers

Physical treatments

Pre-cooling: It was an important operation in post-harvest handling and transport of cut flowers for the fast removal of field heat wherever flowers were held dry pack. All flowers were pre-cooled immediately after harvest by placing them in cold storage without packing or in open boxes until they reach the desired temperature [26]. It varied with the species and cultivars: Cattleya (7-10°C), Cymbidium and Paphiopedilum (0.5 to 4°C), Dendrobium (5-7°C). Precooling lowered respiration rate and decreased the breakdown of nutritional and other stored materials in the stems, leaves and petals; delays bud opening and flower senescence. It also prevented rapid water loss and decreased flower sensitivity to ethylene. Several pre-cooling techniques such as room cooling, forced air cooling, hyder-cooling, vacuum cooling and ice bar cooling etc. were available.

Vase life of cut flowers as affected by stem cut ends: The resistance to water flow through stem segments increased predominantly in the lower most one to two centimeter of the cut stems. Re-cutting stems under water improved longevity due to elimination of air from the conducting vessels.

Chemical treatments

Conditioning: Conditioning or hardening was a simple process where flowers were kept standing loosely in a big container so that air can circulate around the stems. The purpose of the treatment was to restore the turgidity of cut flowers with water stress during storage and transport. Conditioning was achieved by treating the flowers with demineralized water supplemented with germicides and acidified with citric acid to pH 4.5 to 5.0 but with or without sugar. Hydration was improved when water was de-aerated or acidified or when a wetting agent like Tween 20 at the rate of 0.01 to 0.1 % was added. Flower stems were placed in warm water or in a preservative solution in plastic jars at a depth of 2-4 cm and held at room temperature or in cold storage for several hours.

Impregnation: Sometimes, the cut ends of the flower stems were impregnated for a short time with chemicals. This treatment protected the blockage of the water vessel in the stem by microbial growth and stem decay. Impregnation of cut bases of flowers with high concentration (1000 ppm-1500 ppm) of silver nitrate, nickel chloride or cobalt chloride for 10-15 minutes improved the longevity of several flowers such as aster, gerbera, gladiolus, carnation, chrysanthemum, phalaenopsis and snapdragon. In Cymbidium 'Baltic Glaciers Mint Ice', highest longevity was recorded with $CoCl_2$ (1000 ppm) for 15 minutes (46 days) followed by $CoCl_2$ (1500 ppm) for 15 minutes (44 days) over control (39 days).

Pulsing: The absorption of chemical solutions containing sugars and germicides through the lower cut bases of flower stems was known as pulsing. Pulsing could be used by growers, wholesalers or retail florists in order to enhance the cut flowers subsequent vase life in water. Pulsing was employed with higher concentrations of sugar, mainly sucrose, the percentage of which varies with species and cultivars. Other chemicals used in the pulsing treatments were STS, $AgNO_3$, HQ, MH, AOA, $CaCl_2$, $CoCl_2$, nickel sulphate, aluminium sulphate and benzyladenine. Pulsing was found to be of great value in prolonging life, promoting opening and improving the color and petal size of petals through osmo-regulation. In Cymbidium hybrid 'Red Princess' pulsing with 5% sucrose increased vase life (56 days) followed by sucrose @ 8% (54.78 days). In Cymbidium, 'Baltic Glaciers Mint Ice' pulsing of flowers with 5% sucrose followed by 150 ppm 8-HQS increased the vase life of flowers with pollina (49.33 and 46.33 days) and without pollinia (44.00 and 41.67 days), respectively. In Aranda, pulsing with 4 mM STS for 10 minutes and in Dendrobium hybrid 'Pompadour' with 25 ppm $AgNO_3$ + 135 $Na_2S_2O_3$, $5H_20$ for 30 minutes increased vase life of cut flowers [27,28]. In Oncidium 'Goldiana' cut sprays pulsing with $AgNO_3$ for 30 minutes improved vase life [29]. In Phalaenopsis, pulsing with 0.5mM STS for 24 hours blocked the deleterious effect of ethylene. In Dendrobium cv. 'Sonia' pulsing with 4% sucrose + 400 ppm HQ recorded the highest vase life of 21.33 days. Inflorescences pulsed with 6% sucrose + 400 ppm HQ recorded the highest sugar content in the flowers (27.64%) [30].

Bud opening: It was a procedure of harvesting flowers at a stage earlier than normally considered as the cutting stage and then opening the buds off the plant. Such types of post-harvest handling could be applied by growers or wholesalers. Bud opening of flowers increased longevity of cut flowers by reducing the sensitivity of flowers to extreme temperatures, low humidity and ethylene, saving space during shipment and extending the useful storage life. The sugar concentration used was lower than the concentration of pulsing and the optimum temperature is kept lower. In Dendrobium hybrid, 'Thongchai Gold' opened flowers had 29%, half opened flowers had 28.25 % and buds had 16.17% reducing sugars. In Dendrobium hybrids, HQS or $AgNO_3$ (50 ppm) was effective for opening of tight bud cut flowers.

Ketsa et al. [31] reported that a preservative solution containing 225 ppm HQS, 30 ppm $AgNO_3$ and 4% glucose increased bud opening and the time to wilting of the open florets of Dendrobium Cv. 'Ceasar'. In Cymbidium Cv. 'Ensikhan' 4% sucrose + 100 ppm acetyl salicylic acid or 4% sucrose + 100 ppm $Al_2(SO_4)_3$ improved bud opening. In Cymbidium hyb. 'PCMV', sugar(4%) + salicylic acid (200 ppm) showed maximum per cent of flower opening (75%) and vase life (45 days) followed by sugar (4%) + $Al_2(SO_4)_3$(100 ppm) (57% and 44 days) and sugar (4%) + 8-HQS (200 ppm) (53.8% and 44 days) over control (bud drop and senescence on 27th days), respectively (Table 4). Highest content of carbohydrate (140 mg/g) was estimated at bud stage in fresh condition followed by at bud stage (131 mg/g) at senescence in control. Minimum carbohydrate content (60 mg/g) was observed with Sugar 4% + 8-HQS (200 ppm) followed by Sugar 4% + Salicylic acid 200 ppm) (64 mg/g). In Dendrobium hybrid 'Thongchai Gold', per cent of fully opened buds (66%) was recorded maximum with sucrose

(4%) + $Ca(NO_3)_2$ (1%) followed by sucrose (4%) + acetyl acetic acid (100 ppm) (60%). Longest vase life (36 days) was found with sucrose (4%) + $Al_2(SO_4)_3$ (100 ppm) followed by sucrose (4%) + acetyl acetic acid (100 ppm) (33 days).

Table 4: Effect of chemicals on bud opening in Cymbidium hyb 'Pine Clash Moon Venus.

Treatment	Days to first floret opening	Diameter of first floret (cm)	Per cent of half opened buds	Per cent of Fully opened buds	Vase life (Days)
Distilled water	---	----	0	0	27
Sugar 4%	20	5.1	5.8	44	37.8
Sugar 4% + $Al_2(SO_4)_3$(100 ppm)	21	5.4	0	57	44
Sugar 4% + 8-HQS (200 ppm)	18	5.5	30.7	53.8	44
Sugar 4% + Salicylic acid (200 ppm)	21	6.6	0	75	45
Sugar 4% + Ca $(NO_3)_2$ (1%)	25	4.85	2	22.8	37.8
Sugar 4 % + Boric acid 200 ppm + K_2SO_4 (2mM)	20.5	5.5	6.25	25	36.2

Preservatives: Preservatives were used in the holding solutions in the form of tablets containing a mixture of chemicals such as sugars, germicides, salts, growth regulators etc. Besides, the chemicals were employed during conditioning, pulsing and for making bud opening solutions to improve flower shape, size and opening and color of the flowers.

Sugar, biocide, anti-ethylene compounds and hydrated compounds were used for conditioning. The sugar and biocide solutions were effective for opening of bud cut flowers.

The vase solution should contain sugars, acidifying agent and a biocide. Citric acids were mainly used for acidifying agent and hydroxy quinoline as biocide. Metallic salts like silver nitrate, cobalt chloride, aluminium sulphate, zinc sulphate, calcium nitrate and nickel chloride have been found for prolonging post-harvest life of various cut flowers. Among several growth regulators used to increase vase life of cut flowers, BA, IAA, NAA, 2,4,5-T, GA3, B9, CCC are common.

New chemicals that had been found promising as floral preservatives are ethylene inhibitors like amino- oxyacetic acid, 1-amino cyclopropane, aminotriazole, aminoethoxy vinyl glycine, alpha aminoisobutyric acid, diazocyclopentadiene and phenidone.

Different chemicals used in holding solution for improving vase life of orchids are listed in Table 5.

Table 5: Holding solutions for different types of orchids.

Name of orchid	Holding solution
Oncidium	8-HQC (100-200 ppm) + 4% sucrose
	kinetin (50 ppm) + 4% sucrose
Cymbidium	8-HQC 200 ppm + sucrose 2%,
	1-MCP (500ppb)
	1% Sucrose + STS (1 mM)
Arachnis, Aranda, Aranthera, Cattleya	STS (1mM) + 1% sucrose
Paphiopedilum	8-HQC (200 ppm) +2% sucrose
Vanda	$AgNO_3$ (30 ppm) + 1.5% sucrose
Dendrobium	8-HQC (200 ppm) + sucrose (2%),
	0.5 mM AOA + 4% sucrose,
	$AgNO_3$ (30 ppm) + 4% sucrose
	400 ppm HQ + 30 ppm $AgNO_3$ + 2 % sucrose
	200 ppm 8-HQS + 50 ppm $AgNO_3$ + 8% sucrose

In Cymbidium 'PCMV', 2% cane sugar had shown maximum depletion of stored carbohydrates, maximum longevity of first floret (54 days), zero per cent of flower dropping, maximum solution uptake (24ml) and highest vase life (61.2 days) followed by 4% cane sugar. 8% cane sugar had least longevity of first floret (27.2 days) and vase life (36.2 days). Longevity of first floret was recorded maximum with BA (25 ppm) (49.2 days) followed by $Al_2(SO_4)_3$ (100 ppm) (45.2 days) (Table 6). Maximum vase life (54.8 days) was found with $Ca(NO_3)_2$ (100 ppm) followed by $Al_2(SO_4)_3$ (100 ppm) (53.4 days). Solution uptake (33 ml) was observed maximum with $Al_2(SO_4)_3$ (100 ppm) and 8-HQS (200 ppm). In Cymbidium, 'Baltic Glaciers Mint Ice', 8HQS 150 ppm increased the vase life of flowers with pollinia (48.33 days) followed by 8 HQS 200 ppm (46.80 days). 8 HQS 200 ppm increased the vase life (36.67 days) of flowers without pollinia.

In another experiment, in Cymbidium hybrid, 'PCMV', out of six treatment combinations, 2% sucrose + 200 ppm 8-HQS had shown maximum vase life (77.6 days) followed by 2% sucrose + 100 ppm $Al_2(SO_4)_3$ (77.4 days) over control in tap water (65 days).

Packaging in orchids

Cut flowers were inserted in tube containing water or water with preservatives or simply wrapped in wet cotton swab and the same was covered with a piece of plastic and tied with rubber band to keep in its place. Flower spikes were bunched into bunch of 5 or 10 or so. Bunches or individual spikes were placed inside the box in alternate fashion. Ethylene scrubbers with $KMnO_4$ or Purafil could also be kept in the box. For export purpose, packing of flowers in two piece box was better option. In Cymbidium, single flowers backed by a fern leaf were inserted in small flasks containing preservative solution, the flasks were then packed in 3 sided box with a display window. In Dendrobium hybrid 'Sonia-17' a low gauge polyfilm of 100 gauge thickness the cotton dipped in 8-HQS (25 ppm) covering the base of the spike had maximum vase life and flower quality [32]. In Cymbidium, Cilindra - a gift of a glass flute containing a flowering mini Cymbidium and Stylish setting- Festive packaging for special occasions like Birthday were common (Figure 2).

Table 6: Effect of chemicals on post-harvest life of Cym 'PCMV'.

Treatments	Loss in wt (g)	Longevity of first floret (days)	Vase life (days)	Solution uptake (ml)
Distilled water	12.6	37.2	44.4	21
$Al_2(SO_4)_3$ (100 ppm)	18.4	45.2	53.4	33
$Al_2(SO_4)_3$ (500 ppm)	21.4	28.4	36.8	21
Ca $(NO_3)_2$ (100 ppm)	16.4	42.0	54.8	27
Ca $(NO_3)_2$ (500 ppm)	29.6	32	35.6	24
8-HQS (100 ppm)	27.4	33.2	45.6	29
8-HQS (200 ppm)	26.8	42.2	49.8	33
BA (25 ppm)	24.0	49.2	48.0	27

Grading and Packing

Grading

Grading was done based on the basis of parameters like appearance, stage of maturity, blemishes or injuries due to diseases, infestations caused by insect pests, color and size of the bud, straightness, strength as well as length of stem. Flowers were generally grouped into bunches of 5, 10, 12 or 20 stems and loosely tied with rubber bands. Before placing them in the package, individual flower bunches were wrapped with suitable packing materials like cellophane paper, kraft paper, newspaper, tissue paper or corrugated cardboard sheet. For local markets, the bunches were held in buckets containing water or preservative solution. It was advisable that for long distance transport and storage, flower bunches are held in dry cardboard boxes. The minimum length of boxes should be about twice the width and it's width about twice the height. Use of telescope-style boxes made of CFB was ideal.

Grading of cut flowers of orchids was done based upon flower count and spike length of different genera (Tables 7 and 8).

Table 7: Grading of Cymbidium cut flowers.

Cymbidium	Grades	Flower count	Spike length	Other consideration
Cymbidium Standard	AAA	>12	1.25m	• Strong straight stems, uniform length, no marks on flowers. • Bent crooked spikes but with perfect flowers.
	AA	8	90 cm	
Cymbidium Miniature	XL	>15	65+	
	L	12-14	55-64	
	M	8-11	40-54	
	S	< 5	30-39	

Table 8: Grading in Dendrobium Orchids.

Grade	Spike length	No. of opened flowers
Small (S)	30cm	4-5
Medium (M)	40cm	6-8
Large (L)	45cm	8-10
Extra Large	50cm	>10

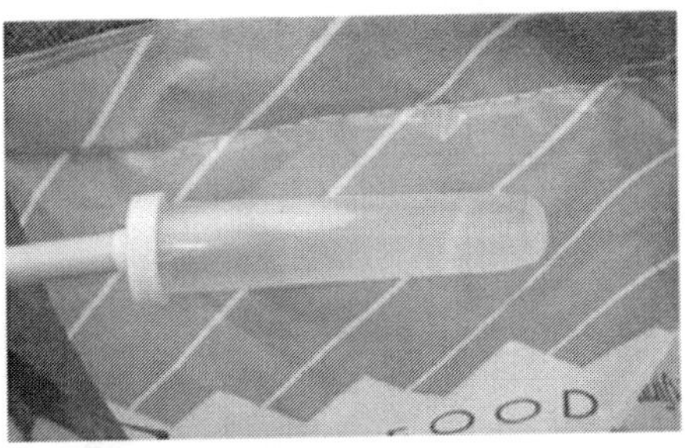

Cut flowers are inserted in tube containing

Water or water with preservatives

Individual spikes are placed inside the box

in alternate fashion

Figure 2: Packaging in orchids.

Storage of Cut Flowers

Low temperature treatment during storage or shipment period reduced the entire metabolism in the tissues, slowed down the respiration, transpiration and ethylene action and retarded the multiplication of bacteria and fungi.

In general, temperate orchids were stored at lower temperature even at 5°C in cold chambers whereas tropical orchids were stored at 7-10°C. A 90-95% relative humidity was necessary during storage to minimize moisture loss and to prevent wilting.

There are two types of cold storage methods, namely 'Wet storage' and 'Dry storage'. In wet storage, flowers were stored with their bases dipped in water or preservative solution for a short time. Dry storage methods were used for long term storage. In this method, fresh flowers were harvested in the morning, graded and sealed in plastic bags or boxes to prevent the loss of moisture.

In Controlled Atmosphere (CA) storage, cut flowers were kept in gas tight cool chambers equipped with cooling systems at a higher level of CO_2 and lower level of 02 to reduce the respiration rate and production and action of ethylene. Generally, the concentration of CO_2 was maintained higher than 4% and not below 0.4% in CA storage (Table 9).

Table 9: Storage of orchid cut flowers.

Name of Orchid	Storage Temperature	Storage period
Oncidium, Phalaenopsis, Odontoglossum, Cattleya	7 - 10°C	2 weeks
Dendrobium	5 - 7°C	10-14 days
Cymbidium	1- 4°C	14 days
Paphiopedilum	-0.5 - 3.0°C	20 days
Arachnis, Aranda, Aranthera Ascocenda, Epidendrum	8-13°C	10-14 days

Transport

Being short lived and perishable in nature, flowers could be delivered to destination as early as possible immediately after harvest. For long distance markets, cut flowers were transported by cargo planes, merchant ships and trucks. Other modes of transportation were head loads, bicycles, two-three wheelers, cars, vans etc. Hence, for long distance transportation, advanced methods of post-harvest handling like cooling, conditioning, impregnation, pulsing, bud opening and packaging were followed.

Short time pulsing of flowers with optimal concentration of sucrose, $AgNO_3$, STS and growth regulators was important for long term truck and sea shipments. Flowers like standard and spray carnations, chrysanthemum with non-hardy stems of full flowers, gerbera, colored bud lily, miniature and floribunda roses are suited for dry transport over a period of several days; while chrysanthemum with hardy stems and single flowered freesias, iris, orchids, narcissus and H.T. roses are unsuited for dry transportation over a long period. Other than tropical flowers, the best method of transport of most of the cut flowers including temperate orchids is under refrigeration from the grower to final consumers.

CONCLUSION

The valuable orchid genera which are highly priced in the international flower trade due to their incredible range of diversity in size, colour, shape, forms, appearance and long lasting qualities of flowers include Cymbidium, Cattleya, Dendrobium, Phalaenopsis, Vanda and Paphiopedilum.

Post –harvest life of orchid cut flowers is influenced by pre-harvest factors like varietal differences, light intensity, sugar level of flowers, temperature and water loss. It is also affected by harvest factors such as time and stage of harvest and post-harvest factors viz. ethylene production, pre-cooling, pulsing, use of preservatives, packaging and storage. Vase life of cut flowers ranges from 18 to 56 days depending upon types of species and hybrids. The optimum harvesting stage of commercial orchids is fully open and mature flowers. Ethylene is the main factor responsible for early senescence. De-capped and emasculated flowers produce more ethylene. There is an obvious increase in flower sensitivity to ethylene following pollination. Precooling of orchids is done by placing the flower boxes in the storage room or in an air conditioned room at 7-10°C in Cattleya; 0.5 to 4°C in Cymbidium and Paphiopedilum and 5-7°C in Dendrobium. In Cymbidium hybrid 'Red Princess' pulsing with 5% sucrose increases vase life upto 56 days followed by sucrose @ 8% (54.78 days). Pulsing with 4 mM STS for 10 mins in Aranda and 0.5 mM STS for 24 hours in Phalaenopsis blocks the deleterious effect of ethylene.

In tropical orchids like Dendrobium and Oncidium, $AgNO_3$ (10- 30 ppm) and HQS (50-100 ppm) extends vase life of cut flowers. In Cymbidium hybrid, 'Red Princess', 75% open flowers with 200 ppm 8-HQS showed highest vase life along with cent percent opening. Acetyl salicylic acid + sucrose or 8-HQS + $AgNO_3$ wss found beneficial on bud opening of Oncidium flowers. In Dendrobium, 'Pompadour', 8-HQS or $AgNO_3$ (50 ppm) gave a high percentage of bud opening. In Cymbidium 'Ensikhan' 4% sucrose + 100 ppm salicylic acid and 4% sucrose + 100 ppm

$Al_2(SO_4)_3$ opened maximum number of buds and maximum longevity. In orchids, cut spikes are inserted in tube containing water or water with preservatives or simply wrapped in wet cotton swabs and then covered with a piece of plastic and tied with rubber band. Flower spikes are then bunched into bunch of 5 or 10 or so and bunch or individual spikes are placed inside the CFB box in alternate fashion. Cool growing orchids like Cymbidiums are stored at lower temperature even at 5°C in cold chambers whereas tropical and subtropical orchids are stored at 7-10°C and 90-95% relative humidity.

Future Research Prospects

- Development of pre-harvest, harvest and post-harvest technologies of major commercially grown orchids for specific target groups like domestic and export market and hybrid/ variety specific technologies.
- Development of packaging for marketing of commercially important fresh and dried produce using locally available materials.
- Development of orchid based technologies for dry flowers and floral arrangement.
- Use of orchid waste for production of phytochemicals including pigments, food, feed, herbal medicines and essential oils.
- Patenting technologies related to species

REFERENCES

1. Chowdhery HJ (2001) Orchid diversity in north-east India. J Orchid Soc India 15: 1-17.
2. De LC, Debnath NG (2011) Vision 2030. NRC for Orchids, Pakyong, Sikkim.
3. Chowdhery HJ (2009) Orchid diversity in northeastern states of India. J Orchid Soc India 23: 19-42.
4. Singh, Foza(1990) Indian orchids. Indian Horticulture 35: 14-15.
5. SarkarIndrajit, MandalT, Naveen Kumar P, Kumar Rajiv , MisraSanyat, et al.(2009) Temperate orchids. AICRP on Floriculture, Technical Bulletin No.28, IARI, Pusa, New Delhi.
6. Bhattacharjee SK, De LC(2005) Post-harvest Technology of Flowers and Ornamental Plants. Pointer Publishers, Jaipur, Rajasthan.
7. Evans RY, Reid MS (1986) Control of petal expansion during diurnal opening of roses. J American SocHort Sci 111: 55-63.
8. Eze JMO, Mayak S, Thompson JE, Dumbroff EB (1986) Senescence in cut carnation flowers, temporal and physiological relationship among water status, ethylene, abscissic acid and membrane permeability. Physiol Plant 68: 323-328.
9. Bieleski RL, Reid MS (1992) Physiological changes accompanying senescence in the ephemeral daylily flower. Plant Physiol 98: 1042-1049.
10. Van Doorn WG (1997) Water relation of cut flowers. Hort Review 18: 1-85.

11. Halevy AH (1986) Flower senescence. In: Processsand Control of Plant senescence. (LeshamYY, HalevyAH and Frenke Cedn), Amsterdam the Netherland: Elsevier.
12. Lindner RC (1946) Studies on packaging and storage of Vanda (Joaquim) flowers. HowaiiAgricExp Stat Progress Notes 49: 1-5.
13. Fischer (1950) Presence of ethylene gas poses problems in cut-flower storage. Florists Rev 106: 31-32.
14. Akamine EK, Goo T(1981) Controlling premature fading in Vanda 'Miss Joaquim flowers with potassium permanganate. University of Hawaii, College of Tropical Agriculture and Human Resources.
15. Arditti J, Flick BH, Jeffrey D (1971) Post-pollination phenomena in orchid flowers-II induction of symptoms by abscissic acid and its interaction with auxin, gibberellic acid and kinetin. New Phytol 70: 333-341.
16. Arditti J, Hogan NM, Chaelwick AV (1973) Post-pollination flowers in orchid flowers-IV, Effects of ethylene. Amer J Bot 60: 883-888.
17. O'Neill SD (1997) Pollination Regulation Of Flower Development. Annu Rev Plant Physiol Plant MolBiol 48: 547-574.
18. Ketsa S, Rugkong A (1999) Senescence of Dendrobium 'Pompadour' flowers following pollination. Journal of Horticultural Science& Biotechnology 74: 608-613.
19. Sheehan (1954) Respiration of cut flowers of Cattleyamossiae. Amer Orchid Soc Bull 23: 241-246.
20. Da Silva, JA Teixeira, (2003)The Cut Flower: Postharvest Considerations. Journal ofBiological Sciences 3: 406-442.
21. Annual Report (2008) NRC for Orchids, Pakyong, Sikkim.
22. ten Have A, Woltering EJ (1997) Ethylene biosynthetic genes are differentially expressed during carnation (Dianthus caryophyllus L.) flower senescence. Plant MolBiol 34: 89-97.
23. Shibuya K1, Yoshioka T, Hashiba T, Satoh S (2000) Role of the gynoecium in natural senescence of carnation (Dianthus caryophy llus L.) flowers. J Exp Bot 51: 2067-2073.
24. De LC, Bhattacharjee SK (2000) Methods for prolonging vase life of cut flowers-a review. Orissa Journal of Horticulture 28: 73-78.
25. De LC, Bhattacharjee SK (2002) Vase life of cut rose 'Queen Elizabeth' as affected by aquatic fungi and bacteria. Indian Rose Annual, 18: 86-90.
26. Bhattacharjee SK (1997) Packaging fresh cut flowers. Indian Horticulture 41:23-27.
27. Hew CS, Wee KH, Lee FY (1987) Factors affecting the longevity of cut Aranda flowers. ActaHorticulturae 205: 195-202.
28. Hew CS, Yong JWH (2004) The Physiology of Tropical Orchids in relation to industry. (2ndedn), World Scientific publishing Co.
29. Ong HT, Lim LL (1983) Use of silver nitrate and citric acid to improve shelf life of Oncidium 'Golden Showers' flowers. Orchid Rev 90: 264-266.
30. Jomy TG, SabinaGT (2002) Effect of conditioning and pulsing on vase life of Dendrobium Sonia inflorescences. Journal of Ornamental Horticulture 5: 80-81.
31. Ketsa S, Uthairatanakij A, Prayurawong A (2001) Senescence of diploid and tetraploid cut inflorescences of Dendrobium 'Caesar'. ScientiaHort 91: 133-141.
32. Jawaharlal M, Dinesh Babu M, IndhumathiK (2006) Postharvest packaging techniques for Dendrobium Hybrid Sonia -17. Journal of Ornamental Horticulture 9: 16-19.

CHAPTER 2

Overview of Growth Pattern and Post-Harvest Techniques for Different Species

Habeeb Pasha*

Department of muthayammal college of Arts and Science, Periyar University, Tamil Nadu, India

ABSTRACT

The world is surrounded with the 3/4th of the water, and one fourth with the land, even though fish production and maintenance has been ignored and this leads to the decrease in the preservation and increase in the mortality rate of the fish. Aquaculture production and research has been increased for the past few years due to unavailability of food in few countries.

As the human population has increased, the extensive research on fisheries started few years to fill the gaps.The gradual increase in the fish industries lead to development of present and future generations. Though the production of consumption and development still lack in few countries due to unawareness of the fish importance and its nutritive value. Fish benefits and the easy way of cultivation can increase the rural development livelihood.

The research in the area of the length weight growth pattern of fish varies in different countries, shown to determine the fish biology, growth and health status of the fishes. A wide variety of the fishes with variations in color, size, structure, growth and life span is still under research and many are unknown based on the growing pattern and adaptation to the environmental conditions, abiotic and biotic factors makes the fishes to grow and survive.

Keywords: Fish, Length Weight Relationship, Preservation, Fish technology, Sperms

INTRODUCTION

In the similar way, to understand the length and weight relationship in fishes can determine the status of fishes fatness and its activity [1,2]. According to the [3,4], helps to determine the conditions of the fishes in Mullet species, whereas according to the Ndimela [5-9], length weight relationship showed nine negative allometric growth pattern from lagos. The six economic fishes showed negative allometric growth pattern in ibeshe waterside in Nigeria reported by Joseph Basmidele Bolarinwa [10-16].

The species of P. daniconius, correlation coefficient found to be high indicating the healthy conditions of fish species [17-21] in upper region of Assam. Whereas growth pattern of length weight relationship for both male and female agrees with isometric growth formula [22-28]. In some fresh water fishes LWR can be determined based on locality, habitat, diet etc according to Esmaeili and Ebrahimi 2006 [29-32].

According to the reservoir habitat, the changes is in the LWR might change with respective to years [33-36]. In species L. senegalensis, the isometric growth shows same results, whereas in L.coubie (male) shown isometric growth compared to female which had negative allometric growth [37-43].

The heavy metals accumulation in the water not only affected the fishes but also human kind. The significant changes occurred with the accumulation of heavy metals with respect to length and weight [44,45].The Hg concentration affected the health condition of Limanda limanda was observed in the water [46-51].The changes in habitat of P.squamosissimus and cichla monoculus to brazil reservoir found the species healthy growth and reproduction in all seasons [52-57]. Fisheries growth can be misleading with the characters including type of sexuality, sex maturity[58-66].

Different ways for Preservation of fishes had different effect on fish

Fish processing and preservation technique includes methods like drying under the sun; salting and chemicals usage etc to aviod the post-harvest losses [67-72].These technologies were helpful to increase the profit and development of aquaculture and to produce the quality of fish according to

Abdi T.G [73]. But according to Sarkar UK, et al. the concept of conservation of aquatic resource and regulating biotic assessment helps in fish protection [74]. Another way of cryopreservation of the fish seed production in Nigeria catfish aquaculture [75]. Gamate preservation for reproduction for the sterile fishes, future prospects to increase young ones and their fertility helps maintaining fish population [76].

To reduce the post-harvest losses the offshore transport of fish increases fisheries value globally [77].The other way to improve the storage of fish by the application of grape and papaya extract in fishes like indian Mackrel has been benificial method [78]. Fermented fish (Raestur Fiskur) in faroe island is significant preservation technique [79]. The storage of the sperms and eggs for different species seems convenient method of preservation to maintain fish fertility [80-88]. Bacteriophage cocktail can be used to preserve smoked salmon [89-99].Vitrification of zebrafish (Marques Ls et al.) found no significant changes [100]. Hence, the length and weight relationship differs among the fish species according to physiological, morphological and external condition, that may include the body shape, fish feed, aquatic environment, fish maturity, sex and seasons.

This review tries to give an overview of situations in this area of research, identifying the challenges and redirecting the reader to more in depth review of papers.

REFERENCES

1. Offem BO et al. Length-weight relationship, condition factor and sex ratio of forty six important fishes in a tropical flood river. Research journal of fisheries and hydrobiology 2009;4:65-72
2. Abowei JFN. The Condition Factor, Length–Weight Relationship and Abundance of Ilishaafricana Block, 1795 from Nkoro River Niger Delta, Nigeria. Advance Journal of Food Science and Technology 2010;2:6-11
3. Adeyemi SO. Length- Weight, Length-Length Relationship and Condition Factor of Synodontisresupinatus At Idah Area of River Niger, Nigeria. 2010;6:85-90
4. Zubia M et al. Length-Weight Relationship, Condition and Relative Condition Factor of Four Mugilid Species Family mugildae from the Karachi Coast of Pakistan. J Coast Dev 2014;17:385
5. Kumolu-Johnson CA, Ndimele PE. Length-Weight relationships and condition factors of twenty-one fish species in Ologe lagoon, Lagos, Nigeria. Asian. Journal of Agricultural Sciences 2010; 2: 174-179
6. Ming-Ming Han et al. Integrins Contributes to Innate Immune Response in Pelteobagrus Fulvidraco.Biochemistry and Physiology 2016
7. Tainá B Andreoli et al. Comb Grouper Mycteroperca acutirostris Information from Catches at Copacabana, Rio de Janeiro, Brazil.J Marine Sci Res Dev 2016;6:200

8. Hema K.Functional Properties of Restructured Surimi Gel Product Prepared from Low Valued Short Nose White Tripod Fish Triacanthus brevirosterus.J Food Process Technol 2016;7:597
9. Waleed M M El-Sayed.Evaluation of Bioethanol Production from Ulva lactuca By Saccharomyces cerevisiae.JBiotechnol Biomater 2016;6:226
10. Bolarinwa JB and Popoola Q. Length-Weight Relationships of Some Economic Fishes of Ibeshe Waterside, Lagos Lagoon, Nigeria. J Aquac Res Development 2013;5:203.
11. Alpina Begossi et al. Collaborative Research on Dusky Grouper Epinephelus marginatus: Catches from the Small-Scale Fishery of Copacabana Beach, Rio De Janeiro, Brazil.J Coast Zone Manag2016
12. Kyuji Watanabe. In-season Forecast of Chum Salmon Return Using Smoothing Spline. Fish Aquac J 2016;7:173
13. Sandipan Gupta and Samir Banerjee. Food, Feeding Habit and Reproductive Biology of Tire-track Spiny Eel Mastacembelus armatus: A Review.J Aquac Res Development 2016;7:429
14. Agmy Beruat.Status of Seagrass Community in Coastal Area in the Kei Besar District of North-East, South-East Maluku Regency.J Aquac Res Development 2016;7:426
15. Assefa Mitike Janko.Organizational Analysis in Value Chain Approach: The Integrated Organizational Model IOM.Neonat Pediatr Med2016
16. Bedaso Kebede et al. Review on Current Status of Vaccines against Parasitic Diseases of Animals.J Veterinar Sci Technol 2015;7:327
17. Dakua Sanjay. Length-Weight relationship and Condition factor of Parluciosoma daniconius Hamilton from the upper Assam, India. J fisheriessciences.com 2015
18. Mohamed S Hassaan and Magdy A Soltan. Evaluation of Essential Oil of Fennel and Garlic Separately or Combined with Bacillus licheniformis on the Growth, Feeding Behaviour, Hemato-biochemical Indices of Oreochromis niloticus L. Fry.J Aquac Res Development2016;7:422
19. Nwabunike MO.Seasonal Variation in Physicochemical Parameters on Fisheries of Ebonyi River System.J Fisheries Livest Prod 2016;4:169
20. Al-Hammady MAM and Mohamed MH.Distribution and Disease Prevalence of Coral Associated Bacteria at Some Impacted Red Sea Reefs.J Biodivers Endanger Species 2016;4:158
21. Ghorab MA and Khalil MS. The Effect of Pesticides Pollution on Our Life and Environment.J Pollut Eff Cont 2016;4:159
22. B Chandrika and N K Balasubramonian. Length-weight relationship of Xenentodon cancila Ham. Teleostei: Belonidae April 1986;95:187-190
23. Pradeepkiran Jangampalli Adi and Ã¯Â»Â¿Bhaskar Matcha.Environmental Acidification Impact on Fisheries by Changing Oxidative Markers of Liver and Intestine of Freshwater Fish Cyprinus Carpio.L.Poult Fish Wildl Sci 2016;4:1
24. Mukunda Goswami et al. Bio-banking: An Emerging Approach for Conservation of Fish Germplasm.Poult Fish Wildl Sci 2016;4:1
25. Abdel Moneim Yones M and Atallah Metwalli A. Influence of Dietary Sorghum Starch on Growth Performance, Digestibility Coefficient and some Hepatic Enzyme Activities in Hybrid Red Tilapia Oreochromis mossambicus ÃƒÂ-Oreochromis niloticus Fingerlings. Fish Aquac J 2016;7:162
26. Rohit Kumar et al. Pathological findings of experimental Aeromonas hydrophila infection in Golden Mahseer Tor putitora.Fish Aquac J 2016;7:160
27. Sandipan Gupta2016Pangasius pangasius Hamilton, 1822, A Threatened Fish of Indian Subcontinent.J Aquac Res Development 2016;7:2
28. Ming-Ming Han et al. Microfibrillar-Associated Protein 4 MFAP4 Genes in Pelteobagrus fulvidraco play a Novel Role in Innate Immune Response.Biochem Physiol 2016;5:198
29. Esmaeili H.R and Ebrahimi M. Length–weight relationships of some freshwater fishes of Iran. Journal ofApplied Ichthyology 2006;22:328-329

30. Asimi OA and Sahu NP. Effect of Antioxidant Rich Spices, Clove and Cardamom Extracts on the Metabolic Enzyme Activity of Labeo rohita.J Fisheries Livest Prod 2016;4:157
31. Stamatopoulos C and Abdallah M. Standardization of Fishing Effort in Qatar Fisheries: Methodology and Case Studies.J Marine Sci Res Dev 2015;5:170
32. Syandri H et al. Social Status of the Fish-farmers of Floating-net-cages in Lake Maninjau, Indonesia.J Aquac Res Development 2015;7:1
33. Şerife Gülsün Kırankaya, F. Güler Ekmekçi, Şükran Yalçın-Özdilek,Baran Yoğurtçuoğlu, Lale Gençoğlu,Condition, Length-Weight And Lengthlength Relationships For Five Fish Species From Hirfanli Reservoir, Turkey. J FisheriesSciences.com 2014;8:208-213
34. Ellaine C Jose et al. Zooplankton Composition and Abundance and its Relationship with Physico-chemical Parameters in Manila Bay.Oceanography 2015;3:136
35. Netti Aryani and Indra Suharman. Effect of Dietary Protein Level on the Reproductive Performance of Female of Green Catfish Hemibagrus nemurus Bagridae.J Aquac Res Development 2015;6:11
36. Elango J et al. Mechanical and Barrier Properties of Multi-Composite Shark Catfish Pangasius fungaseous Skin Gelatin Films with the Addition of Sorbitol, Clay and Chitosan Using Response Surface Methodology.J Mol Genet Med 2015;S4:004
37. Olufeagba SO et al. Some Aspects of the Biology of Labeo Coubie Ruppell, 1832 and Labeo senegalensis Valenciennes, 1842 from Lower River Benue. journal of fisheriessciences.com 2016
38. Hamouda A et al. Acoustic Survey along Heraklieon and East Canopus Ancient Greek Cities, Abu Quir Bay, Alexandria, Egypt.J Earth Sci Clim Change 2015;6:289
39. Machkevskyi Volodymyr K. Marine Parasites as an Object and as a Factor in the Problem of Invasive Species in Marine Ecosystems: Reflections on the Topic.J Biodivers Endanger Species 2015;3:154
40. Md. Nasir Uddin Miah et al. Present Status of Coastal Fisheries in Sitakunda Coast with Special Reference on Climate Change and Fish Catch.J Aquac Res Development 2015;6:9
41. Om AD et al. Molecular Characteristic of Giant Grouper Epinephelus LanceolatusVitellogenin. J Aquac Res Development 2015;6:9
42. Ramachandramohan M and Mamatha P. Impact of Biopesticde Neem Oil for Beneficial to Fisheries Resources, Studies on Skin with Neem Oil Exposure to Fresh Water Fish G. Giuris.J Fisheries Livest Prod 2015;3:132
43. Seinen Chow et al. Universal Primers for Exon-Priming Intron-Crossing EPIC PCR on Ribosomal Protein Genes in Marine Animals.J Marine Sci Res Dev 2015;5:160
44. Eroğlu M et al. Some heavy metals in the muscle of Capoeta trutta: risk assessment for the consumers. Cell Mol Biol Noisy-le-grand. 2016;62:22-26
45. Wei LL et al. [Bioaccumulation and Biomagnification of Heavy Metals in Three Gorges Reservoir and Effect of Biological Factors].Huan Jing Ke Xue. 2016; 3:325-334
46. Lang T et al. Mercury species in dab Limanda limanda from the North Sea, Baltic Sea and Icelandic waters in relation to host-specific variables. Mar Environ Res. 2016
47. Endebu M et al. Fisheries Baseline Survey Describing Status of Fisheries in Lake Zeway, Ethiopia.J Fisheries Livest Prod 2015,3:129
48. Emmanuel HL et al. Amphibian and Benthic Macro Invertebrate Response to Physical and Chemical Properties of Themi River, Arusha, Tanzania 2015
49. Mitchel Andrada. The Dynamics of Philippine Aquaculture.Fish Aquac J 2015;6:e121
50. Ragia Moussa Moussa. Invertebrate Aquaculture.Fish Aquac J 2015;6:130
51. Arabinda Mahanty et al.GC-MS Fingerprinting of Fatty Acids of Freshwater Mollusc Lamellidens Marginalis using Different Columns, TR-Waxms and TR-FAME.J Anal Bioanal Tech 2015;6:238
52. Sousa MM et al. Population structure and reproductive period of two introduced fish species in a Brazilian semiarid region reservoir.Rev Biol Trop 2015;63:727-739

53. Hetisani Chauke and Farai Mudavanhu. An Assessment of the Impacts of the Runde Water Supply on the Life and Business of the Local People. A Case Study of Lundi Business Centre in Mwenezi District, Animal Husbandry, Acoustic Survey, Bioacoustics, Dropline, Fishing Vessel, Fisheries, Gillnet, Jigging, Livestock Production, Marine, Marine Fish, Maritime Policy, Poultry, Sustainable Fishing, Trawling, Pelagic Fish, Livestock Nutrition, Biological Diversity, Fisheries Management, Sustainable fishery.J Fisheries Livest Prod 2014;3:127
54. Muhammad Arifin Dahlan et al. Morphometric And Meristic Comparison Of Decapterus Macrosoma Bleeker, 1851 From Makassar Strait And Bone Bay, South Sulawesi, Indonesia 2014
55. Karl Marx A Quiazon.Updates on Aquatic Parasites in Fisheries: Implications to Food Safety, Food Security and Environmental Protection.J Coast Zone Manag 2015
56. Chhatra M Sharma.Can Bio-Manipulation Be Related to Fisheries and Aquaculture through Environmental Pollution Perspective?Fish Aquac J 2015;6:e116
57. Ashraful MAK et al. Monitoring And Assessment Of Heavy Metals Levels In Littoral Sediments From The North Eastern Part Of The Bay Of Bengal Coast, Bangladesh 2009
58. Renán X et al. Growth stanzas in an Epinephelidae-Lutjanidae complex: considerations to length-weight relationships Rev Biol Trop 2015;63:175-187
59. Athanassios C Tsikliras. Fisheries Mismanagement in the Mediterranean: A Greek Tragedy.Fish Aquac J
60. Mitchel Andrada.Circles of Sustainability: A Good Market Value.Fish Aquac J 2014
61. Seth Mensah Abobi and Elliot Haruna Alhassa.A Review of Fisheries-Related Human Migration in the Gulf of Guinea.J Coast Zone Manag 2015
62. Ali SM et al. Effect of El-Sail Drain Wastewater on Nile Tilapia Oreochromis niloticus from River Nile at Aswan, Egypt.J Aquac Res Development 2015;6:294
63. Jeya Kumari A. Effect of Essential Oil and Aqueous Extract of Ginger Zingiber Officinale on Oxidative Stability of Fish oil-in-Water Emulsion. J Food Process Technol 2015;6:412
64. Mahapatra BK et al. A Review on Status, Potentials, Threats and Challenges of the Fish Biodiversity of West Bengal.J Biodivers Biopros Dev 2014;2:140
65. Volodymyr K Machkevskyi et al. Marine Parasites of Omani Waters: State of knowledge.J Biodivers Endanger Species 2014;2:137.
66. Assefa Mitike Janko. Fish Production, Consumption and Management in Ethiopia.Int J Econ and Manag Sci 2014
67. Oguntimehin GB.Preservation through the application of energy heat processing, Freezing, drying and radiation. Proceedings of Nigerian Food Journal 1985;2:76-81
68. IgeneJO. Notes on the traditional methods of fish preservation in Nigerian sector of Lake Chad. Proceedings of Fish Society of Nigeria 1983;3:123-130
69. Obodai EA et al. Effect of Fuel wood on the Quality of Smoked Freshwater Fish Species Sold in Tamale Central Market, Northern Region,Ghana. Ethiopian journal of environmental studies and management 2009
70. Abera Degebassa. Solar drying of fish fillet using tent in ziway. Proceedings of the12th annual conference of the Ethiopian Society of Animal Production ESAP held in Addis Ababa, Ethiopia 2004
71. Poulter RG et al. Isohalic sorption isotherms. Use in prediction of storage life of dried salted fish. Journal of food technology 1982;17:201-210
72. Daba Tugie. Challenges and Opportunities of Fishery Resource Development of Man-made water bodies in Oromia regional state, Ethiopia. ESAP proceeding 2008
73. Abdi TG. Participatory Evaluation and Verification of Improved Post Harvest Fishery Technologies on Selected Sites of Oromia water bodies. Fish Aquac J 2014;5:90
74. Sarkar UK et al. A Review on the Fish Communities in the Indian Reservoirs and Enhancement of Fisheries and Aquatic Environment. J Aquac Res Development 2015;6:297
75. Olanrewaju AN et al. Cryopreservation: A Viable Tool for Sustainable Catfish Aquaculture Industry in Nigeria. J Fisheries Livest Prod 2015;3:149

76. Asturiano JF. Progress, challenges and perspectives on fish gamete cryopreservation: A mini-review. Gen Comp Endocrinol 2016
77. Asturiano JF et al. First Production of Larvae Using Cryopreserved Sperm: Effects of Preservation Temperature and Cryopreservation on European Eel Sperm Fertilization Capacity. Reprod Domest Anim. 2016; 51:485-491
78. Stringer C et al. Shifting post production patterns: exploring changes in New Zealand's seafood processing industry. N Z Geog 2011;67:161-173
79. Sofi FR et al. All India Coordinated Research Project on Post-Harvest Technology.Antioxidant and antimicrobial properties of grape and papaya seed extracts and their application on the preservation of Indian mackerel Rastrelliger kanagurta during ice storage.J Food Sci Technol 2016;53:104-117
80. Svanberg I and Ræstur fiskur: air-dried fermented fish the Faroese way.J Ethnobiol Ethnomed 2015;11:76
81. Lee S et al. Production of viable trout offspring derived from frozen whole fish.Sci Rep 2015;5:16045
82. Sanches EG et al. Sperm cryopreservation of lane snapper Lutjanus synagris Linnaeus, 1758.Braz J Biol. 2015;75:662-669
83. Golshahi et al. Motility and fertilizing ability of cryopreserved Caspian brown trout Salmo trutta caspius sperm: Effect of post-thaw storage time and different sperm-to-egg ratios.Cryobiology 2015;71:360-363
84. Dietrich MA et al. Proteomic analysis of extracellular medium of cryopreserved carp Cyprinus carpio L. semen.Comp Biochem Physiol Part D Genomics Proteomics. 2015;15:49-57
85. Desai K et al. Use of methanol as cryoprotectant and its effect on sox genes and proteins in chilled zebrafish embryos.Cryobiology 2015;71:1-11
86. Makwana Nayan P et al. Cryopreservation of rainbow trout spermatozoa Onchorhynchus mykiss using different cryodiluents.Cryo Letters. 2015; 36:137-147
87. Maria AN et al. Use of cryotubes for the cryopreservation of tambaqui fish semen Colossoma macropomum.Cryobiology.2015;70:109-114
88. Yildiz C et al. Effect of cholesterol-loaded cyclodextrin on cryosurvival and fertility of cryopreserved carp Cyprinus carpio sperm.Cryobiology 2015;70:190-194
89. Bernáth G et al. Comparison of two different methods in the cryopreservation of Eurasian perch Perca fluviatilis sperm.Cryobiology. 2015;70:76-78
90. Park DH et al. Changes in salmon Oncorhynchus keta flesh quality following ultra-high pressure treatment and 30 d of chilled storage.J Food Sci 2015;80:M142-146
91. Liu QH et al. Summer flounder Paralichthys dentatus sperm cryopreservation and application in interspecific hybridization with olive flounder P olivaceus.Theriogenology 2015;83:703-710
92. Xu XR et al. Detection of DNA damage caused by cryopreservation using a modified SCGE in large yellow croaker, Pseudosciaena crocea.Acta Biol Hung. 2014;65:405-413.
93. Ginson J et al. Effect of high pressure treatment on microbiological quality of Indian white prawn Fenneropenaeus indicus during chilled storage. Food Microbiol. 2015;46:596-603.
94. Öğretmen F and İnanan BE2.Effect of butylated hydroxytoluene BHT on the cryopreservation of common carp Cyprinus carpio spermatozoa.Anim Reprod Sci. 2014;151:269-274
95. Aramli MS and Nazari RM.Motility and fertility of cryopreserved semen in Persian sturgeon, Acipenser persicus, stored for 30-60 min after thawing.Cryobiology. 2014;69:500-502
96. Kutluyer F et al. Cryopreservation of rainbow trout Oncorhynchus mykiss spermatozoa: effects of extender supplemented with different antioxidants on sperm motility, velocity and fertility.Cryobiology 2014;69:462-6
97. Osipova VP et al. Cryoprotective effect of phosphorous-containing phenolic anti-oxidant for the cryopreservation of beluga sperm.Cryobiology 2014;69:467-472

98. Perera MN. Bacteriophage cocktail significantly reduces or eliminates Listeria monocytogenes contamination on lettuce, apples, cheese, smoked salmon and frozen foods.Food Microbiol 2015;52:42-48

99. Marques LS et al. Viability of zebrafish Danio rerio ovarian follicles after vitrification in a metal container. Cryobiology 2015;71:367-373

100. Desai K et al. Short-term chilled storage of zebrafish Danio rerio embryos in cryoprotectant as an alternative to cryopreservation.Zebrafish 2015;12:111-120

CHAPTER 3

Effects of Harvest and Post-Harvest Processing Methods on Quality of Beche-de-mer in Fiji Islands

Ravinesh Ram[1*], Roveena Vandana Chand[2] and Paul C. Southgate[1]

[1]*Centre for Sustainable Tropical Fisheries and Aquaculture, College of Marine & Environmental Sciences, James Cook University, Townsville, Queensland 4811, Australia*

[2]*School of Biological and Chemical Sciences, Faculty of Science Technology and Environment, University of the South Pacific, Suva, Fiji Islands*

ABSTRACT

The sea cucumber fishery and processing of sea cucumbers into bêche-de-mer (BDM) supports livelihoods and generates significant export income in Fiji. Harvesting and processing of sea cucumbers has been done in Fiji for over two centuries and the processing technique has changed little over this time. Processing of sea cucumbers involves a number of steps: first boiling, slitting and gutting, smoke drying and sun drying. This study investigated the effects of harvesting and post-harvest handling techniques on the quality of BDM produced in Fiji and documented difficulties faced by the BDM processors in producing a good quality product. Our results show that to generate income as soon as possible, sea cucumber fishers in Fiji hastened BDM processing by eradicating or shortening necessary processing steps. This resulted in production of lower quality product with lower value and reduced acceptance. The first boiling and smoking steps of processing were particularly problematic for fishers. A well processed BDM product has good odour, appearance, shape, color and size and has moisture content appropriate for storage without spoilage. Processed BDM sold by fishers to the main exporters in Fiji were generally of low quality because of a number of consistent characteristics that greatly impacted on quality and product value. Common problems with product quality included improper cleaning, contamination with sand, peeling skin resulting from over-boiling and burning resulting from an inappropriate smoking technique. The results of our study have prompted immediate awareness and education activities for sea cucumber fishers and processors in Fiji supporting improvements in product quality and improved income from this important coastal livelihood.

Keywords: Bêche-de-mer; Processing; Quality; Harvest; Postharvest processing

INTRODUCTION

Sea cucumbers have been harvested and utilized commercially for at least 1000 years in the Indo – Pacific region [1]. It is a product that is consumed in many tropical and subtropical countries [2,3] and in Fiji was first exploited for commercial purposes by Chinese exporters in the early 1820's [4]. Sea cucumbers are processed into a dried and nonperishable product known as 'bêche-de-mer' (BDM), and processing of BDM offered a source of income for Fijian communities subsequent to the collapse of the sandalwood trade in the early 1800's [4]. Still today, harvesting and producing appropriately processed and presented BDM offers good income opportunities for fishers in Fiji. Consumption of BDM is considered to be beneficial for human health as it is a good source of animal protein, vitamins, minerals, micronutrients and essential fatty acids [5-8].

Literature on the processing methods used for BDM and their effects on resulting quality is very limited. Processing BDM involves a number of steps including boiling, gutting, smoking and drying [9,10]. Each contributes to the resulting quality of the final product which determines the suitability of the processed product for Asian markets [2,9-11] and its value. Small developing nations such as Fiji are reliant on marine resources as their basis for income generation and livelihoods. Nevertheless, due to a lack of equipment required to optimize the quality of BDM, such as kerosene burners, smoking sheds and drying amenities, BDM is produced in Fiji using simple uncomplicated customary technologies principally through fire smoking and air or sun drying [9,12].

Bêche-de-mer processing entails an uncomplicated sequence of actions to accomplish a product that is non – perishable if stored in dry, dark conditions. Post-harvest steps include first boiling, slitting and gutting, second boiling, smoking and finally sun-drying [2,9,13,14]. Although these steps are uncomplicated, processing requires continuous attention to obtain a standard dry product. Failure to do so can result in reduced quality and value of the final product [9,10].

The main objectives of this study were to investigate the effects of harvesting and post-harvest handling techniques of Fijian communities and the main BDM processing agents on the quality of BDM produced in Fiji, and to identify difficulties faced by the BDM processors in producing a good quality product. The economic impacts of inadequate processing of BDM are detailed a related study (Ram et al., in press) [15].

MATERIALS AND METHODS

Study site

This study was conducted in the Republic of the Fiji Islands. The sites of study were eastern and western Viti Levu, northern and southern Vanua Levu and two islands from outer Lau group (Vanua Balavu and Lakeba Island). The latter two islands have two main BDM buyers who also buy product from other islands in the Lau group (**Figure 1**). On this basis the data from Vanua Balavu and Lakeba Island represents the Lau Group as a whole. Catch analyses and interviews with sea cucumber fishers were done when the fishers visited the main buyers on Vanua Balavu and Lakeba Island to sell their products in either fresh or dry form. However, fishers based on the main islands of Fiji (Viti Levu and Vanua Levu) were visited directly in their villages to collect data on harvesting and processing.

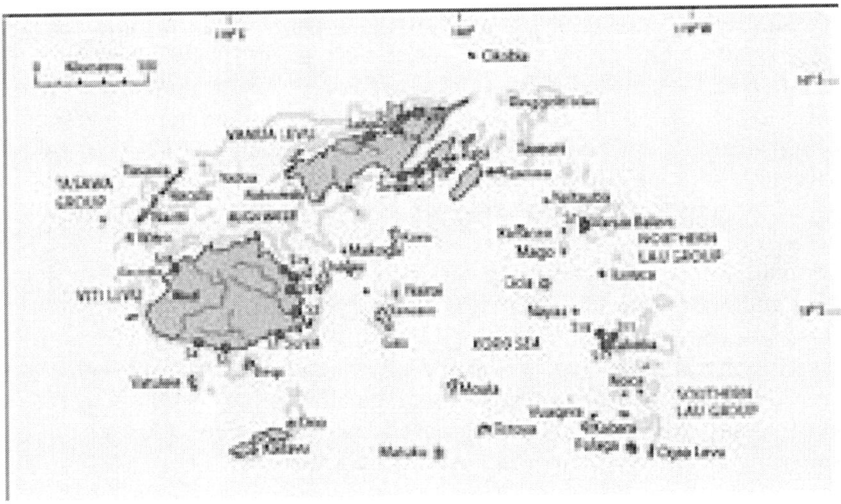

Figure 1: Sites visited in Fiji for collection of data (source: Ram, 2008)

Data collection

Data on processing sea cucumbers was acquired through formal and informal interviews using a structured questionnaire given to fishers at the various study sites around Fiji (n=21; **Figure 1**). The key people targeted for interviews and collection of catch analysis data were artisanal fishers (n=86), middlemen (n=8) who purchase sea cucumbers and process them for a higher return from the exporters, the main agents for BDM exporters (n=5) located in Suva, Fiji and relevant fisheries officers at the Ministry of Fisheries and Forests, Fiji.

The questionnaire was designed to derive information about the species of sea cucumbers that were commonly harvested, total catch of each species harvested, methods used to store sea cucumbers in the sea after harvest, methods used for processing, difficult steps in the processing sequence, and sea cucumber species that were difficult to process or handle. A sea cucumber identification card and pictures of various sea cucumber species were shown to the fishers and buyers to assist with accurate sea cucumber identification [9].

Harvesting and processing of sea cucumber

The daily harvest of the fishers was recorded by species and size. Sea cucumbers were left out of the water for approximately 5 min before length and weight were determined. The length of harvested sea cucumbers was measured using a standard ruler to the nearest 0.1 cm and their weight was measured using an analytical balance to the nearest 0.1 g. Details of the processing techniques (harvesting, boiling, gutting, smoke drying and sun drying) used by the fishers were noted through critically evaluating all steps involved. Measurements (length and weight) of individuals were taken between steps in the processing sequence to determine water loss during processing and to assess whether the processed BDM satisfied Fiji's export size limit of 7.62 cm or 3 inches [16] before shipment. Data on processing methods were analyzed using Logistic regression analysis and a Hosmer and Lemeshow Goodness-of-Fit Test was done to determine the relationships between processing variables (harvesting, smoking, first cook, cutting and gutting) at $P<0.05$. The Hosmer and Lemeshow Goodness-of-Fit Test was performed on the different steps of processing methods such as harvesting, boiling, cutting and gutting, smoking and drying so that the main step of the processing step could be determined leading to major quality losses. All data were expressed as percentages plus or minus standard error and were analyzed using SPSS software.

RESULTS

Problems with harvesting

Collection of all accessible sea cucumbers was done during low tide by communities involved in the BDM trade. A total of 19 species were captured and recorded during this study (**Table 1**). Daily harvests were stored in 20 litre drums that were cut in half. Our results revealed that fishers collected a large number of juvenile sea cucumber that was below the regulated minimum size limit of 7.62 cm (**Table 1**) [16]. The harvested species with the highest proportions of undersized individuals were Curryfish, Greenfish, Lollyfish, Sandfish (currently banned as a target species in Fiji), Curryfish 'D' grade (local name 'Geci' : Stichopus horrens) (no size limit due to its small size), Stonefish, Surf Redfish, Tigerfish and sea cucumber (local name 'Mundra' : Bohadschia marmorata) (**Table 1**). Harvesting of undersized individuals indicates over-exploitation of these sea cucumber species in Fiji. However, the Fisheries Division in Fiji lacks monitoring and a proper management plan for the sea cucumber fishery and this poses a significant threat to the future of Fiji's BDM fishery.

Table 1: The main species of sea cucumbers processed in Fiji and the percentages of each influenced by factors affecting product quality and value.

Common name	Fijian name	Value	% undersized species harvested	% species damaged during harvest	% over cooked	% poorly cutting and gutting	% smoke burnt	% semi dried product	% spoilt
Amberfish	*Dri volavola*	Low	11.2 ± 2.3	14.4 ± 2.6	10.6 ± 2.2	4.3 ± 1.5	8.0 ± 2.0	6.9 ± 1.9	0.0 ± 0.0
Blackfish	*Dri loa*	Medium	12.5 ± 1.4	26.8 ± 1.9	23.0 ± 1.8	50.9 ± 2.1	15.0 ± 1.5	7.9 ± 1.1	17.0 ± 1.6
Black teatfish	*Loaloa*	High	17.4 ± 2.0	14.6 ± 1.9	5.7 ± 1.2	1.7 ± 0.7	7.1 ± 1.4	0.0 ± 0.0	2.3 ± 0.8
Brown sandfish	*Vula*	Low	1.3 ± 0.3	23.1 ± 1.2	35.5 ± 1.4	72.8 ± 1.3	18.9 ± 1.1	31.6 ± 1.3	59.2 ± 1.4
Curryfish	*Lokoloko, Laulevu*	Low – Medium	36.3 ± 1.2	36.0 ± 1.2	30.6 ± 1.1	42.6 ± 1.2	32.4 ± 1.2	26.7 ± 1.1	44.0 ± 1.2
Deepwater redfish	*Dritabua*	Low – Medium	0.0 ± 0.0	13.4 ± 1.9	4.5 ± 1.2	10.5 ± 1.7	10.8 ± 1.8	0.0 ± 0.0	0.0 ± 0.0
Elephants trunkfish	*Dairo ni cokau*	Low	0.0 ± 0.0	8.8 ± 2.2	9.4 ± 2.3	13.1 ± 2.7	8.8 ± 2.2	7.5 ± 2.1	0.0 ± 0.0
Greenfish	*Dri volovolo, Tarase*	Medium - High	46.1 ± 1.1	32.4 ± 1.0	14.5 ± 0.7	11.4 ± 0.7	14.7 ± 0.8	2.1 ± 0.3	18.3 ± 0.8
Lollyfish	*Loliloli*	Low	46.5 ± 1.1	4.7 ± 0.5	14.2 ± 0.8	39.3 ± 1.1	27.6 ± 1.0	29.4 ± 1.0	12.5 ± 0.7
Pinkfish	*Dri damu*	Low	9.0 ± 1.7	0.0 ± 0.0	12.0 ± 1.9	0.0 ± 0.0	19.3 ± 2.3	14.7 ± 2.0	0.0 ± 0.0
Prickly redfish	*Sucudrau*	Medium-High	8.1 ± 1.6	16.3 ± 2.1	0.7 ± 0.5	4.4 ± 1.2	0.0 ± 0.0	6.8 ± 1.5	0.0 ± 0.0
Sandfish **[banned]**	*Dairo*	High	41.2 ± 2.7	0.0 ± 0.0	32.5 ± 2.6	0.0 ± 0.0	45.5 ± 2.8	0.0 ± 0.0	0.0 ± 0.0
Curryfish 'D' grade*	*Laulevu, Geci*	low	100.0 ± 0.0	0.0 ± 0.0	10.7 ± 0.8	0.0 ± 0.0	0.0 ± 0.0	0.0 ± 0.0	0.0 ± 0.0
Stonefish	*Dri vatu*	Medium	34.2 ± 2.3	12.5 ± 1.6	0.0 ± 0.0	2.4 ± 0.7	8.3 ± 1.3	0.5 ± 0.3	0.0 ± 0.0
Surf redfish	*Tarasea*	Medium	21.9 ± 1.9	4.1 ± 0.9	9.0 ± 1.3	16.2 ± 1.7	38.4 ± 2.2	9.4 ± 1.3	3.3 ± 0.8
Tigerfish	*Vula ika*	Low – Medium	32.9 ± 1.5	25.9 ± 1.4	11.8 ± 1.0	38.7 ± 1.6	22.3 ± 1.3	23.9 ± 1.4	19.7 ± 1.3
White teatfish	*sucuwalu*	High	12.6 ± 1.9	45.0 ± 2.8	3.2 ± 1.0	12.9 ± 1.9	4.2 ± 1.1	0.0 ± 0.0	3.9 ± 1.1
Sea cucumber	*Mundra*	Low	31.8 ± 1.2	14.8 ± 0.9	20.1 ± 1.0	16.4 ± 0.9	21.7 ± 1.0	10.3 ± 0.8	12.3 ± 0.8
Snake fish	*Loliloli*	Low	0.0 ± 0.0	0.0 ± 0.0	21.2 ± 2.0	0.0 ± 0.0	6.3 ± 1.2	0.0 ± 0.0	0.0 ± 0.0

*Species with no export size limit of 7.62 cm

During the harvesting of sea cucumbers, villagers also targeted other species for sale or for local consumption including crabs, lobster, clams, fish and other shellfish. With the exception of Sandfish, sea cucumbers usually have a fragile and soft skin that is easily damaged as a result of improper handling [9,10]. Villagers in Fiji tend to mix sea cucumbers with other seafood's after collection and this causes deep scratches in the sea cucumber skin [10] (**Figure 2**). These scratches are sometimes severe and may affect the quality of the product after final drying [9,10,12].

Figure 2: Sea cucumbers damaged during harvesting as a result of mixing with other seafood.

High value species collected by fishers in Fiji are also affected by poor storage in the sea. Instead of being kept separately, White Teatfish is mixed together with other sea cucumber species and usually also with fish and crabs. Proper storage techniques for sea cucumbers have been described previously [9,10], yet our results show that 45% of harvested White Teatfish are damaged due to poor storage (**Table 1** and **Figure 2**). Curryfish (Stichopus hermanni) and Greenfish (S. chloronotos) also showed a high proportion of damage during harvesting (**Table 1**) resulting from skin deterioration (**Figure 3**).

Sea cucumber species with handling issues

Two species of sea cucumber, Curryfish (Stichopus hermanni) and Greenfish (Stichopus cloronotos) were determined to be the most difficult species for handling and storage in the sea by the fishers interviewed. This is because of skin and flesh deteriorated which occurs as soon as they are removed from seawater and mixed with other sea cucumber species (**Figure 3**). The highly fragile skin and flesh of Curryfish and Greenfish discourages fishers from harvesting these two species for processing. This is despite them being ranked as medium to high value species in Fiji and a good revenue earner. Further research on skin and flesh disintegration needs to be done to determine the causes of this problem and to provide a basis for developing more appropriate harvesting and handling methods for these species.

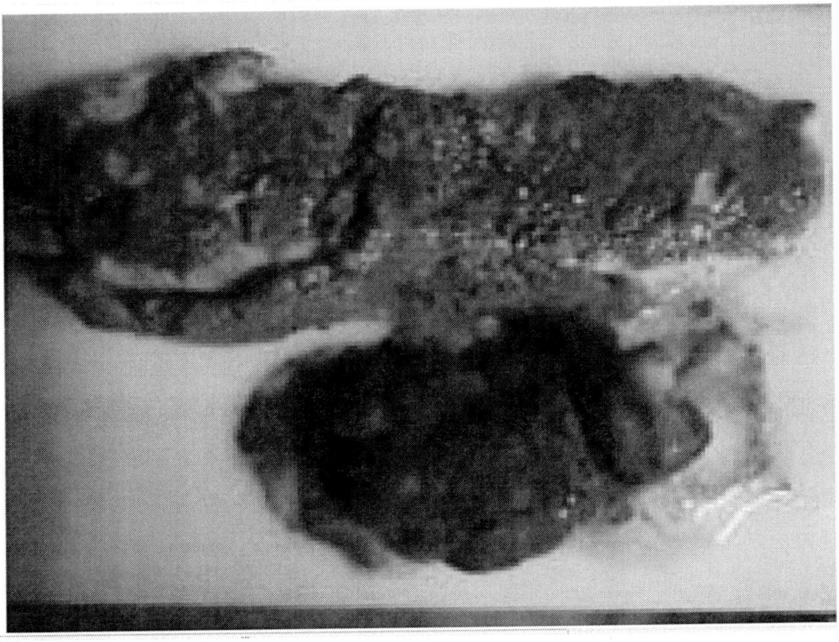

Figure 3: Flesh deterioration observed in Curry fish (Stichopus hermanni) after harvesting

Post harvest processing

Sea cucumber processing is accomplished using two different techniques; the salting and unsalting methods (**Figure 4**). The salting method is generally used by the main agents of bêche-de-mer exporters in Fiji and the middlemen engaged in this trade. The simpler unsalting method is generally used by fishers in Fiji because of their lack of required processing resources. Some major product quality issues relating to production of a 'grade A' BDM are linked to the unsalting method of processing.

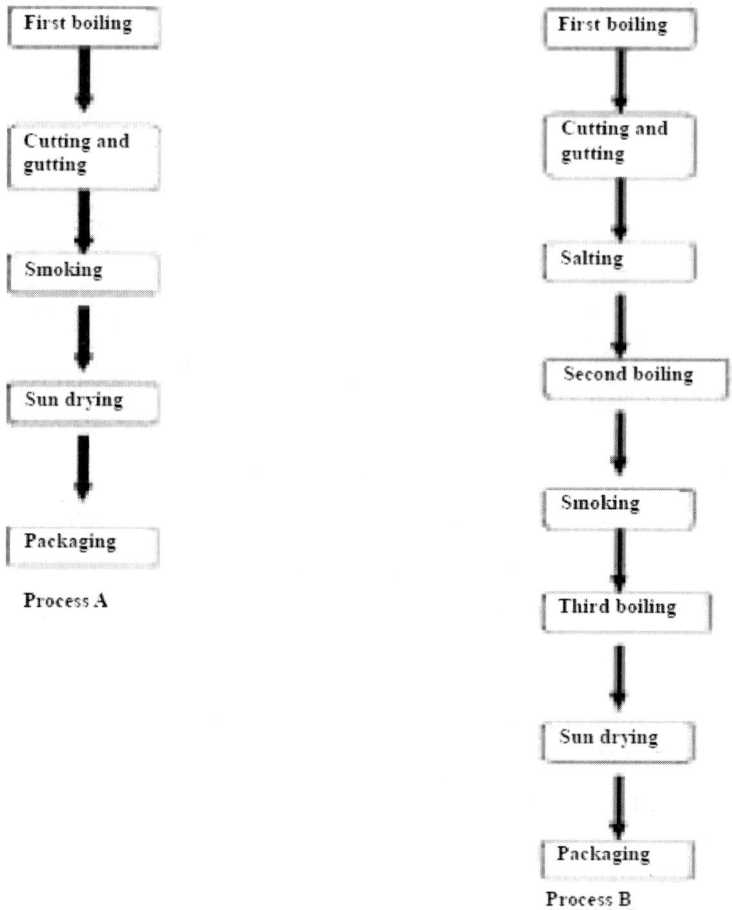

Figure 4: Sea cucumber processing by fishers and the main agents in Fiji Islands. Process A is the 'unsalting' method and process B is the 'salting' method. (Source: Ram, 2008)

Bêche-de-mer processing and problems (post-harvest processing)

Analysis of data using logistic regression analysis and a Hosmer and Lemeshow Goodness-of-Fit Test showed that in addition to first cook ($P<0.05$), cutting and gutting ($P<0.06$) and (smoking) ($P<0.04$) were the major issue leading to loss of BDM quality as a result of poor processing at the fisher's level (**Table 2**). Improperly smoked products have a tendency to deteriorate rapidly due to case hardening. When sold to the marine product agents these products are generally rejected or receive reduced grading with lower value.

Table 2: Logistic regression on likelihood of maturity stages, sex and size of fish

		B	S.E.	Wald	df	Sig.	Exp(B)
Step 1(a)	Smoking	2.132	1.038	4.218	1	.040	8.431
	Harvesting	.690	.426	2.627	1	.105	1.993
	First cook	1.214	.625	3.775	1	.052	3.367
	Cutting and gutting	1.296	.689	3.542	1	.060	3.655
	Constant	-1.652	1.345	1.510	1	.219	.192

The major issue affecting BDM product quality and value in Fiji is that fishers do not follow the standard procedure for processing bêchede- mer [9]. In general they process sea cucumbers with little regard for quality to obtain a quicker source of income [17]. Similar research in Madagascar revealed that, to generate quick income, sea cucumber fishers missed fundamental steps in the processing procedure which badly affected the quality of the final product [18]. If the standard processing steps for BDM are followed, good quality product will result that earns greater income for fishers and communities than poorly processed product [12].

First boiling

Our interviews revealed that 100% of fishers boil all the sea cucumber species of all sizes together without any sorting or grading before processing. Larger sizes of sea cucumber require ample time for cooking during first boil [9,10]. However, attention to this component of processing by Fijian fishers is poor and a high proportion of sea cucumbers are overcooked (**Table 1**). Overcooked sea cucumbers often burst affecting product quality [10].

Inappropriate cooking also results in skin peeling of high value sea cucumber species such as White Teatfish and Black Teatfish, when cooked in boiling water instead of warm water (50–65°C). In addition, Blackfish should be cooked at a high temperature (90–95°C) but not boiled. Introducing commercial sea cucumbers into boiling water results in skin peeling that affects their appearance after drying and reduces value [9,10].

Cutting and gutting

After first boiling sea cucumbers are cut and gutted to remove the viscera. The cutting methods used by fishers in Fiji are generally not of standard quality [9]. Instead of making small cuts in the anal region for species that do not require cuts in the body wall (such as Tigerfish, Stonefish, Deepwater redfish and Brownspotted sandfish) [9], very long cuts are made. These cuts are also generally not straight in the case of high value species. Chinese markets prefer BDM products with neat cuts which enhance their appearance; this is usually not considered during local processing in Fiji. Products may also be incompletely gutted resulting in a product containing gut contents that are contaminated with sand, ash and dust (**Table 1**). The data for cutting and gutting shown for Sandfish in **Table 1** (0%) reflects the fact that this species is generally harvested and sold frozen to the main agents. Processing of Sandfish is currently banned in Fiji and our data are incomplete because the main agents are very secretive regarding Sandfish export data. There could still be significant harvesting of Sandfish in Fiji and, without monitoring and harvesting records, this could have serious consequences for the recovery of Sandfish stocks in Fiji.

Smoking

Smoking is performed after cutting and gutting (**Figure 4**). In the salting procedure (Fiji's current standard procedure), after cutting and gutting, the products are usually salted in grade 11 coarse solar salt for 2-3 days before they are boiled again and smoked (**Figure 4**). Instead of producing a gentle smoke for the smoking step, fishers in Fiji generate fire that either roasts the products (**Figure 5**) or generates a hard skin on the product known as "case – hardening" [19]. "Case – hardening" is described as burnt skin resulting from high temperature that prevents moisture loss from the product resulting in spoilage. The proportion of burnt product resulting from smoke drying during sea cucumber processing by fishers in Fiji is shown for all harvested species in **Table 1**.

Usually the smoking process is done for 24 – 48 hours and the resulting product is sun dried for four to five days to accomplish a 'grade A' product [9,10]. However, to generate rapid income, village processors in Fiji generally perform the smoking and sun drying process together for a day and sell the product to middlemen or the main agents. The products contain 25-35% moisture and due to poor storage, a high proportion of the product may spoil as a result of insufficient drying (**Table 1**). Insufficient drying is a major issue in local processing of BDM in Fiji. This problem results because sea cucumbers are generally piled on top of each other during processing which hinders moisture loss. The products are kept like this for days and this result in spoilage (**Table 1**).

Figure 5: Various examples of smoke burnt sea cucumber species.

Bêche-de-mer quality standard at the main buyer's warehouse before export

The main BDM exporters in Fiji either process sea cucumbers themselves to a marketable standard before exporting to Asian markets, or purchase product processed by local fishers and reprocess prior to export. Reprocessed product is given a low grade and has reduced value. Data generated from the main agent's warehouse revealed that nearly all the bêche-de-mer (all species) exported from Fiji to Asian markets were of poor quality (**Table 3**). Factors that influenced this grading included:

- Undersized (less than 7.62 cm) (**Table 1**)
- Damaged (skin damage, flesh ripped, skin peeled and holes in the tissue)
- Product had poor appearance (shape distorted, gut cavity contaminated with sand, dust and remaining gut contents, poorly cut for gutting) (**Figure 6**)

Table 3: Percentages of poor quality bêche-de-mer products exported by the main buyers in Fiji.

Common name	Grades of product	Percentage undersized species exported	Percentage damaged species exported	Percentage over dried species exported	Percentage discarded as waste
White teatfish	A+	0	4 ± 7.4	0	0
	B	0	17 ± 5.3	0	0
	C	0	26 ± 6.2	0	0
	D	14 ± 3.5	43 ± 5.0	0	0
Black teatfish	A	0	17 ± 5.3	0	0
	B	0	16 ± 5.2	0	0
	C	10 ± 3.0	42 ± 4.9	0	0
Sandfish	Banned	Banned	Banned	Banned	Banned
Prickly redfish	I grade	0	12 ± 4.6	0	0
Stonefish	I grade	82 ± 3.1	0	9.0 ± 2.9	1.3 ± 1.5
Surf Redfish	A	4.0 ± 1.9	12 ± 3.2	0	0
	B	40 ± 5.0	36 ± 4.8	0	1.0 ± 1.3
Blackfish	A+	0	12 ± 4.6	0	0
	B	44 ± 4.9	13 ± 3.4	0	0
	C	86 ± 3.5	29 ± 4.5	7.0 ± 2.6	2.2 ± 1.9
Elephant Trunkfish	I grade	0	0	0	0
Greenfish	1 grade	80 ± 4.0	5 ± 2.2	11 ± 3.1	1.7 ± 1.7
Curryfish	A	0	5 ± 3.1	0	0
	B	12 ± 3.2	10 ± 3.0	3.0 ± 1.7	0
	C	48 ± 5.0	30 ± 4.6	12 ± 3.2	4.0 ± 2.5
	D (Gecr)	100	0	0	0
Amberfish	1 grade	0	15 ± 5.0	0	1.0 ± 1.3
Brown sandfish	1 grade	8.0 ± 2.7	8 ± 2.7	0	0.7 ± 1.1
Deep water redfish	1 grade	0	10 ± 4.2	0	0
Lollyfish	A	36 ± 4.8	0	1.0 ± 1.0	0
	B	80 ± 4.0	0	3.0 ± 1.7	0.5 ± 0.9
Tigerfish	A	0	23 ± 6.0	0	0
	B	16 ± 3.7	44 ± 5.0	2.0 ± 1.4	1.7 ± 1.7
Snake fish	1 grade	17 ± 3.7	0	0	2.5 ± 2.0
Pinkfish	1 grade	26 ± 4.4	4 ± 2.0	0	0
Sea cucumber (*Mundra*)	1 grade	98 ± 1.4	20 ± 4.0	26 ± 4.4	0.8 ± 1.2

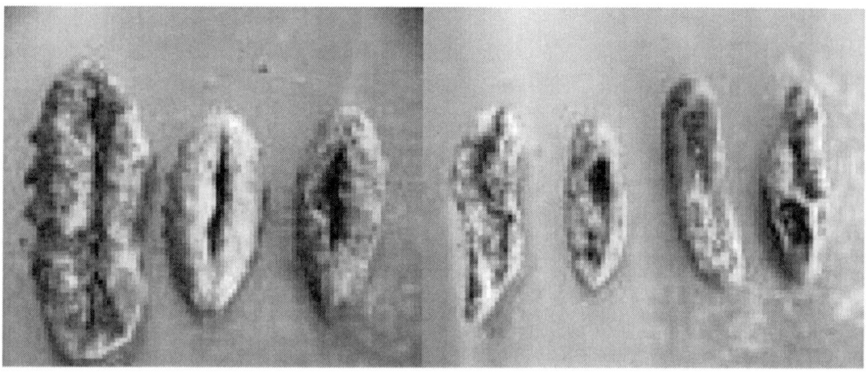

Figure 6: Poorly processed sea cucumbers for export from Fiji

Products were over-dried (**Table 3**) and burnt as a result of smoking at higher temperatures.

When BDM products are traded by village processors to the main agents or exporters, they are sometimes rejected due to extremely bad odour resulting from spoilage, insufficient drying and burning during the smoking process. Buyers however, do purchase the product but when graded by size and quality, extremely poor quality products that cannot be reprocessed for the market are rejected or discarded as waste (**Table 3**). High value species such as White Teatfish, Black Teatfish are not rejected if they are processed by the fishers but instead purchased and reprocessed as a low grade product (e.g. 'Grade D') and exported to Asian markets for low value. Because populations of the high value species of sea cucumbers are declining rapidly in Fiji [20], buyers tend to purchase the product in perishable state from the fishers for little value and process themselves to obtain a higher return from Asian markets (**Table 3**). The results in **Table 3** show that 43% of White Teatfish (a high value species) is exported as damaged product that was processed by fishers. As a result, processors now prevent fishers from processing the higher value species.

The percentages of damaged individuals are high for higher value species (**Table 3**) due to flesh and skin damage caused during harvesting and processing. In addition, high value species are generally purchased raw from the villages and extended exposure of the sea cucumbers to the air or to sun may result death or very poor condition. The skin may also peel as a result of the drying effect from the air or sun [9]. Data for the percentage of damaged individuals shown in **Table 3** also includes smoke burnt products that were processed by village processors.

The proportion of over-dried product for the major exported species is also shown in **Table 3**. Over-dried BDM has reduced nutritive value [21,22] and, especially after over boiling, may have skin and flesh that is fragile and may easily break leading to serious product quality issues [9,10]. Greenfish, Amberfish, Surf Redfish and Curryfish, in particular, show a high proportion of damaged product (**Table 3**) because of the fragility of their skin and flesh and their susceptibility to damage during harvesting.

CONCLUSION

Bêche-de-mer processing in Fiji is generally done rapidly to obtain income as soon as possible. As a result essential processing steps are shortened or missed completely and this has serious consequences for the quality of the final product. Processing steps in Fiji involved first boil after harvest, slitting and gutting, followed by salting (main agents only), smoke drying and finally sun-drying. Fishers found smoke drying and first boiling to be difficult steps during processing and this result in poor quality products being produced in the Fiji. The final dried product generally has poor exterior appearance, distorted or twisted shape, and particulate matter present in the gut cavities and, perhaps of most importance, a high rate of spoilage due to improper storage. These factors reduce the grade of the final product from 'A' to 'B-D' grade. Further research is required into the impacts of processing methods on the quality of BDM produced in the Fiji and to develop ways to add value to BDM exported from Fiji. The results of our study have prompted immediate awareness and education activities for sea cucumber fishers and processors in Fiji supporting improvements in product quality and improved income from this important coastal livelihood.

ACKNOWLEDGEMENTS

The authors wish to thank the main buyers and village people in Fiji for their assistance with data gathering and provision of information. Involvement of officials from the Ministry of Fisheries and Forest of Fiji is also thankfully acknowledged.

REFERENCES

1. Bumrasarinpai R (2006) Overview of issues on concerned aquatic species under the convention on international trade in endangered species of wild fauna and flora (CITES), in Regional technical consultation on international fisheries related issues: WP03.
2. Conand C (1990) The fishing resources of the Pacific Island Countries. Part 2: Holothuria, in Overview of Bêche-de-mer industry in Milne Bay Province, Papua New Guinea, J. Kinch, Editor. SPC Bêche-de-mer information bulletin. p. 2-15.
3. Anon (1994) Sea cucumbers and Bêche-de-mer of the tropical Pacific: a handbook for fishers, no 18, in Bêche-de-mer production in Tuvalu, K. Belhadjali, Editor. 1994, SPC Bêche-de-mer information bulletin. p. 2-3.
4. Ward P (1972) The Pacific Bêche-de-mer with special reference to Fiji. In Pp 2 – 15. , in Bêche-de-mer and sandalwood trade in Fiji. Master's Thesis, M. Dough, Editor. 1972: Macquarie University. Department of History. School of general studies: 2-15.
5. James D (1988) Production, Consumption, and Demand in Fish Drying and Smoking; production and quality, P.E. Doe, Editor Technomic Publishing Co Inc.1-4.
6. Sikorski Z, Haard N, Motohiro T, Pan BS (1998) Quality, in Fish Drying and Smoking; production and quality, P.E. Doe, Editor., Technomic Publishing Co Inc:88-116
7. Ferdouse F (1999) Bêche-de-mer markets and utilisation. SPC Bêche-de-mer information bulletin,11: 3-9.
8. Bordbar S, Anwar F, Saari N (2011) High-value components and bioactives from sea cucumbers for functional foods--a review. Mar Drugs 9: 1761-1805.

9. SPC (1994) Sea cucumbers and beche-de-mer of the tropical pacific. Handbook No. 18 Noumea: SPC. 51.
10. Sachithananthan K, Osman SS, Mlay M, Schoemaker R n (1985) Report on the national workshop on fish handling in Zanzibar, in United Republic of Tanzania and report on the Tanzania/ SWIOP national workshop on Bêche-de-mer processing. 1985, National workshop on fish handling. 45-83.
11. Purcell SW (2014) Value, market preferences and trade of Beche-de-mer from Pacific Island sea cucumbers. PLoS One 9: e95075.
12. Seeto J (1999) Bêche-de-mer processing – a little more effort to get much more money while saving precious resources. SPC Bêche-de-mer information bulletin, 11: 2-3.
13. Schoppe S (2000) Sea cucumber fishery in Philippines. SPC Bêche-de-mer information bulletin, 13:10-12.
14. Jun A (2002) Trepang exploitation in the Philippines: updated information. SPC Bêche-de-mer information bulletin, 17:17-21.
15. Ram R, Chand R.V, Southgate PC (2014) Effects of Processing Methods on the value of bêche-de-mer from the Fiji Islands. Journal of Marine Science: Research and Development ,in press.
16. Bruckner AW, Johnson KA, Field JD (2003) Conservation strategies for sea cucumbers: Can a CITES appendix II listing promote sustainable international trade. SPC Bêche-de-mer information bulletin, 18: 24-33.
17. Kinch J (2002) Overview of Bêche-de-mer industry in Milne Bay Province, Papua New Guinea. SPC Bêche-de-mer information bulletin, 2002. 17:2-15.
18. Irwing MA (1994) Fishing Bêche-de-mer in Madagascar. SPC Bêche-de-mer information bulletin, 6:10-11.
19. Venugopal V (2006) Seafood processing: adding value through quick freezing, retortable packaging, and cook – chilling. United States of America, CRC press, Taylor and Francis group.
20. Pakoa K, Saladrau W, Lalavanua W, Valotu D, Tuinasavusavu I et al, (2013) The status of sea cucumber resources and fisheries management in Fiji. SPC :51.
21. Chen J (2001) Present status and prospects of sea cucumber industry in China, in Advances in sea cucumber aquaculture and management .A. Lovatelli et al, FAO Fisheries Technical Paper. No. 463: Rome; FAO:25-38.
22. Ram R (2008) Impacts of harvest and post-harvest processing methods on quality and value of bêche-de-mer in Fiji Islands. Master of Science Thesis, University of the South Pacific. 191.

CHAPTER 4

Postharvest Technologies of Fresh Horticulture Produce

Alejandro Isabel Luna-Maldonado[1], Clement Vigneault[2] and Kei Nakaji[3]

[1]*Department of Agricultural and Food Engineering, Faculty of Agriculture, Autonomous University of Nuevo Leon, Escobedo, N.L., Mexico*

[2]*Department of Bioresource Engineering, Macdonald Campus, McGill University, Ste-Anne-de-Bellevue, Québec, Canada*

[3]*Kyushu University, Laboratory of Agricultural Ecology, Division of Agricultural Ecology, Department of Plant Resources, Fukuoka, Japan*

1. INTRODUCTION

The actual processing of fresh horticultural produce and the reducing labor force have begun to force the development of robots that are capable of dealing with the variations inherent in the produce being handled, stored and transported. All around the world, there is increasing interest in the use of robots to replace the drastic decreasing number of farmers due to the lower birth rate and the increasing average age of the remaining farmers (Kitamura *et al.*, 2004 & Wang, 2009). In the USA, the supply of workers available for hand harvesting is decreasing steadily and true shortages are occurring. Without enough workers when needed for a few weeks each year, a large amount of hand harvested crops will be lost. Fruit and vegetable crops need new productive harvesting technologies (Sarig *et al.*, 2010)

In general, with horticulture applications, the variations of fresh produce in size and color as well as in the internal structure and defects such as over-ripeness and bruising damage are considered the main barriers to the extension of robotics. Robots have to cope with produce delivered with random orientation and placement, and with environmental issues such as a wide range of humidity and temperature resulting in condensation problems on electronic circuit or dew on sensors of all kinds. It requires manipulator mechanisms, controls and end-manipulators to be designed to interact with the surrounding environment, sensing techniques, mobility, and work cell development. Typical applications where imaging and electro-sensing processes can be used to guide robots include sensing general fruit quality (ElMasry *et al.* 2009; Li *et al.*, 2006 & Lino *et al.*, 2008), measuring internal quality parameters(Chen, 2008 & Li *et al.*, 2009), or detecting mechanical (ElMasry *et al.*, 2008) or physiological(El Masry *et al.*, 2009) damage, for either grading the produce accordingly or involving automatic trimming robots to remove the defect. In other words, processed items are picked from a conveyor belt and presented to a cutting or trimming device to further process the produce in order to maintain a consistency of quality in the finished produce. For example, during the inspection of lettuce heads and their core removal (Key Technology, 2010), a fresh lettuce will be scanned to detect stones and insect damage. At the same time, a two directional measurement device can take images to provide core position and orientation information. Based on these pieces of information, the lettuce will be properly aligned with a blade mounted on a robot to perform the core removal (Applied Sorting Technologies, 2010).

The above example is considered as a typical robot application in horticulture. There is also a totally different type of robot application that is often forgotten due to its relative simplicity which is to fill produce in a standard package unit such as reusable plastic container (RPC) (Vigneault and Émond, 1998). This handling process of horticultural produce is likely comparable to handling any other industrial products such as a TV set or a microwave oven. The main differences are in: the specificity of the environment due to the fact that the horticultural produce is still alive and requires special environmental conditions; the fragility of the produce to mechanical damage and its sensibility to the variations of the environmental conditions; and the length of the time the robot is used for. The particularities of environmental conditions in which the produce are handled and stored require the robot to be able to operate properly at ambient as well as in cold temperatures (close to 0°C) and very high relative humidity (close to 100%), as well as in gas compositions that are hazardous to humans (close to 1% O_2 and up to 20% CO_2) in which the produce is sometimes stored (Leblanc and Vigneault, 2006); as well in a continuous alternating conditions from ambient to any other conditions. In terms of fragility of the produce to mechanical damages, it is easy to understand that horticultural produce are much more sensitive than any industrial products made from metal, plastic or wood. In terms of period of usage, the industry robots are generally used for 12 months/year period, which makes their uses generally economically interesting. However for the horticultural application, each robotic application is of any duration from a few days to the entire year depending on the type of work the robot accomplishes, the produce the robot processes and the produce production period which is produce and localization dependent, and the production system (greenhouse *vs* in-field production, fresh *vs* stored produce, cold climate *vs* subtropical, climacteric *vs* non-climacteric produce, etc). All these horticultural produce particularities must be considered during the design of any robot designated to horticultural applications.

2. HANDLING OF FRESH HORTICULTURAL PRODUCE

Postharvest technologies include the objectives of maintaining the fresh quality of the produce in terms of appearance, texture, flavor, nutritive value, etc, protecting produce, maintaining food safety, and reducing the average losses between harvest and consumption (Saraswathy *et al.,* 2010). The time a produce is exposed to any adverse condition is generally directly proportional to the decrease of quality of any horticultural produce (Leblanc and Vigneault, 2008). Hence, the processing time is one of the most important and controllable factors affecting the quality of the produce. Any delay will generate soon or later important effects on the final quality of the produce (Vigneault *et al.,* 2004 & de Castro *et al.*, 2005) at the marketing step, which may result in a significant loss of marketable value (Leblanc and Vigneault, 2008).

Another controllable factor affecting the quality of the produce is the number of times a product is manipulated or in contact with any fixed or mobile object without being protected (Vigneault et al., 2002 & Vigneault *et al.*, 2009). Packaging systems should be designed for rapid and efficient cooling (Vigneault *et al.,* 2002; Vigneault *et al.,* 2004 b & de Castro *et al.*, 2005b) to prevent physical damage to the produce and to ease handling and storing processes (Vigneault et al., 2009). Finally, storage conditions of refrigerated and controlled atmosphere should be chosen and maintained to retard the deterioration of the perishable produce due to: (1) aging, ripening, softening and color changes (Hoenht *et al.*, 2009); (2) undesirable metabolic changes and respiratory heat production(Leblanc and Vigneault, 2008) ; (3) moisture loss and the resulting wilting (Alvo *et al.,* 2004); (4) spoilage by invasion of bacteria, fungi or yeasts (Kader, 2002); and (5) undesirable physiological processes such as sprouting (Bachmann, 2010).

2.1 Automation and robots for handling of fresh horticultural produce

To maintain the quality of horticultural produce by applying prior knowledge, manufacturers supply a range of solutions based on control, automation and robotics. For example, to reduce processing time a produce is exposed to any adverse conditions, the handling industry proposes to reduce the time between harvest and cold storage by using high speed processing systems such as display container loaders for potato, onion, carrot and other pre-packs, which are capable of obtaining a processing rate up to 80 units of 2.5 kg per minute (Abar Automation, 2011). The automated loaders are generally available for loading 800 × 600 mm footprint containers with options available for the 600 × 400 mm standard RPCs (Abar Automation, 2011). In this process, as all the main movements including package rotation are already carried out in the other industries by a robot, very few conventional technology and mechanical components are required, resulting in a very reliable system. Crate loaders are also available for pre-pack picking and placing systems that load bags, clamshell and other small package units into the 600 × 400 mm North America and Europe standard crates (Vigneault *et al.*, 2009), 400 × 300 mm plastic crates or cases at a processing rate of more than one unit per second (Abar automation, 2009). The robot is housed with a compact space saving structure and space costs, and are fed with bulk or packed produce feeding conveyors. Outer types of packing units such as wooden crates, cardboard boxes, or RPCs are supplied to the system through conveyors either from remote or low level crate de-stackers. The remote de-stacker located on a mezzanine floor or adjacent to the pick-and-place robot separates the packing units from the top of the stack and delivers them over the packing lines. Robot de-stackers and palletizers provide a flexible and relatively low cost solution and high throughput with two or more action spots for single or multiple production lines of pallet loading.

During the picking and placing process in a packinghouse, the multitask operations allow bags, cases, boxes, RPCs and even layer sheets to be placed by a single robotic installation (Abar automation, 2009). Picking and placing is where a robot moves items from one place to another, for example from a conveyor in a production line to another in a packaging line The robot is often guided by a vision system (Piasek, 2010). Pick and place robot cells (Fig. 1) are available for pre-packed bags or trays containing fresh or processed vegetables, salads or mushrooms ranging in weights from a few grams to 50 kg, and are capable of loading plastic crates and roll containers (Abar automation, 2011 & Robots & Robot Controllers Portal, 2010).

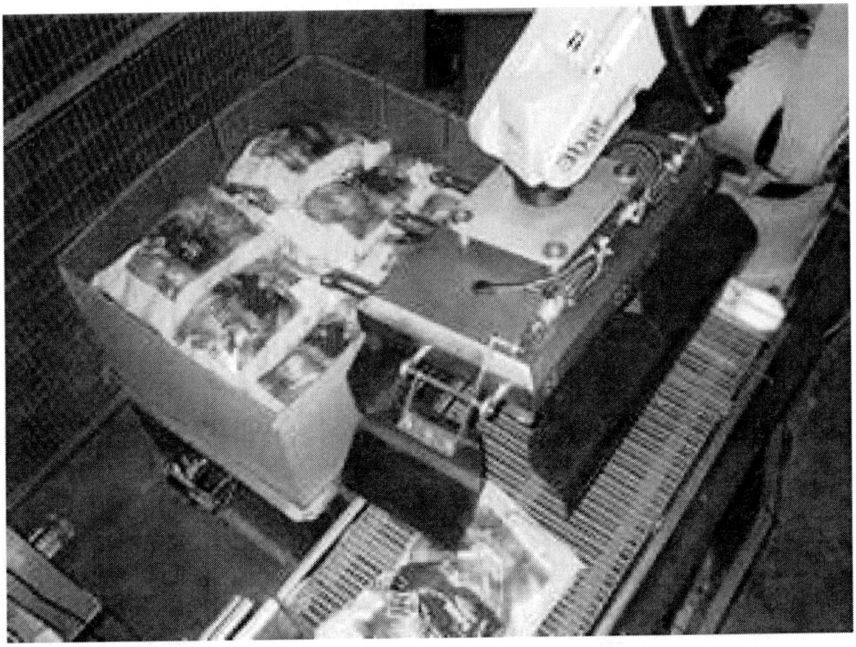

Figure. 1. Pack rotation carry out by an ultra reliable (Courtesy: Abar automation, 2009).

The robotic work cells are becoming more complex. From de-moulding and assembling parts for the car industry to packaging complex objects, the robots are being asked to do faster and more complex moves. This ability to accomplish more sophisticated tasks lies in the capacity to develop more flexible end-of-arm end effecters. A customizable robot has the ability to increase functionality and repeatability as well as allow a higher percentage of automation function within the cell (KUKA Industrial Robots, 2010). For example, the crate sorting system at the Hofbräuhaus Brewery had to deal with different types of beer crates, meaning that the level of mixed packages in the system rose and manual handling became less cost-effective (Raghavan *et al.*, 2005). In addition, the brewery wanted to spare its employees from the heavy physical work of palletizing 2,500 to 3,000 crates per day. The new solution can also be integrated into the existing equipment. The brewery is looking for a customized automation concept. Finally, the system had to be installed in a small space, while at the same time being easily accessible (Raghavan *et al.*, 2005). This brewery example is quite similar in complexity to horticultural packinghouse where robots have to at the same time handle multiple types of fragile and light produce from conveyer to boxes, and heavy boxes to palette with care and precision,

Figure. 2. A robot for vegetable planting (Courtesy: Xinhuanet, Zhu Zheng, 2011)

with a short time to maintain high production rate. Although the similarity in the complexity of the tasks to be accomplished is high which demonstrates the possibility of developing robots for horticulture application, the technologies are not directly applicable to horticultural produce with so many variations. Much modification to the industrial robots is required to adapt these robots for horticultural purposes. On April of this year was presented (Zheng, 2011) in China an intelligent robot (Fig. 2) ables to examine growing conditions such as temperature, humidity, light, detect disease of the vegetables and pick up the ripe ones through identifying the colour.

On the other hand, Dr. Hayashi Shigehiko in the Bio-oriented Technology Research Advancement Institution (Japan) developed the harvesting robot of strawberry in oder t keep the high quality of strawberry freshness (Fig. 3).

Figure. 3. A harvesting robot of strawberry (Courtesy: BRAIN, Shigehiko Hayashi, 2011).

3. STORING OF FRESH HORTICULTURAL PRODUCE

Various storage methods for fresh horticultural produce are used commercially (Liu, 1991 & Li, 2009). The most common and frequently used storage system is the refrigerated storage room (Raghavan *et al.*, 2005). This storage system consists of continuous and uniform control of temperature and humidity inside of well insulated rooms or buildings (Li, 2009). The ambient storage conditions for fresh horticultural produce vary in temperature from 0 to 15° C and in relative humidity from 65 to 98%

[Kader, 2002; Raghavan *et al.*, 2005; Liu, 1991 & Li, 2009). These conditions are generally controlled by mechanical refrigeration [FAO, 2010 & Hollingum, 1999] and electric thermostat and humidistat, or using an electronic automated system based on adjusted enthalpy of the ambient air (Markarian *et al.*, 2006).

The controlled atmosphere (CA) storage frequently used in postharvest technology of fruits and vegetables consists of controls on the concentrations of O_2, CO_2 and N_2, within a perfectly sealed refrigerated storage room [38]. Modified atmosphere (MA) storage or packaging (MAP) is also used for horticultural produce. It consists generally of modifying gas composition of the internal atmosphere of a package unit that could be as small as 200 g (eg, for leafy vegetable) or a large transportable unit such as a pallet size of 250 to 1,000 kg (eg, for strawberry). The gas concentration encountered in CA and MAP are not different generally except the size of the units which is extremely different and use completely different atmosphere generation and controlling systems. The O_2 gas concentration used for CA and MAP is from 21% (ambient air) to 0%, while the CO_2 concentration increases from 0.003% to values sometimes reaching 15 to 20% de-pending on the produce stored. The N_2 is normally used as a filling gas for both MAP and CA technologies (Raghavan et al., 2010). The ethylene (C_2H_4) is sometimes controlled by scrubbing such as potassium permanganate or O_3 (ozone), or by flushing for better quality control of produce responding to it. The control of temperature, humidity and atmospheric composition may be performed manually, but automated systems frequently measuring the gas compositions and performing the adequate adjustment of the gas processing system also exist to achieve more accurate and uniform conditions (Raghavan *et al.*, 2010).

3.1 Robots for storing of fresh horticultural produce

For storage of fresh produce, the robot handling system must be able to operate within the ambient conditions generally encountered in the storage rooms. At the same time, the robots have to be able to work within the conditions encountered outside of the storage room since they are continuously required to handle the produce to and from the storage room. They must then be resistant to heavy ambient humidity condensation due to this continuously changing environment. Besides the humidity condition, the basic principles used for developing industrial storage robots are basically the same for horticultural purposes. The robot must be able to circulate in a specific environment recognising fixed and moving objects consisting of rows and columns of produce, walls, doors, humans and other robots working individually or in collaboration with human and other robots.

There are two great advantages of robots used for horticultural storage purposes. The first is the robot's capacity for managing extremely large quantities of information associated with any specific produce. These pieces of information could be related to the type of produce, the origin, the harvesting date, the processing date, the initial quality, and its eventual destination, etc (Leblanc and Vigneault ,2006). The information could be easily transferred from and to the robots for managing purposes and improvement of efficiency. For example, maintaining the link between the information and the produce is essential for choosing and executing any managing types such as "the first in first out", "consumer preference", "quality based" or "shelf live" objectives. The second advantage is that in a great facility, a robot can be considered as an economical aspect in the handling the produce such as optimisation of space occupancy, organisation of traffic circulation and managing picking order (Piasek, 2010) to maximise handling efficiency and minimise produce waiting time on docks or truck loading process duration.

Another advantage of a robot, which has not been explored yet in horticulture, is the robot's capacity to work in hazardous environment such as CA room or high O_3 concentration atmosphere for C_2H_4 control. In the case of C_2H_4 control, one has to understand that the optimal O_3 level for the best control of C_2H_4 is generally too high for worker safety, which considerably limits the use of this interesting technique. For a CA room, the main problem is the fact that the quality of the produce is highly related to the time delay between harvest and storage and between storage and commercialization. It is frequent to encounter produce that has been harvested at the optimum maturity level but has to wait for days and even weeks before being under CA due to the time required to fill the CA room. The same way, it is very common to encounter CA rooms that were degassed for marketing a first load of produce and had to wait weeks and months before the last load quits the room. An airlock system has been developed for CA apple storage which can maintain the gas composition inside of the room continuously (Vigneault, 2009), reduce the waiting time before and after storage to a few hours only. However, workers have to use autonomous breading systems to enter in the airlock and transfer the produce from and to the CA room, which has to be performed with a very high precaution level. Human life can be in danger if any mistake happens (Vigneault, 2009).

4. TRANSPORTATION OF FRESH HORTICULTURAL PRODUCE

Ground transportation of fresh horticultural produce is usually by trucks and occasionally by railway (Leblanc and Hui, 2005). Overseas transportation is by ship (Tanner and Smale,2005) or plane (Thompson et al., 2004). A limited amount of high-valued produce is sometimes transported overland by air (Vigneault et al., 2009). The basic requirements for conditions during transportation are similar to those needed for storage (Vigneault, 2005), including proper control of temperature and humidity, and adequate ventilation (Hui et al., 2006 & Hui et al., 2008a). In addition to ambient condition require-ment, the produce must be immobilised by proper packaging, stacking, and bracing methods to avoid excessive movement or vibrations (Vigneault et al., 2009) without being detrimental to air circulation to maintain a uniform temperature distribution within a load (Hui et al., 2008b).

4.1 Robots for transportation of fresh horticultural produce

Loading trucks with produce can be done robots with a vision machine and integrated sensors able to detect objects or people obstructing their way and find and follow a specific path (Belforte *et al.*, 2006). For the purpose of this handling associated with transportation, the robots should be compact enough to perform this task within a restricted work area (Garcia Ruiz *et al.*, 2007), robust but able to perform delicate and very precise operations since the load placement within the truck greatly affects the air circulation (Vigneault , 2005; Hui *et al.*, 2006 & Hui *et al.,* 2008a). To perform these tasks, transportation robots coupled with forks and elevator can carry packed industrial products as well as horticultural produce (Logistic Systems Design, 2010).

5. CONCLUSION

Packaging and refrigeration prevent physical damage and decrease the physiological and pathological damages to the produce. Automation and robots are capable of speeding up the postharvest processes, decrease the period during which the perishable produce is exposed to undesirable conditions, and enhancing the facility and the safety level of human tasks for handling fresh horticultural produce. Robot palletisers could provide more than one course at the time and work within multi line production.

Picking and placing robot cells adapted from industrial uses for pre-packed bags, trays or boxes containing fresh fruits or vegetables are becoming more available and able to accomplish complex tasks, but they have to be redesigned to meet with the specific requirements and environmental conditions of horticulture.

6. ACKNOWLEDGMENT

The authors would like to thank to the Mexican, Canadian and Japanese Ministries of Education for the financial support. Moreover, we thank the robot companies and institutions for the photographs.

7. REFERENCES

1. Kitamura S.; Oka K and Takeda F. (2004). Development of Picking Robot in Greenhouse Horticulture. SICE Annual Conference, pp. 3176–3179, ISBN 0780320247 Okayama, Japan
2. Wang M. What will modern agriculture bring to people? Sciences by Experts, Series 3. (2009) China Publishing Group, ISBN/ISSN 9787500084785 pp. 145– 160, Beijing, China
3. Sarig Y, Thompson JF and Brown GK. (2010). Alternatives to Immigrant Labor? The Status of Fruit and Vegetable Harvest Mechanization in the United States. 2000: 29.07.2010. Available from http://www.cis.org/articles/2000/back1200.html
4. ElMasry G, Nassar A, Wang N and Vigneault C. (2008) Spectral methods for measuring quality changes of fresh fruits and vegetables. Stewart Postharvest Review, Vol.4 , No. 4, pp. 1–3.13, ISSN 1745-9656
5. Li Z, Wang N and Vigneault C. (2006) Electronic nose and electronic tongue in food production and processing. Stewart Postharvest Review, Vol.2, No.4, pp. 7.1–7.6., ISSN 1745-9656
6. Lino ACL, Sanchez J and Dalfabro I. M. (2006). Image processing techniques for lemons and tomatoes classification. Bragantia, Vol.3, pp. 85–789, ISSN 0006-8705
7. Chen M. (2008). Non-destructive Measurement of Tomato Quality using Visible and Near- infrared Reflectance Spectroscopy. Master degree thesis. G.S.V. Raghavan and C. Vigneault (Supervisors). Department of Bioresource Engineering. Macdonald Campus, McGill, University. Sainte-Anne de Bellevue, Québec, Canada
8. Li Z, Wang N, Vigneault C and Raghavan GSV. (2009). Ripeness and rot evaluation of 'Tommy Atkins' mango fruit through volatiles detection. Journal of Food Engineering, Vol. 91, pp. 319–324. ISSN 0260-8774
9. ElMasry G, Wang N, Qiao J, Vigneault C and ElSayed A. (2008). Early detection of apple bruises on different background colors using hyperspectral imaging. LWT – Food science and technology, Vol. 41, No.2, pp, 337–345, ISSN 0023-6438
10. ElMasry G, Wang N and Vigneault C. (2009). Detecting Chilling Injury in "Red Delicious" Apple Using Hyperspectral Imaging and Neural Network. Postharvest Biology and Technology, Vol. 52, No.1, pp. 1–8, ISSN 0925-5214
11. Key Technology. Lettuce Core Removal System. (2010). 25.07.2010. Available from http://www.key.net/ products/oncore-lettuce-core-removal-system/default.html Applied Sorting Technologies. (2010). 24.07.2010. Available from http://www. appliedsorting.com.au/ index.php?~iframe1

12. Vigneault C and Émond JP. (1998). Reusable container for the preservation of fresh fruits and vegetables. Agriculture and Agro-Food Canada and Laval University. United States Patent Office, Washington, Washington DC, USA
13. Leblanc D.I. and Vigneault C. (2006). Traceability of environmental conditions for maintaining horticultural produce quality. Stewart Postharvest Review,Vol.2 , No. 2 pp. 4.1–4.10, ISSN 1745-9656
14. Saraswathy S, Preethi TL, Balasubramanyan S, Suresh J, Revathy N and Natarajan S. (2010). Postharvest Management of Horticultural Crops. Agrobios, India
15. Leblanc DI and Vigneault C. (2008). Predicting quality changes of fresh fruits and vegetables during postharvest handling and distribution. Stewart Postharvest Review,Vol.4 , No. 4, pp. 1.1–1.3., ISSN 1745-9656
16. Vigneault C, Gariépy Y, Roussel D and Goyette B. (2004a). The effect of precooling delay on the quality of stored sweet corn. International Journal of Food, Agriculture and Environment,Vol.2, No. 2, pp. 71–73. ISSN (printed) 1459-0255. ISSN (electronic) 1459-0263
17. de Castro LR, Vigneault C, Charles MT and Cortez LAB. (2005a). Effect of cool-ing delay and cold-chain breakage on Santa Clara tomato. International Journal of Food, Agriculture and Environment Vol.3 , No. 1, pp. 49–54. ISSN (printed) 1459-0255. ISSN (electronic) 1459-0263
18. Vigneault C, Bordint MR and Abrahão RF. (2002). Embalagem para frutas e hortaliças. In : Cortez LAB, Honório SL and Moretti CL (Eds). Resfriamento de frutas e hortaliças. Embrapa Informaçã Tecnológica, Brasília, DF, Brazil, ISSN 1678-3921
19. Vigneault C, Thompson J and Wu S. (2009). Designing container for handling fresh horticultural produce. In Benkeblia N (ed). Postharvest Technologies for Horticultural Crops. Research Signpost, India, Vol. 2, pp. 25–47. ISBN 978-81-308-0356-2
20. Vigneault C, Markarian NR, da Silva A and Goyette B. (2004b).Pressure drop during forced-air ventilation of various horticultural produce in containers with different opening configuration. Transaction of the ASAE, Vol.47, No.3, pp. 807– 814, ISSN 0001-2351
21. de Castro LR, Vigneault C and Cortez LAB. (2005b). Effect of container openings and airflow on energy required for forced air cooling of horticultural produce. Canadian Biosystems Engineering, Vol.47, No., pp. 3.1–3.9., ISSN 1492-9058
22. Hoehn E, Prange R and Vigneault C. (2009). Storage technology and applications. In: Yahia EM (ed). Modified and controlled atmospheres for the storage, transportation, and packaging of horticultural commodities. CRC Press (Taylor & Francis Group), pp. 17– 50, ISBN 9781420069570.ISBN10 1420069578
23. Alvo P, Vigneault C, DeEll JR and Gariépy Y. (2004). Texture characteristics of carrots asaffected by storage temperature and duration. International Journal of Food, Agriculture and Environment 2004: 2(2): 33–37, ISBN 9781420069570
24. Kader A (ed). (2002). Postharvest technology of horticultural crops. University of California, 2002: Publication 3311. Davis, CA, USA
25. Bachmann J and Earles R. (2010). Postharvest Handling of Fruits and Vegetables. 27.07.2010. Available from http://attra.ncat.org/attra-pub/postharvest.html.
26. Abar automation. (2011). Pick and Place Robots for Fresh Produce, Netherlands. 18.10.2011. Available from http://www.abar.nl/pick_and_place_fresh_produce.htm
27. Piasek D. (2010). Order Picking: Methods and Equipment for Piece Pick, Case Pick, and Pallet Pick Operations. Logistics Handling, Dematic, Le Dôme, Luxembourg. 28.07.2010.Available from http://www.logisticshandling.com/absolutenm/templates/article-conveyors_sortation.aspx?articleid=36&zoneid=29
28. Robots & Robot Controllers Portal. (2010). Specify Multi-functioning End-of-Arm Tooling for Robotic Workcells. 26.07.2010. Available from http://www.automation.com/robots-robot-controllers/preview-articles/more

29. KUKA Industrial Robots. (2010). Handling of beverage crates with 180 PA robot. 25.07.2010. Available from http://www.kuka-robotics.com/en/solutions/solutions_search/ L_R171_Handling_of_beverage_crates_with_180_PA_robot.htm.
30. Raghavan GSV, Vigneault C, Gariépy Y, Markarian NR and Alvo P. (2005). Refrigerated and controlled/modified atmosphere storage. In Barrett D, Somogyi L and Ramaswamy H (eds). Processing Fruits, Science and technology. 2nd edition. CRC Press. ISBN 9781420069570 ISBN10-1420069578
31. Zheng Zhu (2011). Robots for vegetable planting works at 12th China Int'l Vegetable Sci-tech Fair in Jinan, China's Shandong. 20.04.2011. Available from http://news.xinhuanet.com/english2010/photo/2011-04/20/c_13838241_2.htm
32. Shigehiko H. (2011). Specific Project Research Team (Robot) Bio-oriented Technology Research Advancement Institution. Japan. 30.11.2011 Available from http://brain.naro.affrc.go.jp/iam/Team/iam_team_robo00.html
33. Liu FW. (1991). Precooling of horticultural crops. In: Yahia EM and Higuera IHC. (eds.) Memorias Simposio Nacional Fisiologia y Tecnologia Post-cosecha de Productos Horticolas en Mexico. Noriega Editores, Mexico, pp.235–240.
34. Li FW. (2009). Postharvest Handling in Asia 2. Horticultural Crops Food and Fertilizer 21.12.2009. Available from http://www.agnet.org/library/eb/465b/
35. FAO. (2010). Small-scale post-harvest handling practices. Section 6: Temperature and relative humidity control. 27.07.2010. Available from http://www.fao.org/WAIRdocs/ x5403e/x5403e08.htm.
36. Hollingum J. (1999). Robots in agriculture. Industrial Robot: An International Journal, Vol. 26, No. 6, pp. 438–446, ISSN 0143-991X
37. Markarian NR, Landry JA and Vigneault C. (2006). Development of a model for simulating ambient conditions in fresh fruit and vegetable storage facility. International Journal of Food, Agriculture and Environment, Vol.4, No.1, pp. 34–40, ISSN (printed) 1459- 0255. ISSN (electronic) 1459-0263
38. Raghavan GSV, Gariépy Y and Vigneault C. (2010). Controlled Atmosphere Storage. Encyclopedia of Agriculture, Food, and Biological Engineering, 2nd Edition. Taylor & Francis 2010: (In press).
39. Vigneault C. (2009). La combinaison d'un sas AC et un système de refroidisse-ment à l'eau pour une meilleure gestion de la qualité des pommes. Agriculture and Agri-Food Canada, Saint-Jean-sur-Richelieu, Québec, Canada, 3pp
40. LeBlanc DI and Hui KPC. (2005). Land transportation of fresh fruits and vegetables: An update. Stewart Postharvest Review, pp. 1.4.1–1.4.13. ISSN 1745-9656
41. Tanner D and Smale N. (2005). Sea transportation of fruits and vegetables: An update. Stewart Postharvest Review,pp. 1.1.1–1.1.9, ISSN 1745-9656
42. Thompson JF, Bishop CFH and Brecht PE. (2004). Air transport of perishable products. Agriculture and Natural Resources, University of California, 2004: Publication 21618. 22 pp. Davis, CA, USA
43. Vigneault C, Thompson J, Wu S, Hui KPC and LeBlanc DL. (2009).Transportation of fresh horticultural produce. In Benkeblia N (ed). Postharvest Technologies for Horticultural Crops. Research Signpost, Kerale, India. Vol. 2, pp. 1–24, ISBN 978-81-308-0356-2
44. Vigneault C. (2005). Transport of fruits and vegetables. Stewart Postharvest Review,Vol.1, No.1, pp. 6.1–6.4, ISSN 1745-9656
45. Hui KPC, Raghavan GSV, Vigneault C and de Castro LR. (2006) Evaluation of the air circulation uniformity inside refrigerated semi-trailer transporting fresh horticultural produce. Journal of Food, Agriculture and Environment ,Vol.4, No. 1, pp. 109–114, ISSN 1459-0255
46. Hui KPC, Vigneault C, de Castro LR and Raghavan GSV. (2008a). Effect of different accessories on airflow pattern inside refrigerated semi-trailers transporting fresh produce. Applied Engineering in Agriculture, Vol. 24, No. 3, pp. 337–343, ISSN 0883-8542
47. Hui KPC, Vigneault C, Sotocinal SA, de Castro LR and Raghavan GSV. (2008b). Effects of loading and air bag bracing patterns on correlated relative air distribution inside

refrigerated semi-trailers transporting fresh horticultural produce. Canadian Biosystems Engineering, pp. 3.27–3.35, ISSN 1492-9058

48. Belforte G, Deboli R, Gay P, Piccarolo P and Ricauda Aimonino D. (2006). Robot Design and Testing for Greenhouse Applications. Biosystems Engineering,Vol. 95, No. 3, pp. 309– 321,ISSN 1537-5110

49. García Ruiz MA, Gutiérrez Díaz S, López Hernández HC, Rivera Ibáñez S and Ruiz Tadeo AC. (2007). Estado del arte de la tecnología de robots aplicada a invernaderos state of the art of robot technology applied to greenhouses. Avances de Investigación Agropecuaria, Universidad de Colima, México, Vol. 11, pp. 53–61.

50. Logistic Systems Design. (2010). Handling Systems. 21.07.2010. Available from http:/www2.isye.gatech.edu/~mgoetsch/cali/logistics_systems_design/material_handling_systems/ material_handling_systems.pdf.

CHAPTER 5

Postharvest Handling Practices and Treatment Methods for Tomato Handlers in Developing Countries: A Mini Review

Isaac Kojo Arah,[1] Gerald K. Ahorbo,[1] Etornam Kosi Anku,[1] Ernest Kodzo Kumah,[1] and Harrison Amaglo[2]

[1]*Department of Agro-Enterprise Development, Ho Polytechnic, P.O. Box HP 217, Ho, Ghana*

[2]*Department of Agricultural Engineering, Ho Polytechnic, P.O. Box HP 217, Ho, Ghana*

ABSTRACT

Tomato production has increased in recent years due to the economic and nutritional importance of the crop. This increase is made possible by the numerous research advances made along the entire value chain. However, scientific research has been focussed mainly on production whilst neglecting postharvest issues. Tomato producers have therefore enjoyed good harvests in recent times, though the good harvests of those from developing countries do not translate into profit as most are lost after harvest. From this study, it was revealed that the postharvest quality and shelf life of the fruit in part will depend on some postharvest handling practices and treatments carried out after harvest. Handling practices like harvesting, precooling, cleaning and disinfecting, sorting and grading, packaging, storing, and transportation played an important role in maintaining quality and extending shelf life. Using appropriate postharvest treatments like refrigeration, heat treatment, modified atmosphere packaging (MAP), and 1-methylcyclopropene (1-MCP) and calcium chloride ($CaCl_2$) application was also vital. It was concluded by this study that the quality of the harvested fruit can be maintained and shelf life extended by using appropriate postharvest handling practices and treatment methods. Failure to adhere to these practices will result in a high amount of loss.

1. INTRODUCTION

Tomato (Solanum lycopersicum L.) is one of the most widely cultivated and extensively consumed horticultural crops globally [1]. Although the crop is believed to have been originated from the wild in Peru, Ecuador, and other parts of tropical Americas [2], the nutritional and economic importance of the crop has led to its global production. By weight, tomatoes rank second only to potatoes in global production of all horticultural produce [3]. Tomato can be eaten in various ways and in a countless number of dishes. It can be eaten raw in salads or as an extract or sauce in many dishes and in drinks [4]. Tomato and tomato-based foods provide a wide variety of nutrients and many health-related benefits to the body [5]. Tomato contains higher amounts of lycopene, a type of carotenoid with antioxidant properties [6] which is beneficial in reducing the incidence of some chronic diseases [7] like cancer and many other cardiovascular disorders [8]. In regions where it is being cultivated and consumed, it constitutes a very essential part of the people's diet. The numerous uses of tomatoes can be a contributing factor to its widespread production. Tomato production has increased in recent years to about 163 million tonnes by 2013 [9] due to the economic and nutritional importance of the crop. This increase according to Alonso et al. [10] is made possible by the numerous research advances along the entire value chain. However, scientific research has focussed on production mainly whilst neglecting postharvest issues. Kader [11] reported that less than 5% of resource allocation in agricultural research in developing countries is on postharvest. This means more than 95% of resource allocation is on production. Research in the production aspect of the entire value chain in tomato production has resulted in improved cultivars or varieties which are high yielding and resistant to both diseases and drought. This has made tomato producers enjoy good harvests in recent years; however, the good harvest of many producers in developing countries does not translate into profit as most are lost after harvest. Postharvest loss is a major challenge hampering tomatoes production in most developing countries [12]. Tomato being a perishable crop as a result of its high moisture content has short shelf life of about 48 hours [13] under tropical conditions. Specialised postharvest handling practices and treatment methods are needed in order to extend the shelf life of the crop after harvest. Failure to adhere to these specialised handling practices and treatment methods will result in high amount of loss. Losses of up to 50% can be recorded in tomatoes between the harvesting and consumption stages of the distribution chain in tropical countries [11, 14] which is in line with estimates by Gustavo et al. [15] that between 49 and 80% of all agricultural commodities end up with the consumer whilst the remainder is lost. It is therefore important to know the appropriate handling practices and treatment methods needed for harvested tomatoes in order to reduce postharvest losses thereby increasing profitability for handlers in developing countries. The purpose of this paper therefore is to look at some postharvest handling practices and treatment methods that can be used by handlers of developing countries and how they can affect the postharvest qualities and shelf life of harvested tomatoes.

2. SOME POSTHARVEST HANDLING PRACTICES FOR TOMATOES

Physical handling can have a drastic effect on the postharvest quality and shelf life of most harvested fruits and vegetables. For instance, rough handling during harvesting and after harvesting can cause mechanical injuries which can affect the postharvest quality and shelf life of harvested fruit like tomatoes [16]. It is therefore important to know suitable postharvest handling practices needed to maintain the quality and extend the shelf life of harvested tomatoes for producers in developing countries. Some of the handling practices which include harvesting, precooling, cleaning and disinfecting, sorting and grading, packaging, transportation, and storage are discussed below.

2.1. Harvesting

The physiological maturity of any fruit at harvest has an important effect on postharvest quality of that fruit [17]. Therefore, care must be taken as to when to harvest the fruit in order to attain the best quality. The shelf life of fruits and vegetables is described by postharvest physiologists in three stages: the maturation, ripening, and senescence stages. The maturation stage gives an indication of the fruit being ready for harvest [18]. Tomatoes can be harvested in either matured green, partially ripe, or ripe state. Tomato being a climacteric fruit can be harvested at the matured green state allowing ripening and senescence to occur during the postharvest period of the fruit. Producers targeting distant markets must harvest their tomatoes in a matured green state [19]. Harvesting tomatoes in matured green state will not only give producers ample time to prepare the fruit for the market but also prevent mechanical injuries during harvesting. Unfortunately, most producers from developing countries especially those in Africa harvest tomatoes when they are partially or fully ripened [20]. Fully ripened tomatoes are susceptible to mechanical injuries during harvesting resulting in shorter shelf life [21, 22]. Care must therefore be taken when harvesting tomatoes in ripe state to avoid these injuries which will hasten deterioration. Also, the use of harvesting and packaging containers with sharp edges must be discouraged to prevent bruising and puncturing of the fruits. Harvesting of fruits should be done in either early or late hours of the day to avoid excessive field heat generation. The inability of producers to follow these simple but vital harvesting procedures coupled with some inefficiencies (like lack of ready market and processing facilities) in the entire value chain may explain the reasons why there are lots of losses in tomatoes harvested at fully ripened state in most developing countries. Access to ready market is a big challenge when dealing with highly perishable crops like tomatoes in most developing countries. This challenge can be attributed to many factors but the major factor is the pattern of production resulting in high gluts. In most developing countries, a greater proportion of tomato production is rain fed. This causes high peaks in production during the raining seasons which is always more than the local consumption of the crop. The problem is further compounded by the lack of processing facilities or equipment which can be used to process the crop into a more durable form for later use. Processing tomatoes into a more durable form will also serve as a means of value addition which will finally give a better market value for the crop. Producers from developed countries always have supply contracts with multinational supermarkets to supply tomatoes. For instance, Blush Tomatoes, a tomato producing company in Australia, supplies Coles and Woolworth supermarkets with tomatoes making access to market already predetermined [12]. In the case of producers in many developing countries, there is lack of communication between producers and consumers and also lack of market information [11]. This could be the main reason for the mismatch between production and available markets. Producers therefore have to sell their harvest at very low cost to prevent total loss.

2.2. Precooling after Harvest

Field heat is usually high and undesirable at harvesting stage of many fruits and vegetables and should be removed as quickly as possible before any postharvest handling activity [23]. Excessive field heat gives rise to an undesirable increase in metabolic activity and immediate cooling after harvest is therefore important [24]. Precooling minimises the effect of microbial activity, metabolic activity, respiration rate, and ethylene production [25], whilst reducing the ripening rate, water loss, and decay, thereby preserving quality and extending shelf life of harvested tomatoes [26]. The suitable temperature range of about 13–20°C for tomato handling can be attained either in the early hours of the morning or late in the evening [27, 28]. Harvested fruit must be precooled to remove excessive field heat if harvested at times other than the recommended periods. A cheap but effective way of precooling harvested tomatoes for producers of developing countries can be by dipping fruits in cold water (hydrocooling) mixed with disinfectants such as thiabendazole and sodium hypochlorite if availability of clean water is not a challenge. This method is effective in removing field heat whilst reducing microbial loads on the harvested fruits. Tomato producers in developing countries especially those from Africa assemble their harvested produce under tree shade in an attempt to reduce field heat [29]. Tree shade, however, is not a reliable and effective way of reducing field heat in harvested produce. Arah et al. [12] therefore suggested that the adoption of a simple on-farm structure like a small hut made of thatch can be very beneficial in precooling of harvested tomatoes.

2.3. Cleaning or Disinfecting

Proper hygiene is a major concern to all produce handlers, because of not only postharvest diseases, but also incidence of food-borne illnesses that can be transmitted to consumers. According to a report by the Government of India [30], Salmonella, Cryptosporidium, Cyclospora, and hepatitis A virus are some examples of disease causing organisms that have been transmitted to consumers through fresh fruits and vegetables. Unfortunately, cleaning or disinfecting tomatoes after harvest is not a common practice for most tomatoes handlers in developing countries especially those from Africa. This practice may be attributed to either the unavailability of portable water at the production sites or the sheer ignorance of the practice. However, in places where water is not a constraint, the use of disinfectants in water either for washing or for cooling can reduce both postharvest and food-borne diseases in fruits and vegetables. The use of various disinfectants during postharvest treatment of tomatoes is well documented. For instance, sodium hypochlorite solution has been used to sterilise tomato fruits in order to reduce the incidence of fungal infection before any postharvest treatment was applied [31]. Dipping of tomato fruits in thiabendazole solution reduced the microbial load on the fruits [32]. Fruits and vegetables are usually treated with chlorinated water after washing to reduce the microbial load prior to packaging. Workneh et al. [33] indicate that anolyte water dipping disinfection of tomatoes not only reduced the microbial loads on the fruits but also maintained superior quality of tomatoes during storage. Disinfection can be used in conjunction with hydrocooling to achieve the purpose of reducing excessive field heat and reducing microbial infection at the same time.

2.4. Sorting and Grading

One of the most important processes in packaging and marketing of fruit and vegetables is sorting [34] and grading. Sorting is the removal of rotten, damaged, or diseased fruits from the healthy and clean ones. The damaged or diseased fruits can produce ethylene in substantial amounts which can affect the adjacent fruits [35]. Grading is also the process of categorising fruits and vegetables on the basis of colour, size, stage of maturity, or degree of ripening. The two processes are vital in maintaining postharvest shelf life and quality of harvested tomatoes. Sorting limits the spread of infectious microorganisms from bad fruits to other healthy fruits during postharvest handling of tomatoes. Grading also helps handlers to categorise fruits and vegetables in a given common parameter which enables easy handling. For instance, grading on the basis of colour or maturity stage will help eliminate overripe fruits which will easily produce ethylene to hasten the ripening process in the whole batch. Commercial tomato producers normally use sophisticated systems that require precise sorting and grading standards for their produce. Small-scale producers and retailers in developing countries in contrast may not use written down grading and sorting standards; however, the produce must still be sorted and sized to some degree before selling or processing it.

2.5. Packaging

Packaging is also one of the important aspects to consider in addressing postharvest losses in fruits and vegetables. It is enclosing food produce or product to protect it from mechanical injuries, tampering, and contamination from physical, chemical, and biological sources [36]. Packaging as a postharvest handling practice in tomato production is essential in putting the produce into sizeable portions for easy handling. However, using unsuitable packaging can cause fruit damage resulting in losses [37]. Some common packaging materials used in most developing countries include wooden crates, cardboard boxes, woven palm baskets, plastic crates, nylon sacks, jute sacks, and polythene bags [37]. Most of the abovementioned packaging materials do not give all the protection needed by the commodity. Whilst the majority of these packaging materials like the nylon sacks do not allow good aeration within the packaged commodity causing a build-up of heat due to respiration, others like the woven basket have rough surfaces and edges which cause mechanical injuries to the produce. The wooden crate and the woven palm basket are some of the common packaging materials used in many developing countries especially those in Africa for packaging tomatoes. The major shortcoming of the wooden crate is in its height which creates a lot of compressive forces on fruits located at the base of the crate [38]. These undesirable compressive forces cause internal injuries which finally result in reduced postharvest quality of the tomatoes [20]. There have been suggestions of modifying the wooden crate to make it more suitable for packaging tomatoes. Kitinoja [39] has therefore suggested that the depth of the crate should be reduced considerably to reduce the build-up of compressive forces which can cause mechanical injuries to fruits at the base of the crate after packaging. The palm woven baskets used by tomato handlers have sharp edges lining the inside which puncture or bruise the fruit when they are used. It has also been recommended by Idah et al. [37] that woven palm baskets should be woven with the smooth side of the material turned inward.

2.6. Storage

Tomato has very high moisture content and therefore is very difficult to store at ambient temperatures for a long time. Meanwhile, storage in the value chain is usually required to ensure uninterrupted supply of raw materials for processors. Storage extends the length of the processing season and helps provide continuity of product supply throughout the seasons. For short-term storage (up to a week), tomato fruits can be stored at ambient conditions [40] if there is enough ventilation to reduce the accumulation of heat from respiration. For longer-term storage, ripe tomatoes can be stored at temperatures of about 10–15°C and 85–95% relative humidity [41]. At these temperatures, both ripening and chilling injuries are reduced to the minimal levels. These conditions are also difficult to obtain in most tropical countries and therefore losses of appreciable quantities of harvested tomatoes have been reported [11, 14]. This is consistent with the claim that the quality of tomato is compromised when exposed to high temperatures and high relative humidity [42]. Very low temperature storage too is detrimental to the shelf life and quality of many tropical fruits like tomatoes. For instance, refrigerating a tomato will reduce its flavour, a quality trait of tomatoes which is largely determined by the total soluble solids (TSS) and pH of the fruit [43]. An understanding of the correct temperature management during storage of tomatoes is vital in extending the shelf life of the fruit whilst maintaining fruit qualities. Tomatoes handlers in tropical countries can store tomatoes for short to intermediate time by using evaporative cooling system made from woven jute sacks.

2.7. Transportation

In most developing countries, the production sites for many tomato producers are far from the marketing centres and also inaccessible by road. Transporting harvested tomatoes to the market on such bad road network and the lack of proper transportation like refrigerated vans become a big challenge for both producers and distributors [44]. This challenge therefore causes unnecessary delays in getting the produce to the market. Meanwhile, any delay between harvest and consumption of tomatoes can result in losses [45]. Losses of up to about 20% are incurred by producers due to transportation delays [46]. Producers will therefore make use of any available means of transport for their produce without considering its appropriateness in order to avoid delays. Some modes of transportation include human labour, donkeys, public transport, rented trucks, busses, lorries, fuel tankers, articulator trucks, and pick-up vans [37, 47]. However, the use of appropriate transportation for tomatoes is a major factor to consider in postharvest handling of the fruit. During transportation, the produce should be immobilised by proper packaging and stacking to avoid excessive movement or vibration. Vibration and impact during transportation as a result of undulations on roads are one of the major causes of postharvest losses to most fruits and vegetables especially tomatoes [37]. The bad nature of road networks in most developing countries therefore provides these unfavourable factors during transportation resulting in great losses. The wobbling nature of most of the vehicles coupled with the bad nature of roads causes a lot of mechanical damage to the produce before it reaches its destination. Handlers from developed countries on the other hand use refrigerated containers and trailers which travel on reasonably good roads. Transporting tomatoes in refrigerated trucks is not only convenient, but also effective in preserving the quality of fruits. However, both the initial investment and the operation costs of these vehicles are very high and beyond the affordable reach of most producers in developing countries. Handlers of developing countries therefore transport their produce using the most affordable mode of transport without considering the effect it will have on the postharvest quality of the produce. Even though handlers from developing countries may not have the capacity to use refrigerated trucks, they should be well educated on the consequences that any other transportation option they use may have on their produce.

3. POSTHARVEST TREATMENT TECHNIQUES FOR TOMATOES

After harvesting, the tomato fruit still remains living and performs all functions of a living tissue [48]. However, the postharvest quality of the fruit at harvest cannot be enhanced by any postharvest technology but can only be maintained [49]. In order to maintain this quality, there are some postharvest treatment methods that have to be adhered to in order to achieve this goal. Below are some of the treatments methods that can be used for harvested tomatoes.

3.1. Refrigeration Storage

Refrigeration is one of the most effective methods of preserving the quality of many fruits and vegetables for several days [32, 50]. Low temperature storage can protect nonappearance quality attributes like texture, nutrition, aroma, and flavour in many harvested fruits [51]. Tomato handlers have also used refrigeration storage for tomatoes in attempt to extend shelf life. However, some fruits and vegetables of tropical origin, like tomatoes, are sensitive to chilling injury when they are stored below their critical temperature of 10°C [52]. This shortcoming of refrigeration storage was reported by Lee et al. [53] and Babitha et al. [54] where low temperatures from refrigeration storage caused chilling injuries which resulted in pitting, uneven ripening, and fungal infestation of stored fruits. This gives an indication that refrigeration storage may not be the most effective method of storing tomatoes for a long period. Another challenge in using refrigeration storage in tomato handling in most developing countries is the huge initial cost which is beyond the reach of most of underresourced handlers. However, in situations where handlers can afford refrigeration storage and temperature regulation is possible, temperatures of about 10–15°C should be maintained to avoid chilling injuries [41]. In spite of the high cost of refrigeration, it is very important to control storage temperatures and relative humidity during storage, as these two parameters are the main causes of deterioration in fruits and vegetables. The required optimum temperatures of about 10–15°C and 85–95% relative humidity [41] can be achieved by using less expensive methods of cooling such as evaporative cooling system as suggested by Workneh and Woldetsadik [55]. In such cooling system, air temperatures can be decreased to about 16°C, whilst relative humidity can be increased to about 91%, which is appropriate for reducing deterioration of harvested tomatoes due to physiological weight loss [56]. Evaporative coolers can be manufactured locally using low cost materials like jute sacks, wooden planks, and basins.

3.2. Postharvest Heat Treatment of Tomatoes

Postharvest heat treatment of fruits and vegetables is receiving more attention as a way of reducing the incidence of chilling injuries in temperature sensitive fruits of tropical origin. It is one of the approaches that can be used to avoid or reduce chilling injuries in stored fruits [57]. Postharvest heat treatments using hot air and heated water have been reported to reduce chilling injuries in fruits like mangoes, oranges, zucchini, and tomatoes [50]. Heat treating of tomato fruit at temperatures of about 37–42°C prior to cold storage can slow down ripening whilst increasing pathogenic resistance when in storage [58]. Some studies have shown that heat treatment prior to storage enhanced or caused no change in some quality traits of stored tomatoes. For instance, TSS of heat treated tomatoes was unaffected when tomatoes ripened at ambient temperatures [59] or when they ripened in a modified atmosphere storage system [60]. However, uniform heat treatment before cold storage at 14°C actually increased TSS and titratable acids (TA) when fruits ripened as compared to the untreated fruits [61]. In situations where refrigeration storage is possible, postharvest heat treatment of tomatoes can be used in conjunction with refrigeration storage in extending shelf life of harvested tomatoes.

3.3. Modified Atmosphere Packaging (MAP)

Modified atmosphere packaging (MAP) refers to a packaging technique of using specialised materials in packaging products in a predetermined composition of gases which are mainly oxygen (O_2) and carbon dioxide (CO_2) after which there is no active effort of modifying the storage space [17]. The packaging materials used in MAP allow for diffusion of gases through them until a stable equilibrium is reached between the external gases and those inside the package [62]. The most commonly used MAP materials are polyethylene terephthalate (PET), low density polyethylene (LDP), high density polyethylene (HDP), polyvinyl chloride (PVC), polypropylene [63], polystyrene [64, 65], and some chemically modified derivatives [17]. The benefit of using MAP is not only in providing a modified atmosphere to control ripening [66], but also in reducing water loss in stored products [67], reducing mechanical injuries, and enhancing better hygiene which reduces the spread of food-borne diseases [66]. MAP creates water saturated or near-saturated atmosphere (high relative humidity) around the fruit which reduces water loss and shrinkage [32]. Water loss and subsequent shriveling of tomatoes in tropical regions are one of the causes of their deterioration. Fruit shriveling may become evident with any small percentage of moisture loss. The use of MAP by tomatoes handlers in developing countries will therefore prevent or reduce the problem of water loss in harvested tomatoes. However, maintaining excessively high level of relative humidity inside the package can result in moisture condensation on the commodity, which will in turn create conducive environment favourable for pathogenic activities thereby increasing the risk of fruit deterioration [60]. Tomato handlers must therefore be trained in the proper use of MAP for tomatoes to avoid moisture condensation which will result in fruit deterioration.

3.4. 1-Methylcyclopropene (1-MCP)

The use of 1-methylcyclopropene (1-MCP) has been shown to suppress the action of ethylene in many fruits and vegetables [68]. The rate of ethylene production in harvested climacteric fruit like tomato is indicative of the metabolic activities within the fruit. The higher the metabolic activities within the harvested fruit, the shorter its shelf life. However, the aim of every postharvest technology is to slow down the metabolism in the harvested produce thereby increasing shelf life. The use of 1-MCP by handlers in developing countries is therefore essential in extending shelf life of harvested tomatoes. The use of 1-MCP has been shown to slow down many of the metabolic activities associated with the ripening process such as colour change, cell wall breakdown, and respiration rates making it a useful technique in extending storage life of fruits [69]. Although 1-MCP treatment preserves quality of fruits at advanced ripened stage, its application in green fruit may result in nonuniform ripening [70, 71]. Storing green fruits using 1-MCP must therefore be done with caution in order to achieve full ripeness [72]. In postharvest tomato treatment in particular, 1-MCP has been shown to prolong the shelf life of fruits by retaining fruit firmness and delaying both lycopene accumulation and external colour development [68]. Another benefit of using 1-MCP in tomatoes is the prevention of abscission in fruits that are sold on the vine [73]. This will have a desirable photosynthetic effect in the vegetative tissues which could lead to uninterrupted supply of certain vital substances or nutrients to the fruit thereby increasing the consumption quality. Educating and training handlers on how use 1-MCP in tomatoes will help reduce some amount of postharvest loss associated with the crop.

3.5. Calcium Chloride ($CaCl_2$) Application

Postharvest calcium chloride ($CaCl_2$) application is receiving considerable attention in recent times due to its positive effects on shelf life whilst maintaining quality of many fruits and vegetables [74]. It has been found that calcium chloride delays ripening and senescence, reduces respiration, extends shelf life, maintains firmness, and reduces physiological disorders of many fruits and vegetables [75, 76]. Lester and Grusak [77] also noted that both pre- and postharvest calcium application delayed senescence in many fruits without any negative effect on consumer acceptability. For instance, a 1% $CaCl_2$ treatment was found to have reduced fungal attack, slowed down fruit ripening, and maintained structural integrity of cell walls of strawberry [78], whilst the same application also delayed softening and increased storage life by almost 3 months in Kiwi fruits stored at 0°C [79]. In loquat fruit, calcium chloride ($CaCl_2$) dip extended shelf life by 4-5 weeks [76]. In tomatoes, calcium chloride treatment is vital for maintaining quality of fruits by reducing the physiological disorders, increasing the fruit firmness, delaying ripening process, and prolonging the shelf life [80]. $CaCl_2$ has been found to have delayed fruit colour development in tomatoes and slowed down ethylene production thereby extending shelf life by 92% [74]. Also, fruits treated with $CaCl_2$ have been shown to have reduced physiological weight loss and maintain higher firmness levels during storage [81]. Bhattarai and Gautam [82] also reported a reduction of physiological weight loss in tomatoes from 19% to 17% by using 0.25% $CaCl_2$ application for 10-day storage. However, $CaCl_2$ is a very cheap and soluble salt which can be dissolved into any concentration for use by producers. The affordable cost of the $CaCl_2$ salt and the relatively easy preparation of the solution and its application therefore make it a favourable choice for adoption by underresourced handlers of developing countries in reducing postharvest losses in tomatoes. A simple and cost-effective way of using $CaCl_2$ is by adding the required dosage of the salt in water used for precooling or cleaning of the fruits after harvesting.

4. CONCLUSION

The postharvest quality status and shelf life of the fruits in part will depend on some postharvest handling practices and treatments carried out after harvest. Even though the quality of any fruit after harvest cannot be improved by the use of any postharvest handling practices or treatment methods, it can however be maintained. Shelf life of the fruit can also be extended when appropriate postharvest handling practices and treatment methods are employed. Failure to adhere to these best practices has resulted in high amount of loss especially in developing countries. Although most tomatoes handlers from developing countries may not have high-tech postharvest technologies in addressing postharvest losses in tomatoes, understanding simple and the best postharvest practices has been found to be beneficial. Postharvest handling practices like harvesting, precooling, cleaning or disinfecting, sorting and grading, packaging, storage, and transporting played an important role in maintaining quality and extending shelf life of the tomato fruits after harvest. Also, the use of appropriate postharvest treatment methods like refrigeration, postharvest heat treatment, modified atmosphere packaging (MAP), and 1-

methylcyclopropene (1-MCP) and calcium chloride ($CaCl_2$) application was also vital. It is concluded by this study that the quality of the harvested fruit can be maintained and shelf life extended by simply using appropriate postharvest handling practices and treatment methods. Until these simple postharvest practices are followed, postharvest losses in tomatoes will continue to be a major challenge for tomatoes handlers of developing countries.

REFERENCES

1. S. Grandillo, D. Zamir, and S. D. Tanksley, "Genetic improvement of processing tomatoes: a 20 years perspective," Euphytica, vol. 110, no. 2, pp. 85–97, 1999.
2. C. M. Rick and L. Butler, "Cytogenetics of the tomato," Advances in Genetics, vol. 8, no. 1, pp. 267–382, 1956.
3. H.-L. Tan, J. M. Thomas-Ahner, E. M. Grainger et al., "Tomato-based food products for prostate cancer prevention: what have we learned?" Cancer and Metastasis Reviews, vol. 29, no. 3, pp. 553–568, 2010.
4. T. Alam and G. K. Goyal, "Packaging and storage of tomato puree and paste," Stewart Postharvest Review, vol. 3, no. 5, pp. 1–8, 2007.
5. J. K. Willcox, G. L. Catignani, and S. Lazarus, "Tomatoes and cardiovascular health," Critical Reviews in Food Science and Nutrition, vol. 43, no. 1, pp. 1–18, 2003.
6. L. Arab and S. Steck, "Lycopene and cardiovascular disease," The American Journal of Clinical Nutrition, vol. 71, no. 6, pp. 1691S–1695S, 2000.
7. Basu and V. Imrhan, "Tomatoes versus lycopene in oxidative stress and carcinogenesis: conclusions from clinical trials," European Journal of Clinical Nutrition, vol. 61, no. 3, pp. 295–303, 2007.
8. Burton-Freeman and K. Reimers, "Tomato consumption and health: emerging benefits," American Journal of Lifestyle Medicine, vol. 5, no. 2, pp. 182–191, 2011.
9. FAOSTAT, Global Tomato Production in 2013, FAO, Rome, Italy, 2014.
10. Alonso, S. García-Martínez, L. Vázquez-Araújo, J. J. Ruiz, and Á. A. Carbonell-Barrachina, "Comparative post-harvest behaviour of traditional and virus-resistant Muchamiel tomatoes," Journal of the Science of Food and Agriculture, vol. 90, no. 6, pp. 1056–1062, 2010.
11. A. Kader, "Increasing food availability by reducing postharvest losses of fresh produce," Acta Horticulturae, vol. 682, no. 1, pp. 2169–2176, 2005.
12. K. Arah, E. K. Kumah, E. K. Anku, and H. Amaglo, "An overview of post-harvest losses in tomato production in Africa: causes and possible prevention strategies," Journal of Biology, Agriculture and Healthcare, vol. 5, no. 16, pp. 78–88, 2015.
13. R. H. Muhammad, E. Bamisheyi, and F. F. Olayemi, "The effect of stage of ripening on the shelf life of tomatoes (Lycopersicon esculentum) stored in the evaporative cooling system (E.C.S)," Journal of Dairying, Foods & Home Sciences, vol. 30, no. 4, pp. 299–301, 2011.
14. N. Pila, N. B. Gol, and T. V. R. Rao, "Effect of post harvest treatments on physicochemical characteristics and shelf life of tomato (Lycopersicon esculentum Mill.) fruits during storage," American-Eurasian Journal of Agricultural & Environmental Science, vol. 9, no. 5, pp. 470–479, 2010.
15. C. V. Gustavo, F. M. J. Juan, M. Stella, S. T. Maria, L. M. Aurelio, and W. C. Jorge, "Handling and preservation of fruits and vegetables by combined methods for rural areas," Technical Manual FAO Agricultural Services Bulletin 149, FAO, Rome, Italy, 2003.

16. K. Arah, H. Amaglo, E. K. Kumah, and H. Ofori, "Preharvest and postharvest factors affecting the quality and shelf life of harvested tomatoes: a mini review," International Journal of Agronomy, vol. 2015, Article ID 478041, 6 pages, 2015.
17. M. Beckles, "Factors affecting the postharvest soluble solids and sugar content of tomato (Solanum lycopersicum L.) fruit," Postharvest Biology and Technology, vol. 63, no. 1, pp. 129–140, 2012.
18. FAO, Basic Harvest and Post-Harvest Handling Considerations for Fresh Fruits and Vegetableshandling and Preservation, FAO, Rome, Italy, 2008.
19. M. Moneruzzaman, A. B. M. S. Hossain, W. Sani, M. Saifuddin, and M. Alenazi, "Effect of harvesting and storage conditions on the post harvest quality of tomato," Australian Journal of Crop Science, vol. 3, no. 2, pp. 113–121, 2009.
20. K. Arah, "An overview of post-harvest challenges facing tomato production in Africa," in Africa: Diversity and Development, 37th AFSAAP Conference Proceedings, AFSAAP, pp. 1–21, The African Studies Association of Australasia and the Pacific, 2015, http://afsaap.org.au/assets/Arah-Isaac-Kojo.-An-overview-of-post-harvest-challenges.pdf.
21. B. Watkins, "The use of 1-methylcyclopropene (1-MCP) on fruits and vegetables," Biotechnology Advances, vol. 24, no. 4, pp. 389–409, 2006.
22. P. M. A. Toivonen, "Fruit maturation and ripening and their relationship to quality," Stewart Postharvest Review, vol. 3, no. 2, 5 pages, 2007.
23. Bachmann and R. Earles, Postharvest Handling of Fruits and Vegetables, Appropriate Technology Transfer for Rural Areas, Fayetteville, NC, USA, 2000.
24. B. Akbudak, N. Akbudak, V. Seniz, and A. Eris, "Effect of pre-harvest harpin and modified atmosphere packaging on quality of cherry tomato cultivars 'Alona' and 'Cluster'," British Food Journal, vol. 114, no. 2, pp. 180–196, 2012.
25. N. C. Shahi, U. C. Lohani, K. Chand, and A. Singh, "Effect of pre-cooling treatments on shelf life of tomato in ambient condition," International Journal of Food, Agriculture and Veterinary Sciences, vol. 2, no. 3, pp. 50–56, 2012.
26. M. D. Ferreira, J. K. Brecht, S. A. Sargent, and J. J. Aracena, "Physiological responses of strawberry to film wrapping and precooling methods," Proceedings of the Florida State Horticultural Society, vol. 107, pp. 265–269, 1994.
27. A. Kader, "Effects of postharvest handling procedures on tomato quality," in Symposium on Tomato Production on Arid Land 190, pp. 209–222, 1984.
28. A. Risse, W. R. Miller, and R. E. McDonald, "Effects of film wrapping on mature green tomatoes before and after ethylene treatment," Proceedings of the Florida State Horticultural Society, vol. 97, pp. 112–114, 1984.
29. F. Olayemi, J. A. Adegbola, E. I. Bamishaiye, and A. M. Daura, "Assessment of post-harvest challenges of small scale farm holders of tomotoes, bell and hot pepper in some local government areas of Kano State, Nigeria," Bayero Journal of Pure and Applied Sciences, vol. 3, no. 2, pp. 39–42, 2011.
30. Government of India, "Estimation loss of horticulture produce due to non-availability of post harvest & food processing facilities in Bihar & Uttar Pradesh," Socio-Economic Research Planning Commission, ASET, New Delhi, India.
31. T. Genanew, "Effect of post harvest treatments on storage behavior and quality of tomato fruits," World Journal of Agricultural Sciences, vol. 9, no. 1, pp. 29–37, 2013.
32. Batu and A. Keith Thompson, "Effects of modified atmosphere packaging on post harvest qualities of pink tomatoes," Turkish Journal of Agriculture and Forestry, vol. 22, no. 4, pp. 365–372, 1998.
33. T. S. Workneh, G. Osthoff, and M. Steyn, "Effects of preharvest treatment, disinfections, packaging and storage environment on quality of tomato," Journal of Food Science and Technology, vol. 49, no. 6, pp. 685–694, 2012.
34. O. O. Arjenaki, P. A. Moghaddam, and A. M. Motlagh, "Online tomato sorting based on shape, maturity, size, and surface defects using machine vision," Turkish Journal of Agriculture and Forestry, vol. 37, no. 1, pp. 62–68, 2013.

35. E. Saltveit, "Effect of ethylene on quality of fresh fruits and vegetables," Postharvest Biology and Technology, vol. 15, no. 3, pp. 279–292, 1999.
36. P. Prasad and A. Kochhar, "Active packaging in food industry: a review," Journal of Environmental Science, Toxicology and Food Technology, vol. 8, no. 5, pp. 1–7, 2014.
37. P. A. Idah, E. S. A. Ajisegiri, and M. G. Yisa, "Fruits and vegetables handling and transportation in Nigeria," AU Journal of Technology, vol. 10, no. 3, pp. 175–183, 2007.
38. W. C. Hurst, "Harvest, handling and sanitation," in Commercial Tomato Production Handbook B 1312, CAES Publications, University of Georgia, 2010.
39. Kitinoja, "Causes and sources of postharvest problems," Postharvest Training CDRom\Sample Presentations from Ghana, pp. 1–19, 2008.
40. D. Znidarcic and T. Pozrl, "Comparative study of quality changes in tomato cv. Malike (Lycopersicon esculentum Mill) while stored at different temperatures," Acta Agriculturae Slovenica, vol. 87, no. 2, pp. 235–243, 2006.
41. R. Castro, C. Vigneault, M. T. Charles, and L. A. B. Cortez, "Effect of cooling delay and cold-chain breakage on 'Santa Clara' tomato," Journal of Food, Agriculture & Environment, vol. 3, no. 1, pp. 49–54, 2005.
42. R. Parker and B. Maalekuu, "The effect of harvesting stage on fruit quality and shelf-life of four tomato cultivars (Lycopersicon esculentum Mill)," Agriculture and Biology Journal of North America, vol. 4, no. 3, pp. 252–259, 2013.
43. C. L. Moretti, S. A. Sargent, D. J. Huber, A. G. Calbo, and R. Puschmann, "Chemical composition and physical properties of pericarp, locule, and placental tissues of tomatoes with internal bruising," Journal of the American Society for Horticultural Science, vol. 123, no. 4, pp. 656–660, 1998.
44. A. Abimbola, "Post-harvest losses and welfare of tomato farmers in Ogbomosho, Osun state, Nigeria," Journal of Stored Products and Postharvest Research, vol. 5, no. 2, pp. 8–13, 2014.
45. A. Kader, "Effects of postharvest handling procedures on tomato quality," Acta Horticulturae, vol. 190, pp. 209–222, 1986.
46. A. Babatola, D. O. Ojo, and O. I. Lawal, "Effect of storage conditions on tomato (Lycopersicon esculentum Mili.) quality and shelf life," Journal of Biological Sciences, vol. 8, no. 2, pp. 490–493, 2008.
47. H. Getinet, T. S. Workneh, and K. Woldetsadik, "Effect of maturity stages, variety and storage environment on sugar content of tomato stored in multiple pads evaporative cooler," African Journal of Biotechnology, vol. 10, no. 80, pp. 18481–18492, 2011.
48. Joas and M. Léchaudel, "A comprehensive integrated approach for more effective control of tropical fruit quality," Stewart Postharvest Review, vol. 4, no. 2, pp. 1–14, 2008.
49. Tigist, T. S. Workneh, and K. Woldetsadik, "Effects of variety on the quality of tomato stored under ambient conditions," Journal of Food Science and Technology, vol. 50, no. 3, pp. 477–486, 2013.
50. S. D. C. Rodriguez, B. López, and A. R. Chaves, "Effect of different treatments on the evolution of polyamines during refrigerated storage of eggplants," Journal of Agricultural and Food Chemistry, vol. 49, no. 10, pp. 4700–4705, 2001.
51. R. E. Paull, "Effect of temperature and relative humidity on fresh commodity quality," Postharvest Biology and Technology, vol. 15, no. 3, pp. 263–277, 1999.
52. K. Raison and J. M. Lyons, "Chilling injury: a plea for uniform terminology," Plant, Cell & Environment, vol. 9, no. 9, pp. 685–686, 1986.
53. Y. S. Lee, J. N. Kim, and D. S. Chung, "The effect of 1-MCP (1-methylcyclopropene) treatment methods on the ripening process of tomato fruit during postharvest storage," Journal of Horticulture, Environment & Biotechnology, vol. 4, no. 3, pp. 175–181, 2008.
54. C. Babitha, B. T. Ninganur, M. B. Chetti, D. I. Jirali, and S. I. Harlapur, "Effect harvest treatment and atorage conditions on physio-chemical parameters in extending shelf life in tomato," Plant Archhives, vol. 10, no. 1, pp. 37–47, 2010.

55. T. S. Workneh and K. Woldetsadik, "Forced ventilation evaporative cooling: a case study on banana, papaya, orange, mandarin, and lemon," Tropical Agirculture, vol. 81, no. 1, pp. 1–6, 2004.
56. T. S. Workneh, "Feasibility and economic evaluation of low-costevaporative cooling system in fruit and vegetables storage," African Journal of Food, Agriculture, Nutrition and Development, vol. 10, no. 8, pp. 2984–2997, 2010.
57. S. Lurie and J. D. Klein, "Ripening characteristics of tomatoes stored at 12°C and 2°C following a prestorage heat treatment," Journal of Horticultural Science, vol. 51, no. 1-2, pp. 55–64, 1992.
58. B. Akbudak, N. Akbudak, V. Seniz, and A. Eris, "Sequential treatments of hot water and modified atmosphere packaging in cherry tomatoes," Journal of Food Quality, vol. 30, no. 6, pp. 896–910, 2007.
59. R. E. McDonald, T. G. McCollum, and E. A. Baldwin, "Temperature of water heat treatments influences tomato fruit quality following low-temperature storage," Postharvest Biology and Technology, vol. 16, no. 2, pp. 147–155, 1999.
60. Suparlan and K. Itoh, "Combined Effects of Hot Water Treatment (HWT) and Modified Atmosphere Packaging (MAP) on quality of tomatoes," Packaging Technology and Science, vol. 16, no. 4, pp. 171–178, 2003.
61. J. Lu, M. T. Charles, C. Vigneault, B. Goyette, and G. S. V. Raghavan, "Effect of heat treatment uniformity on tomato ripening and chilling injury," Postharvest Biology and Technology, vol. 56, no. 2, pp. 155–162, 2010.
62. C. A. Phillips, "Review: modified atmosphere packaging and its effects on the microbiological quality and safety of produce," International Journal of Food Science and Technology, vol. 31, no. 6, pp. 463–479, 1996.
63. H. P. J. De Wild, E. C. Otma, and H. W. Peppelenbos, "Carbon dioxide action on ethylene biosynthesis of preclimacteric and climacteric pear fruit," Journal of Experimental Botany, vol. 54, no. 387, pp. 1537–1544, 2003.
64. Artés, P. A. Gómez, and F. Artés-Hernández, "Modified atmosphere packaging of fruits and vegetables," Stewart Postharvest Review, vol. 2, no. 5, pp. 1–13, 2006.
65. Sandhya, "Modified atmosphere packaging of fresh produce: current status and future needs," LWT—Food Science and Technology, vol. 43, no. 3, pp. 381–392, 2010.
66. A. Kader and C. B. Watkins, "Modified atmosphere packaging—toward 2000 and beyond," HortTechnology, vol. 10, no. 3, pp. 483–486, 2000.
67. I. Cantwell, X. Nie, and G. Hong, "Impact of storage conditions on grape tomato quality," in Proceedings of the 6th ISHS Postharvest Symposium. International Society of Horticultural Sciences, Antalya, Turkey, 2009.
68. Cliff, S. Lok, C. Lu, and P. M. A. Toivonen, "Effect of 1-methylcyclopropene on the sensory, visual, and analytical quality of greenhouse tomatoes," Postharvest Biology and Technology, vol. 53, no. 1-2, pp. 11–15, 2009.
69. B. Watkins, "Overview of 1-methylcyclopropene trials and uses for edible horticultural crops," HortScience, vol. 43, no. 1, pp. 86–94, 2008.
70. Y. Mostofi, P. M. A. Toivonen, H. Lessani, M. Babalar, and C. Lu, "Effects of 1-methylcyclopropene on ripening of greenhouse tomatoes at three storage temperatures," Postharvest Biology and Technology, vol. 27, no. 3, pp. 285–292, 2003.
71. B. M. Hurr, D. J. Huber, and J. H. Lee, "Differential responses in color changes and softening of 'Florida 47' tomato fruit treated at green and advanced ripening stages with the ethylene antagonist 1-methylcyclopropene," HortTechnology, vol. 15, no. 3, pp. 617–622, 2005.
72. D. J. Huber, "Suppression of ethylene responses through application of 1-methylcyclopropene: a powerful tool for elucidating ripening and senescence mechanisms in climacteric and nonclimacteric fruits and vegetables," HortScience, vol. 43, no. 1, pp. 106–111, 2008.
73. C. Passam, I. C. Karapanos, P. J. Bebeli, and D. Savvas, "A review of recent research on tomato nutrition, breeding and post-harvest technology with refrence to fruit

quality," European Journal of Plant Science and Biotechnology, vol. 1, no. 1, pp. 1–21, 2007.

74. Senevirathna and W. Daundasekera, "Effect of postharvest calcium chloride vacuum infiltration on the shelf life and quality of tomato (cv. 'Thilina')," Ceylon Journal of Science, vol. 39, no. 1, pp. 35–44, 2010.

75. N. Hong, B. C. Lee, S. Mendonca, M. V. E. Grossmann, and R. Verhe, "Effect of infiltrated calcium on ripening of tomato fruits," LWT Journal of Food Science, vol. 33, pp. 2–8, 1999.

76. Akhtar, N. A. Abbasi, and A. Hussain, "Effect of calcium chloride treatments on quality characteristics of loquat fruit during storage," Pakistan Journal of Botany, vol. 42, no. 1, pp. 181–188, 2010.

77. E. Lester and M. A. Grusak, "Field application of chelated calcium: postharvest effects on cantaloupe and honeydew fruit quality," HortTechnology, vol. 14, no. 1, pp. 29–38, 2004.

78. Lara, P. García, and M. Vendrell, "Modifications in cell wall composition after cold storage of calcium-treated strawberry (Fragaria x ananassa Duch.) fruit," Postharvest Biology and Technology, vol. 34, no. 3, pp. 331–339, 2004.

79. D. Gerasopoulos and P. D. Drogoudi, "Summer-pruning and preharvest calcium chloride sprays affect storability and low temperature breakdown incidence in kiwifruit," Postharvest Biology and Technology, vol. 36, no. 3, pp. 303–308, 2005.

80. A. Abbasi, L. Zafar, H. A. Khan, and A. Ahad Qureshi, "Effects of naphthalene acetic acid and calcium chloride application on nutrient uptake, growth, yield and post harvest performance of tomato fruit," Pakistan Journal of Botany, vol. 45, no. 5, pp. 1581–1587, 2013.

81. M. Gharezi and M. Gharezi, "Effect of post harvest treatment on stored cherry tomatoes," Journal of Nutrition & Food Sciences, vol. 2, no. 8, pp. 1–10, 2012.

82. D. R. Bhattarai and D. M. Gautam, "Effect of harvesting method and calcium on post harvest physiology of tomato," Nepal Agriculture Research Journal, vol. 7, pp. 37–41, 2009.

CHAPTER 6

Characterization of Pre- and Postharvest Losses of Tomato Supply Chain in Ethiopia

Bezabih Emana[1], Victor Afari-Sefa[2*], Ngoni Nenguwo[3], Amsalu Ayana1, Dereje Kebede[1] and Hedija Mohammed[1]

[1]*HEDBEZ Business & Consultancy PLC, P. O. Box 15805, Addis Ababa, Ethiopia.*

[2]*World Vegetable Center, West and Central Africa, Samanko Research Station, PB 320, Bamako, Mali*

[3]*World Vegetable Center, Eastern and Southern Africa, P. O. Box 10, Duluti, Arusha, Tanzania.*

ABSTRACT

Background

Tomato has significant economic importance in Ethiopia. Although quantitative evidence is limited, postharvest loss in tomato is considerably high. This study presents qualitative and quantitative postharvest losses of tomato. The study was conducted in 2015 in two districts (Bora and Dugda) located in East Shewa Zone of Oromia National Regional State, Ethiopia, located southeast of Addis Ababa. Primary data were collected from producers (smallholder farmers) and traders via household survey, focus group discussions and key informant interviews as well as estimation of losses based on samples. Secondary data and information were collected from published and unpublished sources.

Results

Results indicate that tomato production is being done by relatively young married individuals who have at least primary level education. About three-quarter of land holding is allocated for vegetable production, which largely takes place under irrigation during dry season. All sample producers sold more than 95% of their tomato produce, mostly to wholesalers as compared to collectors and retailers. Postharvest losses occur at collectors, wholesalers and retailers level. The finding indicates that more than 16% of respondents encounter produce losses due to high incidence of diseases, insect pest and mechanical injuries, each of them accounting for more than 20% of postharvest losses.

Conclusions

The findings from our study underscore that the ability of actors to mitigate postharvest losses is limited due to lack of technical know-how. They also lack necessary support and complementary resources to improve postharvest handling practices and technology.

Keywords: Postharvest management, Produce shelf life, Value chain analysis, Vegetable marketing, Vegetable production

BACKGROUND

In Ethiopia, the vegetable subsector has a vital role in human nutrition and health, farm income generation, poverty alleviation and foreign currency earnings through export and foreign direct investment [2, 5, 10]. Processed products such as tomato paste and tomato juice are produced for export to Somalia, Djibouti and Saudi Arabia, making a significant contribution to the national economy [3, 6]. Ethiopia's wide range of agro-climatic conditions and soil types makes it suitable for the production of both warm and cool season vegetables [7, 29]. Vegetable crops are suitable for production under intensive systems, where some farmers produce two to three times within a calendar year in Ethiopia [8]. However, vegetable production in the country is constrained by several challenges [5, 9]. Among them, postharvest loss of vegetables such as tomato is of critical importance [30, 37, 38, 39].

Postharvest losses (PHL) refer to the losses that occur along the food supply chain, from the farm gate through till it gets on the table of the final consumer. Losses are encountered along the chain in the handling, storage, transportation and processing, thereby resulting in a reduction in the quantity, quality and market value of agricultural commodities [4, 19, 27]. Within developing countries' context and in Ethiopia particularly, concerns about reduction of quantitative losses (i.e., weight, volume or total wastage of agricultural produce) are of higher priority than qualitative losses such as loss in edibility, nutritional quality, caloric value and consumer acceptability of the produce. It is also known that, in general, qualitative losses are much more difficult to assess than quantitative losses.

The main causes of postharvest losses include mechanical damage, physiological deterioration and biological (i.e., postharvest diseases and insect pests) [1, 11, 19, 26, 27, 32]. Rodents and birds also cause postharvest losses, especially in fruits such as tomatoes [17], although such losses tend to be relatively small for vegetables compared to damages due to rough handling, poor packaging and quality losses caused by temperature stress [35]. In some cases, postharvest losses of vegetables such as tomato are also attributed to socioeconomic and institutional factors, viz. inadequate marketing information and support systems, inappropriate transportation facilities, unfavorable government policies, inability to implement regulations and legislations, lack of appropriate tools and equipment, lack of technical know-how and poor maintenance culture for existing facilities and infrastructure [19, 28, 35]. In most developing countries such as Ethiopia, roads are not adequate for proper transport of horticultural crops, while transport vehicles and other modes, especially those suited for fresh horticultural perishables, are in short supply. Moreover, the extent of losses is significantly influenced by preharvest conditions and field operations such as cultivar and soil types, crop management practices, poor weather conditions, insect pest control programs and harvesting as well as packaging and handling practices [21].

Moreover, postharvest losses vary greatly with commodities, production areas and seasons as well as the level of development of infrastructure and technology for postharvest management and market system [15, 18, 19, 20, 22, 23]. Vegetable postharvest loss in Ethiopia could be as high as 40%. However, quantitative and qualitative assessment information of postharvest losses of vegetables in Ethiopia remains scarce and is mostly based on guess estimates [30, 37, 38, 39] as opposed to formal quantitative field surveys. Despite the rich knowledge of postharvest losses in tomato production and marketing in the world, there is huge knowledge gap in postharvest handling and management in Ethiopia. The extent of losses is also not known with reasonable accuracy. There exists only little recent quantitative and systematic evidence on the magnitude of postharvest losses in tomato production and marketing in Ethiopia [16, 30]. Even so, these studies mostly considered a range of horticultural crops that exclude tomato. The objective of this paper therefore is to generate country-specific knowledge in tomato regarding the production system, postharvest management and losses and analyze the factors responsible for such observed losses. The remaining part of the article describes the study area and methodology, key findings of the study and conclusions that trigger policy interventions.

METHODS

Description of the study area

The study was conducted in two districts (Bora and Dugda) located in East Shewa Zone of Oromia Regional State, located southeast of Addis Ababa, Ethiopia (Fig. 1). In 2013, the population of the two districts was 236,794 of which 48.4% was female. About 25% of the population lives in towns. Geographically, the study districts are located from 8°00′00″ to 8°25′00″ east and from 38°30′00″ to 39°50′00″ north. The districts are located in the Rift Valley, where major vegetables such as tomato, onion and leafy vegetables like kales are produced primarily for the market. The districts are known for their good road network and other basic infrastructure such as transport, telecommunication and electricity. They are also well connected to major market areas like Addis Ababa, Adama, Hawassa, Shashemene and Bishoftu. Intensive small-scale farming is practiced whereby farmlands are used for 2–3 times per year for vegetable production under both rain-fed and irrigated conditions.

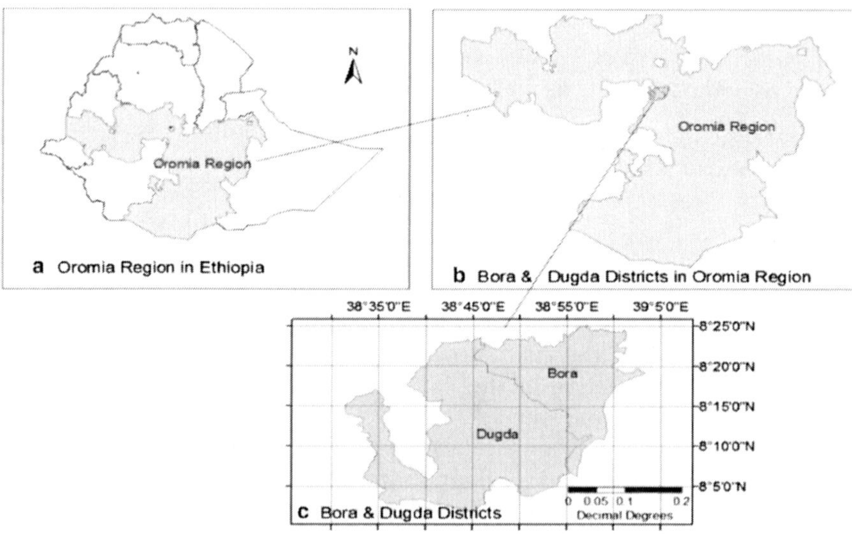

Figure. 1 Map of the study districts

The study area also shares major challenges faced by vegetable producers in the country. The challenges include lack of capacity in terms of knowledge and skill gap for postharvest loss management and limited access to viable produce markets.

Data collection and analysis

Data for the study were collected from primary and secondary sources in December, 2014. Secondary data and information were collected from desk reviews and unpublished raw data collected from the Central Statistical Agency (CSA). Primary data were collected from producers (smallholder farmers) and traders via one-on-one survey (producers and traders), focus group discussions (FGD) and key informant interviews. The producers' survey covered five sampled *kebeles* [1] involving 155 sample households drawn from both districts, while the traders' survey covered 56 traders—29 wholesalers, 5 collectors and 22 retailers. The sampled producers were equally distributed among the *kebeles* (31 from each *kebele*), while the proportion of female farmers interviewed was 11%. A structured questionnaire was designed and used for the one-on-one quantitative interview. The questionnaire covered diverse issues such as tomato production, produce utilization, marketing, postharvest handling, transportation, postharvest losses and its management and factors causing postharvest losses.

Moreover, a physical fresh produce quality survey was also conducted to determine the type, extent and causes of postharvest losses. In total, tomato samples were randomly collected from 45 sampled producers and traders. Two kilograms of tomato was purchased from each of the sampled respondents on a random basis. Digital weighing scales were used to measure the weight of tomatoes purchased. Damaged tomato was sorted based on physical damage, damage by diseases and insect pests, wilting, damage due to deformity or color. Sorting was made with replacement after the weight of tomato classified under a particular damage category was recorded. In addition, an MM laboratory thermometer was used to measure pulp temperature of tomato and corresponding ambient temperatures of the produce.

Moreover, information that complements quantitative data was collected through focus group discussion (FGD) with group of producers and key informants interview (KII). Two FGDs were conducted in two of the sample *kebeles* (i.e., one for each district with 10 producers in each *kebele*). Key informant interviews were also conducted with relevant district-level agriculture offices and farmers' cooperative union. In total, eight district agriculture office experts and two farmers' cooperative union representatives were interviewed as key informants in both target districts. In-depth discussion on production system, postharvest losses, and production and marketing constraints and opportunities were undertaken during the FGDs and KIIs. Moreover, horticulture experts at district and Farmers' Training Centers were interviewed as key informants on extent of tomato postharvest losses and possible causes of such losses. The primary data were augmented with secondary data collected from various publications based on national and international studies. The data collected were encoded and entered into Microsoft Excel spreadsheets and Statistical Package for the Social Science (SPSS) Software version 21 package. Data obtained from various sources were also triangulated, checked for consistency and analyzed.

RESULTS AND DISCUSSION

Demographic characteristics of producers

The demographic profiles of the sampled tomato producers are presented in Table 1. About 47% of the sample producers (64 respondents) were involved in tomato production, of which 89% are males. Approximately 78% of tomato producers are married and have an average household size of 5 persons per household. Producers' age range from 20 to 67 years, with an average of 33 years implying that young cohort of farmers are more involved than elder ones in tomato production. The majority of them have less than 20 years of experiences in farming. As tomato is largely produced for the market, younger farmers tend to establish good marketing network, use market information and bear risk involved in tomato production than the elderly. Younger farmers also have better access to education with 75% of the sampled tomato producers having received formal education (primary school to college level). Education gives the farmers also to analyze the benefits, costs and risks involved in tomato production.

Demographic profile of tomato traders

Close to two-third of the respondents are male tomato traders (Table 2). Female traders are mostly involved in small vegetable retailing in urban settings. This presupposes that women do not have enough capital to engage in wholesaling of vegetables, and such business activities also involve a lot of engagement and networking, which is hard to be attained for most women traders. With regard to marital status, most of the vegetable traders are married and have an average household size of five. Vegetable collectors are few (2% of the sampled traders), indicating that the wholesalers are also involved in purchasing vegetables at farm gate. Interestingly, 75% of the sampled vegetable traders have attained formal education (primary school to college level) showing that the vegetable trading is quite attractive to educated people, partly as it provides opportunities for educated people and partly because educated farmers see the attractiveness of trade business.

Table 1 Proportion of producers by demographic profile

Category	Mean
Average age ($N = 64$)	33
Average household size	5.5
	Percent
Gender ($N = 64$)	
Female	11
Male	89
Marital status ($N = 64$)	
Married	78.3
Unmarried	18.3
Divorced	0
Widow	3.3
Educational level	
Primary	51.7
Secondary	20.0
College	3.3
No formal education	25
Experiences (year) ($N = 56$)	
<10	39
11–20	47.5
>21	13.6

Source: Own Survey, 2015

Tomato production

Vegetable production including tomato is largely practiced under irrigation using water pumped from nearby lakes and from over 5000 shallow wells dug for irrigation. Based on information elicited from the KIIs, 15,547 rural households are engaged in irrigated production of vegetables in the study districts. More than 90% of these households use irrigation for vegetable production. Similarly, the survey findings indicate that 99% of the vegetable producers practice irrigated farming, while 34% produce vegetables under rain-fed conditions, with no significant difference between the two districts. The FGD results also show that tomato is preferably produced during the dry season under irrigation mainly to reduce risks of diseases and insect pests. Irrigated production of vegetables including tomato is undertaken twice a year, from January to June with irrigation only and from September to December with supplementary irrigation.

Table 2 Proportion of traders by demographic profile

Category	Percent of traders		Total
	Wholesaler	Retailer	
Gender ($N = 56$)			
Female	3.4	86.4	37.5
Male	96.6	13.6	62.5
Marital status ($N = 56$)			
Married	65.5	59.1	66.1
Unmarried	34.5	13.6	23.2
Divorced	–	4.5	1.8
Widow	–	22.7	8.9
Educational level of traders ($N = 56$)			
Primary	24.1	36.4	30.4
Secondary	51.7	18.2	35.7
College	13.8	–	8.9
No formal education	10.3	45.5	25
Experiences (in year) ($N = 56$)			
<5	64.3	40	56
5–10	21.4	30	23
>10	14.3	30	21
Average household size (mean value)			
	4.5	4.5	5

Source: Own Survey, 2015

Tomato is a major vegetable grown in the study area with the survey results indicating that about 47% of the sampled farmers produce tomato. The survey result also indicates that households on the average allocate around 44% of their farmland (0.74 ha) for tomato production. The total output per household is about 24 tons per season. The average yield of tomato is 39.8 tons/ha showing considerable variability between sampled farmers. The productivity of tomato at farm-gate level is fairly comparable with yield potential under research field for tomato in Ethiopia [12]. Based on the area allocated for tomato production and the average yields, the total supply of the produce from the two districts was estimated at 387,567 tons per annum.

Marketing of tomato

As it is indicated earlier, a significant proportion of the tomato harvest are meant for sales in various cities/towns within Ethiopia, implying that production of the crop is commercially driven and access to market is crucial to improve household incomes. The survey results indicate that about 98% of tomato was sold during the last production season. The remaining 2% was used for different purposes, including family consumption and gift to neighbors. The tomato supplied to the market is also subjected to quality and quantity loss, as also discussed in the subsequent sections.

Farmers have various marketing channels through which they sell their produce. As shown in Fig. 2, farmers can sell tomato to village collectors, wholesalers, retailers, cooperatives and consumers. Sales to wholesalers account for significant share of tomato, accounting for 93% of tomato sold by the producers.

Village collectors and cooperatives buy very small proportion of the produce. Some farmers especially women also sell small quantities of tomato in the open market to consumers. However, the FGDs results indicate that producers have limited access to wholesalers. The transaction to wholesalers is exclusively through brokers who play a dominant role in facilitating trade between producers and wholesalers. They link the wholesalers to the producers at farm gate or local wholesalers who operate at smaller scale to larger wholesalers in other cities/towns. Only 10–15% of producers sell the produce through retailers, and fewer farmers sell through other lesser known channels. Not surprisingly, the quantity of tomato sold through these channels is smaller (5%).

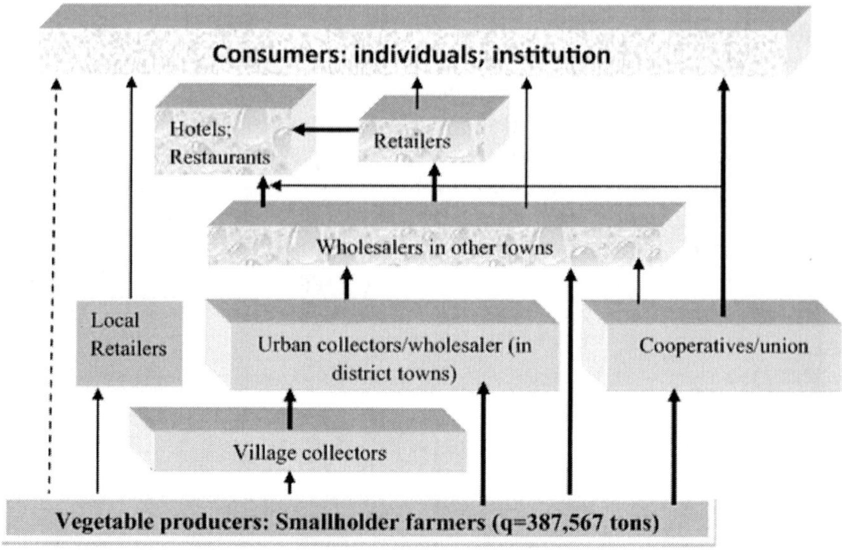

Figure. 2 Tomato marketing channel

Moreover, the survey results indicate that the majority of the urban collectors or local wholesalers supply to the larger wholesalers coming from other towns (83%) and are often called gap fillers. More than 40% of them also sell to urban small retailers. The result indicates that all wholesalers in main cities are registered companies, while the majority of the urban collectors/local wholesalers are informal and not registered as a trade company. This implies that the majorities of these traders are not recognized and lack access to requisite institutional support.

Unlike most producers, the majority of fruits and vegetable traders are simultaneously involved in handling different types of vegetables and rarely specialize in a particular commodity (Table 3) with the aim to diversify the fresh produce they handle so as to maximize their returns and reduce losses. This approach to vegetable trading is practiced by wholesalers and retailers alike though retailers diversify more than wholesalers. The survey result shows that tomato traders also handle onion and some of them handle kales and hot peppers. Some retailers also trade in potato and garlic, so that the consumers who purchase these produce in smaller quantities are able to source all their supplies from one location. It is, however, important to note that vegetable and fruit traders do usually handle other food commodities such as cereals which have distinct features.

Table 3 Proportion of traders by their main trading activities (%)

Fresh produce	Collectors	Wholesaler	Retailer	Total
Tomato	100	79.3	59.1	64.3
Onion	40	79.3	68.2	67.9
Kale		6.9	40.9	19.6
Hot pepper		13.8	18.2	14.3
Green bean		6.9	–	3.6
Potato		–	18.2	7.1
Garlic		–	4.5	1.8
N	5	34	22	56

Source: Own Survey, 2015

Vegetable trading in the study area is largely informal. Despite the policy that requires registration of business and the need for a valid trade license, only 17.2% of the sampled wholesalers reported to have business license, while all the retailers do not possess one. Tomato marketing/trading is done in collaboration with brokers and other traders, who are not also licensed, hence making accountability for their marketing or service behavior more challenging. Collaboration can be between traders of the same category where wholesalers collaborate with other wholesalers or with other categories where wholesalers collaborate with collectors, broker or retailers. However, such collaboration involved each a maximum of two brokers/commission agents, wholesalers or retailers. The means of such collaboration largely include sharing with each other market information and even to some extent supply of vegetables to the collaborator when there is supply shortage (Fig. 3).

Membership into an association or traders group also helps for systematic networking of the trade business by improving social capital. However, only 9% of the traders belong to trade associations or groups relevant to their business undertakings (Table 4). This may limit their access to necessary services such as credit and joint transportation services as well as voicing their needs to policy and decision makers in the government administration. The size of associations to which the wholesalers belong to is significantly larger than that of the retailers.

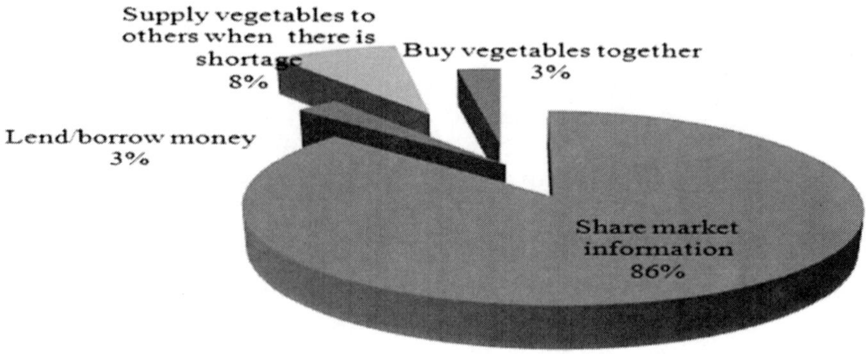

Figure. 3 Means of collaboration of tomato traders with each other

Table 4 Traders by some features of their tomato business

Trader	Collectors	Wholesaler	Retailer
Licensed (%)	0	17.2	0
Member of associations (%)	20	7.4	9.1
Size of association (# of members)	4	118	22
Collaborating with brokers (%)	80	82.8	18.2
Collaborating with retailers (%)	–	72.4	45.5
Collaborating with wholesalers (%)	60	79.3	22.7
N	5	34	22

Source: Own Survey, 2015

Tomato price determination

Tomato traders buy tomato from different sources and sell to different buyers. Just considering the producer, collector, wholesaler, retailer and consumer produce flow channel, the buying and selling prices of the traders are given in Table 5. The buying price of collectors represents the selling price of the producers, while the selling price of the retailer represents the consumer price. Brokers press producers to reduce price. When the tomato crop is ready for harvest in many fields at the same time, the producers become price takers and sell it at whatever price could be offered by the traders rather than negotiating for higher prices, due to the perishable nature of the produce. The survey result reveals that actors along the value chain handle tomato of different qualities. The traders sort tomato every time they handle it. Good-quality tomato is sold to larger wholesalers coming from other cities/towns, while low-quality tomato is sold to retailers at lower prices. The wholesalers face less price reduction for tomato because they trade in relatively high-quality fresh produce in comparison with retailers and hence attract higher margins.

Table 5 Average buying and selling prices of tomato in the study districts (ETB/100 kg)

Trader	Buying	Selling	Trader's margin
Collector ($N = 5$)	440	560	120
Wholesaler ($N = 34$)	518	678	159
Retailer ($N = 22$)	365	463	98

Source: Own Survey, 2015

Postharvest handling of tomato

With regard to transport, some vegetable traders use cart to transport the produce directly from the field/farm to their business center, while others engage the services of suppliers to bring the produce to their place. Collectors used only vehicles and animal-drawn carts, while retailers used animal-drawn carts and human labor to transport tomato. Wholesalers use different means of transportation (Table 6).

Vegetable traders, who do not transport the product directly from field/farm to their business center, receive the produce from their suppliers/partners transported to their business area. Most of the collectors and retailers receive the produce transported by the sellers. The suppliers used different modes of transport to bring tomato to the buyers. Vehicles (own and rented) are the most common means of transport followed by animal-drawn cart used by suppliers to deliver tomato to collectors and wholesalers. Retailers receive tomato transported by human and carts only.

Table 6 Proportion of traders by how they transport vegetables (%)

Means of transporting purchased tomato by buyer	Collectors	Wholesaler	Retailer
Hand cart	–	11.1	–
Vehicle	40.0	61.1	–
Pickup truck	–	11.1	–
Animal-drawn cart	60.0	11.1	66.7
Public transport	–	5.6	–
Manual labor	–	–	33.3
N	5	34	22

Source: Own Survey, 2015

Produce from the farm gate is mostly packed and delivered in wooden boxes (Table 7). Insulated Styrofoam boxes and airtight plastic packing materials were not known as containers used for delivery by the respondents involved in this study.

Table 7 Traders stating the form the incoming produce is delivered to their business center (%)

Form of deliver	Collector	Wholesaler	Retailer
Plastic bag	–	–	9.1
Sacks	–	–	18.2
Wooden boxes	100.0	88.9	81.8
Bamboo basket	–	–	9.1
Loose	–	5.6	–
N	5	24	22

Source: Own Survey, 2015

In terms of storage, about 37% of the traders used rented premises to store tomato and other vegetables they handle (Table 8). However, in most cases, tomato is highly perishable and about 42% of the traders sell it immediately as they received.

Table 8 Proportion of traders by where they store tomatoes (%)

Place of storage	No	Tomato
Own storage	3	9.7
Rented storage	14	38.7
Cooperative's storage place	3	9.7
I don't store	14	41.9

Source: Own Survey, 2015

Several studies have reported high postharvest losses due to poor packaging, inadequate storage facilities and poor means of transportation using human labor, donkeys and mules, public transport and rented trucks [14]. Such losses can be reduced by harvesting produce at optimal maturity, through grading, packaging and careful handling of the produce, maintaining higher sanitation standards, decreasing injury incidence and maintaining good storage and environmental conditions [14, 37, 38, 39]. Moreover, treatment combinations such as low temperature, waxing, low oxygen and high carbon dioxide storage and ethylene inhibitor such as calcium chloride treatment have been reported to have the potential to extend the storage life of fresh produce such as tomatoes [13]. Moreover, some literature also claims that tomatoes could be marketed at premium quality if lower storage temperatures were accessible and encouraged private sector to provide such facilities particularly in urban markets where retail prices will merit such investments. In general, maintaining appropriate storage of vegetables can minimize moisture loss and wilting, slow down respiration rate, prolong shelf life and inhibit development of decay-causing pathogens.

Postharvest loss is a key concern of the tomato value chain actors at different levels. Sample tomato producers were asked whether they faced losses during the last production season, while the traders were asked about losses encountered on a weekly basis. The results indicate that 76% of the producers and more than 60% of traders encountered physical postharvest losses in tomato marketing (Table 9).

Table 9 Proportion of tomato value chain actors stating of tomato traded (%)

Type of trading activity	No	% of respondents	Average quantity traded (ton)	% loss (measured)	% loss (perceived)
Producers	64	75.9	21.46**	3.7	5.3
Traders					
Collector	5	80.0	2.31*	2.8	12.7
Wholesaler	34	65.5	22.48*	3.0	7.2
Retailer	22	59.1	0.13*	6.7	8.2

* Quantity traded per week and ** Quantity sold per season

Source: Own Survey, 2015

The sample producers sold an average of about 22 tons of tomato per single production season of which 3.7% was lost due to spoilage. The result indicates about 21.46 tons of tomato was sold by the producer household to different market actors, while the remaining balance was used in different ways (consumption, gift, etc.). As it is shown in Table 9, wholesalers traded about 22.48 tons of tomato per week, while the percentage loss is more than double in the case of retailers who handle about 0.13 tons of tomato per week. Retailers buy and sell smaller quantities of tomato, hold the produce for relatively longer time and encounter larger percentage losses. Samples collected from producers and traders measured the actual level of loss due to damage and poor quality. As can be seen in Table 9, these estimated measured levels are lower than the perceived losses of producers and traders.

In order to determine the extent of losses and characterize the different types of postharvest damage on tomato, postharvest losses were measured based on real production and marketing contexts along the value chain. In total, 40 tomato samples were collected from producers and traders, characterized and measured for damages. The assessment result indicates the prevalence of various forms of damages including mechanical damage, disease and insect attack; poor shape and sun burn among others are observed along the value chain. At farm level, significant proportion of tomato damage occurred due to insect attack (26%) and disease (25%) (Table 10). Limited access to appropriate harvesting tools and skills as well as poor disease and insect pest control measures are the notable factors causing damage at farm level.

Table 10 Percent of tomato observed having undesirable quality (physical) characteristics

Causes of poor quality	Producer	Wholesaler	Retailer
Mechanical damage	16.0	15.4	13.0
Disease	25.0	41.1	28.0
Insect pests	26.0	14.7	31.9
Deformed shape	14.8	13.5	13.7
Sun burn	9.5	8.6	15.8
Undesired color	24.1	31.5	31.6
No	64	34	22

Source: Own Survey, 2015

On the other hand, among the wholesalers, over 40% of tomato was affected by disease. Retailers were also engaged in marketing of low-quality fresh produce due to varied factors. Undesirable color and deformed shape of the produce are serious problems across the value chain. The respondents partly attributed to the poor-quality tomato sold in the market along the value chain to nutrient deficiency during the plant growth.

The findings are also consistent with the findings of other authors. Hussen et al. [16] for example found out that inadequate storage and transportation facilities, diseases and insect pests are found to be significant factors contributing to postharvest losses in horticultural crops. Lack of storage facilities forces the value chain actors to dispose of tomato even if the prices are low. According to Teka [31], the farmers store tomato when the price gets low for an average of 3 days. Due to lack of well-ventilated storage facilities and adequate, reliable and timely market information, farmers are forced to dispose of their produce within a short period even if the selling price is low. Efforts made to apply modern postharvest handling practice (such as curing, grading and storing) to increase the shelf life of tomato are low [25]. Knowing this lack of ability and facilities to store tomato for long, wholesalers put pressure on producers to sell at low price.

High temperature is one of the factors for enhanced postharvest loss of tomato. The climate of most zones of Ethiopia is characterized by high ambient temperature combined with low relative humidity, which has a negative effect on tomato quality during harvesting, transportation, storage and marketing. Under such conditions, using methods such as the forced ventilation evaporative cooling system is expected to effectively reduce postharvest loss and prolong shelf life of tomato [36].

The measured pulp temperature of tomato was less than that of the corresponding average ambient temperatures (Table 11). Although the pulp temperatures for tomato were found to be lower than the corresponding ambient temperature, produce temperatures are far above the recommended temperature (e.g., 15 °C for tomato pulp temperature) [34]. It is interesting to note that the temperature does decrease from producer through to retailer (Table 11). High temperatures are well known to result in increased rates of respiration, deterioration and water loss in fresh produce, leading to reduced market value and decreased nutritional value [24]. It is to be noted that fresh produces such as tomato need low temperature and high relative humidity during the storage and transportation. Reduced temperature decreases physiological, biochemical and microbiological activities, which are the causes of quality deterioration such as flavor and color [14, 37, 38, 39, 40].

Table 11 Average ambient pulp temperature of tomato taken from actors (°C)

Temperature	Producer	Wholesaler	Retailer
Ambient temp	30.0	26.8	25.2
Pulp temperature	26.0	24.3	23.0
N	9	17	3

Source: Own Survey, 2015

Various causes of postharvest losses were identified along the value chain. The majority of producers reported diseases and insect pests as the major problems affecting their produce (Table 12), while wholesalers indicated that damage during transportation was a concern. Due to its perishable nature, tomato harvesting is time sensitive and requires intensive engagement over a short period of time. Workers that do not have adequate skills and tools are engaged in harvesting, and this has been causing significant damage to produce. Unfavorable weather conditions, during and after harvest, affects the produce quality. Damage during transport was also reported by the majority of village collectors and wholesalers. Lack of market is also reported by sizeable number of producers, collectors and retailers as a reason for produce quality deterioration. Brokers' practice of creating information symmetry aggravates the situation.

Table 12 Proportion of value chain actors stating reasons for loss of tomato (%)

Reason	Producers ($N = 44$)	Wholesaler ($N = 29$)	Retailer ($N = 22$)
Unfavorable weather condition	9.1	20.6	13.6
Disease and pests	61.4	20.7	22.7
Damage during harvest	54.5	30	22.7
Delay of harvesting	11.4		
Damage during transport	13.6	48.3	18.2
Lack of market	11.4	20.7	27.3
Damage during packing	2.3	10.3	9.1

Source: Own Survey, 2015

Practices to reduce produce loss

The tomato value chain actors take different measures to reduce postharvest loss or produce damage (Table 13). Producers focus on appropriate harvesting practices and identification of potential buyers before harvesting, while village collectors and wholesalers give due attention to better transportation infrastructure, and produce handling and management. Retailers on the other hand manage the situation through storing the produce under shade/cool areas. The FGDs and KIIs results indicate that produce cooling and storing under shade are insignificant in terms of addressing postharvest losses mainly because of limited capacities and access to necessary facilities and services. A recent report by [33] also indicates that farmers mainly sell vegetables immediately after harvesting as a means of reducing postharvest loss and costs associated with storage and transportation of such perishable fresh produce. Consequently, the information generated through this study has wide applications for other regions in Ethiopia.

Table 13 Proportion of value actors using different means to reduce spoilage (%)

Practice to reduce spoilage	Producers ($N = 60$)	Wholesaler ($N = 29$)	Retailer ($N = 22$)
Collect during cool weather	23.3	10.3	0
Careful handling during harvest	48.3	34.5	18.75
Store under shade	10.0	3.4	37.5
Store in a cool place	1.7	17.2	12.5
Take care during transport	6.7	31.0	6.25
Use padding/cushioning material during transport		–	0
Harvest after buyers identified	23.3		
Sell at lower price		10.3	6.25
Nothing	8.3	3.4	12.5

Source: Own Survey, 2015

SUMMARY AND CONCLUSIONS

Significant postharvest losses occur along the tomato value chain with detrimental effects on the incomes of smallholder farmers and traders. About a quarter of tomato produced is damaged and puts out of normal use. The forms of damage include physical or mechanical damage, disease and/or insect pest infection, and/or poor shape, color and size of produce or combinations of these factors thereof. At farm level, postharvest loss is a continuum of disease and pest attack, lack of access to appropriate tools and skills during harvesting, poor postharvest handling and lack of market to sell the produce immediately after harvest. Postharvest loss which occurred at one value chain node extends to the other chain actors and also aggravates along the value chain due to poor handling, transporting, storage and ambient temperature which deteriorate the produce quality.

Fresh produce handling and marketing facilities such as cooling facilities are lacking. To reduce the postharvest losses of tomato, serious interventions are needed, including skill building to improve preharvest crop management practices, capacity for postharvest handling including cool storage, improving market information, facilities and services.

FOOTNOTES

1 Kebeles are the lowest administrative unit below district in Ethiopia.

ABBREVIATIONS

EHDA: Ethiopian Horticulture Development Agency

FAO: Food and Agricultural Organization of the United Nations

FGD: focus group discussion

KII: key informant interview

USDA: United States Department of Agriculture

WFLO: World Food Logistics Organization

WorldVeg: World Vegetable Center

ACKNOWLEDGEMENTS

The support for this research work was provided by the Bureau for Food Security, US Agency for International Development, under the terms of Award No. AID-BFS-IO-1200004 through: The World Vegetable Center, Eastern and Southern Africa in Arusha, Tanzania. The opinions expressed in this publication are those of the author(s) and do not necessarily reflect the views of the US Agency for International Development.

Ethical approval and consent to participate

Appropriate ethical clearance was sort verbally from the sector Ministry of Agriculture and local government authorities and respective communities prior to the study. All respondents who participated in the field study were duly informed about the purpose of the study and their right to decline participation in the study, and verbal consent was obtained from all participants prior to interviews.

Funding

Funding for this study was provided by the Bureau for Food Security of the US Agency for International Development, Washington, D.C., USA via the World Vegetable Center (WorldVeg).

REFERENCES

1. Adeoye IB, Odeleye OMO, Babalola SO, Afolayan SO. Economic analysis of tomato losses in Ibadan Metropolis, Oyo State, Nigeria. Afr J Basic Appl Sci. 2009;1(5–6):87–92.
2. Adish A. Micronutrient deficiencies in Ethiopia: present situation and way forward. 2012. http://www.epseth.org/a/files/Micronutrient%20Deficiencies%20in%20Ethiopia.pdf. Accessed 11 Nov 2014.
3. Aklilu S. Research achievement on variety development and seed production of vegetable crops in Ethiopia. In: Chadha ML, Altoveros EC, Nono-Womdim R, Mndiga H editors. AVRDC Africa Regional Program 2000. Varietal evaluation and seed production of vegetable crops held at AVRDC Africa Regional Program, Arusha, Tanzania, 29th September to 5th October 1997. Asian Vegetable Research and Development Center—Africa regional Program, Arusha, Tanzania. AVRDC ARP Publication No. 2000-2; 2000; p. 6–11.
4. Aramyan LH, van Gouh JB. Reducing postharvest food losses in developing economies by using a Network of Excellence as an intervention tool, paper presented at the 2014 IFAMA and CCA Food and Agribusiness World Forum 'People Feed the World' in Cape Town, South Africa, June 15–19; 2014.
5. Ayana A, Afari-Sefa V, Emana B, Dinssa FF, Balemi F, Temesgen M. Analysis of vegetable seed systems and implications for vegetable development in the Humid Tropics of Ethiopia. Int J Agric For. 2014;4(4):325–37.
6. Baredo Y. Gamo Gofa zone diagnosis and planning document, Livestock and Irrigation Value Chains for Ethiopian Smallholders (LIVES) project. Ethiopia: Addis Ababa; 2012.
7. EHDA, Ethiopian Horticulture Development Agency. Ethiopian Horticulture Sector Statistical Bulletin. Issue 01, Addis Ababa Ethiopia; 2012. Available online at: http://ehda.gov.et/Downloads/Statistical_Bulletin%282012%29.pdf. Accessed 17 Nov 2013.
8. Emana B, Gebremedhin H. Constraints and opportunities of horticulture production and marketing in eastern Ethiopia. DCG Report No. 46, Addis Ababa, Ethiopia; 2007.
9. Emana B, Afari-Sefa V, Dinssa FF, Ayana A, Balemi F, Temesgen M. Characterization and assessment of vegetable production and marketing systems in the Humid Tropics of Ethiopia. Q J Int Agric. 2015;54(2):163–85.
10. Ethiopian Investment Agency. Investment opportunity profile for production of fruits and vegetables in Ethiopia; 2012.
11. Etebu E, Nwauzoma AB, Bawo DDS. Postharvest spoilage of tomato (Lycopersicon esculentum Mill.) and control strategies in Nigeria. J Biol Agric Healthc. 2013;3(10):51–61.
12. Gebre E. An overview of horticultural crops with emphasis on vegetables production in Ethiopia, paper presented at Tuta Absoluta Workshop—Meeting the challenge of

the tomato leaf miner, Nov 26–28, 2013, Intercontinental Hotel, Addis Ababa, Ethiopia; 2013.
13. Genanew T. Effect of post harvest treatments on storage behavior and quality of tomato fruits. World J Agric Sci. 2013;9(1):29–37.
14. Getinet H, Workneh TS, Woldetsadik K. Effect of maturity stages, variety and storage environment on sugar content of tomato stored in multiple pads evaporative cooler. Afr J Biotechnol. 2011;10(80):18481–92.
15. FAO. Role of agro-industry in reducing food losses in the middle East and North Africa region; 2012.
16. Hussen S, Beshir H, Yitbarek Woldehawariyat Y. Postharvest loss assessment of commercial horticultural crops in South Wollo, Ethiopia "Challenges and Opportunities". Food Sci Qual Manag. 2013;17:34–9.
17. Helvetas. Consequences of post-harvest losses to vegetable farmers in Tanzania: a case of Singida region. HELVETAS Swiss Intercooperation Survey Report 2014. 2014; p. 32.
18. Kader AA. A perspective on postharvest horticulture (1978–2003). HortSciience. 2003;38(5):1004–8.
19. Kader AA. Increasing food availability by reducing postharvest losses of fresh produce. In: Proceedings of 5th international Postharvest symposium. Acta Hort 682, 2169–78; 2005.
20. Kader AA. The return on investment in postharvest technology for assuring quality and safety of horticultural crops. J Agric Invest. 2006;4:45–52.
21. Kahsay Y, Abay F, Belew D. Intra row spacing effect on shelf life of onion varieties (*Allium cepa* L.) at Aksum, northern Ethiopia. World J Agric Res Food Saf. 2013;1(2):34–42.
22. Kaminski J, Christiaensen L. Post-harvest loss in Sub-Saharan Africa—What Do Farmers Say? Policy Research Working Paper 6831, World Bank; 2014.
23. Kitinoja L. Identification of appropriate postharvest technologies for improving market access and incomes for small horticultural farmers in sub-Saharan Africa and south Asia. WFLO Grant Final Report, 2010; p. 323.
24. Kitinoja L, AlHassan HY. Identification of appropriate postharvest technologies for small scale horticultural farmers and marketers in Sub-Saharan Africa and South Asia: Part 1. Postharvest losses and quality assessments. Acta Hort. 2012;934:31–40.
25. Negasi TA, Nigussie DR, Kebede W, Lemma D, Abuhay T. Characterization of soil nutrient management and postharvest handling practices for onion production in the Central Rift Valley region of Ethiopia. Agric For Fish. 2013;2(5):184–95.
26. Olayemi FF, Adegbola JA, Bamishaiye EI, Daura AM. Assessment of post- harvest challenges of small scale farm holders of tomatoes, bell and hot pepper in some local government areas of Kano State, Nigeria. Bayero J Pure Appl Sci. 2010;3(2):39–42.
27. Parfitt J, Barthel M, Macnaughton S. Food waste within food supply chains: quantification and potential for change to 2050. Philos Trans R Soc. 2010;365:3065–81.
28. Prigojin I, Fallik E, Qat Y, Ajalin I, Allam H, Ezzat M, Bader M. Middle East regional agricultural program: survey on postharvest losses of tomato fruit and table grapes. In: Proceedings of the 5th international postharvest symposium, June 6–11 in Verona, Italy. Acta horticulture; 2005.
29. Tabor G, Yesuf M. Mapping the current knowledge of carrot cultivation in Ethiopia. Charlottenlund: Technical Report Submitted to Carrot Aid; 2012.
30. Tadesse F. Postharvest losses of fruits and vegetables. J Hortic. 1991;270:261–70.
31. Teka TA. Analysis of the effect of maturity stage on the postharvest biochemical quality characteristics of tomato (*Lycopersicon esculentum* MILL.) fruit. Int Res J Pharm Appl Sci. 2013;3(5):180–6.
32. Ukech JA, Chiejina NV. Preliminary investigations of the cause of post-harvest fungal rot of tomato. IOSR J Pharm Biol Sci. 2012;4(5):36–9.

33. USAID. Cost-benefit analysis of the potatoes, onions, and tomatoes value chains in Ethiopia: graduation with resilience to achieve sustainable development (GRAD) project final report; 2013.
34. USDA. The commercial storage of fruits, vegetables and nursery stocks; 2004. Agric. Handbook 66. http://www.ba.ars.usda.gov/hb66/contents.html. Accessed 2 April 2014.
35. WFLO, World Food Logistics Organization. Identification of appropriate postharvest technologies for improving market access and incomes for small horticultural farmers in Sub-Saharan Africa and South Asia. 2010; WFLO Grant Final Report, University of California, UC, Davis, USA.
36. Workneh TS. Feasibility and economic evaluation of low-cost evaporative cooling system in fruit and vegetables storage. Afr J Food Agric Nutr Dev. 2010;10(8):2984–97.
37. Workneh TS, Osthoff G, Steyn MS. Preharvest treatment, disinfections and storage environment on quality of tomato. J Food Sci Technol. 2011. doi:10.1007/s13197-011-0391-3.
38. Workneh TS, Osthoff G, Steyn MS. Influence of pre-harvest and postharvest treatments on stored tomato quality. Afr J Agric Res. 2011;6(12):2725–36.
39. Workneh TS, Osthoff G, Steyn MS. Physiological and chemical quality of carrots subjected to pre-and postharvest treatments. Afr J Agric Res. 2011;6(12):2715–24.
40. Workneh TS, Woldetsadik K. Forced ventilation evaporative cooling: a case study on banana, papaya, orange, mandarin, and lemon. Trop Agric. 2004;81(3):179–85.

CHAPTER 7

Comparative Proteomic and Metabolomic Profiling of Citrus Fruit With Enhancement of Disease Resistance by Postharvest Heat Treatment

Ze Yun†, Huijun Gao†, Ping Liu, Shuzhen Liu, Tao Luo, Shuai Jin, Qiang Xu, Juan Xu, Yunjiang Cheng* and Xiuxin Deng

Key Laboratory of Horticultural Plant Biology of Ministry of Education, Huazhong Agricultural University, Wuhan, Hubei 430070, P.R. China

ABSTRACT

Background

From field harvest to the consumer's table, fresh citrus fruit spends a considerable amount of time in shipment and storage. During these processes, physiological disorders and pathological diseases are the main causes of fruit loss. Heat treatment (HT) has been widely used to maintain fruit quality during postharvest storage; however, limited molecular information related to this treatment is currently available at a systemic biological level.

Results

Mature 'Kamei' Satsuma mandarin (*Citrus unshiu* Marc.) fruits were selected for exploring the disease resistance mechanisms induced by HT during postharvest storage. Proteomic analyses based on two-dimensional gel electrophoresis (2-DE), and metabolomic research based on gas chromatography coupled to mass spectrometry (GC-MS), and liquid chromatography quadrupole time-of-flight mass spectrometry (LC-QToF-MS) were conducted. The results show resistance associated proteins were up-regulated in heat treated pericarp, such as beta-1, 3-glucanase, Class III chitinase, 17.7 kDa heat shock protein and low molecular weight heat-shock protein. Also, redox metabolism enzymes were down-regulated in heat treated pericarp, including isoflavone reductase, oxidoreductase and superoxide dismutase. Primary metabolic profiling revealed organic acids and amino acids were down-regulated in heat treated pericarp; but significant accumulation of metabolites, including tetradecanoic acid, oleic acid, ornithine, 2-keto-d-gluconic acid, succinic acid, turanose, sucrose, galactose, myo-inositol, glucose and fructose were detected. Noticeably, H_2O_2 content decreased, while, lignin content increased in heat treated pericarp compared to the control, which might increase fruit resistibility in response to external stress. Also, flavonoids, substances which are well-known to be effective in reducing external stress, were up-regulated in heat treated pericarp.

CONCLUSIONS

This study provides a broad picture of differential accumulation of proteins and metabolites in postharvest citrus fruit, and gives new insights into HT improved fruit disease resistance during subsequent storage of 'Kamei' Satsuma mandarin. Interpretation of the data for the proteins and metabolites revealed reactive oxygen species (ROS) and lignin play important roles in heat treatment induced fruit resistance to pathogens and physiological disorders.

Keywords: Disease resistance, H_2O_2, Heat treatment, Lignin postharvest storage, Metabolomics, Proteomics

BACKGROUND

Fresh citrus fruits, from harvest to human consumption, require a lengthy period for shipping, storing and marketing. During that time, the main damage to fruit comes from biological diseases and physical damage. Storage pre-treatments, such as biotic or abiotic typological treatments, can be used to inhibit disease and extend the shelf-life. Among the various types of treatment available, heat treatment has previously been reported to induce disease resistance in apples, citrus, grapes, peaches, pears and mangos against a multitude of pathogens [1, 2, 3].

In the fruit industry, hot water treatment is an alternative physical treatment, which has been used in citrus fruit storage since 1922 [4]. More recently, hot air and hot water systems have been widely adopted in several countries to inhibit postharvest decay of fruits and to alleviate chilling injuries during storage. Heat treatment (HT) has the advantage of minimizing surface infection in pericarp without leaving any chemical residues [5]. HT can prevent the development of plant pathogens and disease by heat-killing disease organisms or by inhibiting pathogen mycelial growth, by removing surface insects and by structurally altering epicuticular wax [6]. HT can also enhance fruit resistance to chilling injury in cold-sensitive cultivars, and help the fruit retain quality during cold storage and help maintain shelf life [7]. Aside from controlling pathogenic organisms, HT may also reduce disease by changing the physical characteristics of the interior pericarp. For example, HT not only delayed fruit ripening and coloration, inhibited fruit softening, and slowed ethylene production [3], it also increased fruit hardness, as well as increasing the content of polyphenolics, flavonoids and dopamine [6, 7, 8]. However, less is understood about the regulatory genes which are active in the response to microbial attack, and there is no exhaustive metabolomics profile detailing the changes in heat treated pericarp.

In the past few decades, researchers have given increasing attention to the study of gene expression in heat treated fruits. Sapinitskaya *et al.*[9] reported HT activates stress genes and lipid modification genes in response to chilling stress in grapefruit, such as glutathione-dependent formaldehyde dehydrogenase, dehydrin, heat shock protein (HSP), ascorbate peroxidase, superoxide dismutase, fatty acid desaturase and lipid transfer protein. Some scientists discovered HT induces phenylalanine ammonia lyase and accumulates phenylpropanoid compounds, which play a vital role in heat-induced fruit resistibility [2]. Dixon *et al.*[10] studied the functions of phenylpropanoid compounds in plants involved in local and systemic signalling for defence-gene induction. However, little information is available related to which particular class of phenylpropanoid compounds plays a core role in defensive functions, and even less is known about the regulatory genes which orchestrate the rapid induction of phenylpropanoid defences in response to HT. Proteome analysis reveals the induction of HSPs, allergen proteins, dehydrin, and other proteins involved in the stress response. Repression of polyphenol oxidase caused by HT may constitute the molecular basis for the protection against chilling stress in peach fruit [3]. However, few proteins involved in HT were reported to have induced fruit resistance to deterioration, especially in citrus fruit.

Also, proteomics analysis and metabolic profiling approaches make possible comparative parallel analyses of global changes at the proteome and metabolome levels, facilitating an understanding of the relationships between changes in specific proteins and subsequent alterations in metabolites in response to HT. In the present study, fruits of 'Kamei' Satsuma mandarin (*Citrus unshiu* Marc.) received HT. The comparative proteomics analysis and metabolic profiling (the primary metabolome and the secondary metabolome) analysis were performed based on two-dimensional electrophoresis (2-DE), gas chromatography and mass spectrometry (GC-MS) and high-performance liquid chromatography/electrospray ionization-time of flight-mass spectrometry (HPLC-QTOF-MS). Comprehensive physiological and biochemical analysis were completed. Results will provide additional theoretical evidence to improvement of disease resistance in heat treated pericarp.

RESULTS

Satsuma mandarin is one of the main citrus crops in East Asia, including China, South Korea and Japan. However, this cultivar has poor storability mainly because it is susceptible to biotic stress and physical damage. HT improved fruit resistance to stress and damage effectively and inhibited postharvest decay of fruits. In present study, HT fruits were dipped in a 52°C warm water bath for 2 min, and control fruits were dipped in a 25°C water bath for 2 min. After treatments, fruits were stored in a ventilated warehouse (storage temperature: 12–16°C; relative humidity: 90–95%). Samples were collected at 2 h, and 1, 2, 3, 4, 6, 9, 12, 16, 20, 24, 28, 32, 38, 44, 50 and 60 days after treatments (DAT) for further analysis.

HT increased fruit weight loss slightly, but did not affect fruit quality

Fruit quality directly determines its commercial value. Parameters related to fruit quality, including weight loss, respiration rate, and soluble solids content, were evaluated to investigate the influence of HT on fruit quality. Weight loss increased significantly in heat treated pericarp compared to control pericarp during the storage period (Figure 1A). However, no obvious changes were observed in the respiration rate and soluble solids content between HT and control fruits during the entire storage period

(Figures 1B, C). Since soluble solids content is a well-known standard in measuring fruit internal quality and respiration acts as an important standard for fruit storability, the minor changes in soluble solids content and respiration indicated that heat treatment did not affect fruit flesh quality.

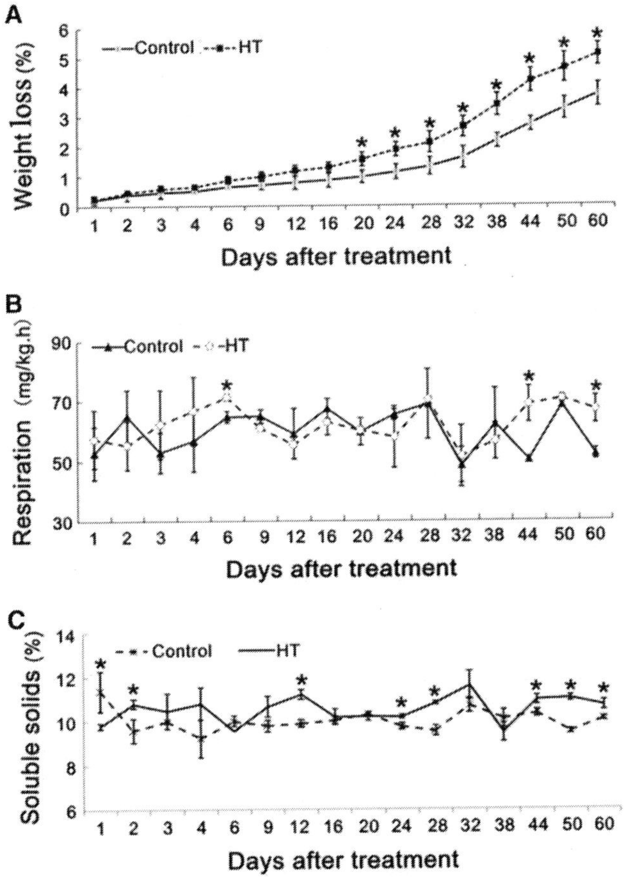

Figure 1 The influence of heat treatment on 'Kamei' Satsuma mandarin fruit quality during the entire period of storage at ambient temperature. Parameters related to fruit quality, including weight loss, respiration rate, and soluble solids content, were evaluated. 30 fruits (each sample) and three replications were applied to quality analysis. Control: fruits dipped in a water bath at 25°C for 2 min; HT: fruits dipped in a water bath at 52°C for 2 min. An analysis of statistically significant differences was conducted between HT and control pericarp at the same period using Student's t-test. *: significant difference ($P < 0.05$). Mean values and SE bars are provided.

HT suppression of *Penicillium* germination on pericarp

Artificial inoculation tests were conducted with the blue mould *Penicillium italicum* to assess the effects of HT on the fruit's response to biotic stresses. Few spores had germinated on the peel of HT fruits 4 d after inoculation, but about 90% of the control fruits showed typical symptoms of blue mould development at the inoculation sites (Figure 2A). As for the rate of disease incidence, the incidence of disease on control fruit was significantly higher than on HT fruit from 4–7 d after inoculation (Figure 2B); lesion diameters were significantly higher in control fruit than HT fruit from 4–7 d after inoculation (Figure 2C). Based on the above experiments, HT effectively inhibited the germination of *Penicillium* on pericarp.

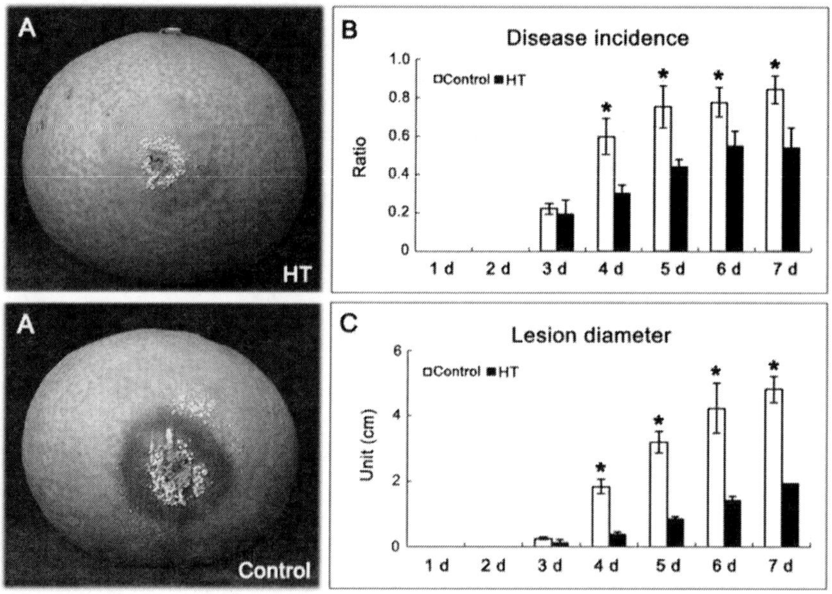

Figure 2 Effects of HT on disease development in 'Kamei' Satsuma mandarin fruit during storage at ambient temperature. 150 fruits were used for fungal infection. A uniform lesion (4 mm deep, 3 mm wide) was made at the equator of the fruit using a sterile nail. Aliquots of 20 μl suspension of *Penicillium italicum* at 1×10^5 spores ml^{-1} were inoculated into each wound site. After fungal inoculation, the incidence of disease in fruit was detected at 1, 2, 3, 4, 5, 6 and 7 days after *Penicillium* inoculation. Statistically significant differences analysis was conducted between HT and control pericarp during the same period using Student's t-test. *: significant difference ($P < 0.05$). Mean values and SE bars are provided. **A**: Fruit status at 3 d after *Penicillium* inoculation; **B**: Disease incidence ratio; **C**: Fruit lesion diameter.

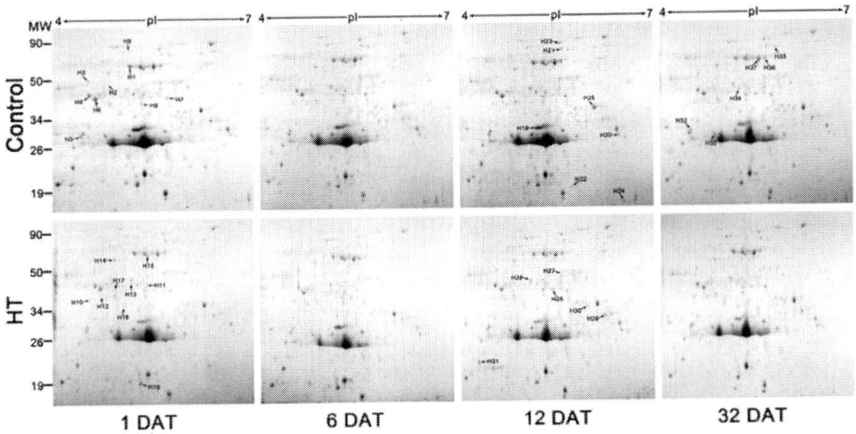

Figure 3 Representative 2-DE profiles of proteins from different treated Satsuma mandarin fruits during postharvest storage at ambient temperature. Total proteins were extracted from the fruit at 1, 6, 12 and 32 DAT. Proteins (l mg) were separated in the first dimension on IPG strip (17 cm, pH 4–7) and in the second dimension on a 15% SDS-PAGE gel. More than 600 spots were observed. 37 of 50 differential accumulated proteins were identified using MALDI TOF/TOF. The proteins were classified into 5 categories, including glycolysis and TCA cycle (9 proteins, H1-9), redox reactions (9 proteins, H10-18), stress response (7 proteins, H19-25), other metabolism (6 proteins, H26-31), protein folding (6 proteins, H32-37). Table one shows detailed information related to the proteins.

Differentially accumulated proteins in HT citrus pericarp

To explore the differential expression of proteins between HT and control at the same period, the pericarp of 'Kamei' Satsuma mandarin fruit were analysed with samples collected at 1, 6, 12 and 32 DAT. After 2-DE analysing, more than 600 protein spots were reproducibly detected (Figure 3). The protein relative abundance was obtained with the PDQuest 2-D software version 7.4. Protein spots were scored only when they were reproducibly observed in four independent replicates. Data were normalized as a percentage of the total density of all spots on the corresponding gel. Comparative analysis was carried out using 1.5-fold comparative analysis and one-way analysis of variance (ANOVA). Result showed there were only 50 differentially displayed protein spots whose abundance was altered at least 1.5-fold when comparing the HT and control during the same period of storage. Of these, 37 were identified using Applied Biosystems 4800 matrix-assisted laser desorption/ionization-time-of-flight tandem mass spectrometry (MALDI-TOF MS) based on tryptic peptide sequences (Figure 3, Table 1). Based on previously reported functions of these proteins, they were classified into five categories (Table 1): glycolysis and TCA cycle (nine proteins, 24%), redox regulation (nine proteins, 24%), stress response (seven proteins, 19%), other metabolism (six proteins, 16%), and protein folding (six proteins, 16%).

Table 1 Identities of HT induced/decreased proteins in the peels of Satsuma mandarin fruit

Glycolysis and TCA cycle: 9								
No.	**Protein name**	**Protein accumulation** □CK ■HT	**Source organism**	**GI no.**	**Theo Mr/pI**	**Expt Mr/pI**	**PC**	**MS**
H1	Enolase	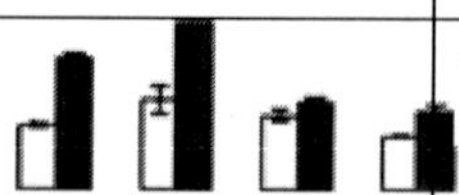	*Glycine max*	42521309	47.97/5.31	50.84/5.23	8	443
H2	Phosphoglycerate kinase, putative	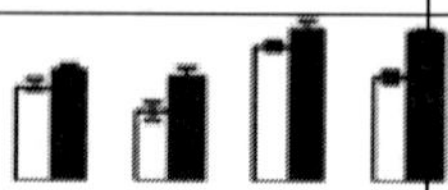	*Arabidopsis thaliana*	15223484	50.02/8.27	41.88/4.90	9	108
H3	Putative 2-oxoglutarate dehydrogenase E2 subunit	□CK ■HT	*Oryza sativa*	48716382	49.46/6.78	43.64/4.62	4	105
H4	Malate dehydrogenase, cytoplasmic	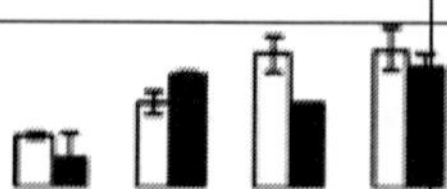	*Beta vulgaris*	11133601	35.81/5.89	37.61/4.61	7	347
H5	Triosephosphate isomerase, cytosolic (TIM)	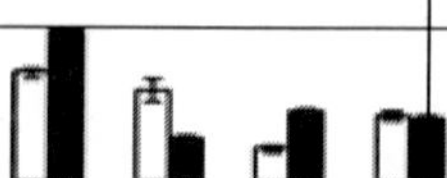	*Zea mays*	136063	27.23/5.52	24.97/4.47	9	310

H6	Malate dehydrogenase, mitochondrial precursor	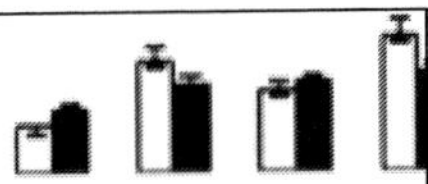	*Citrullus lanatus*	126896	36.40/8.88	36.36/4.71	8	376
H7	Pyruvate dehydrogenase E1 beta subunit isoform 2	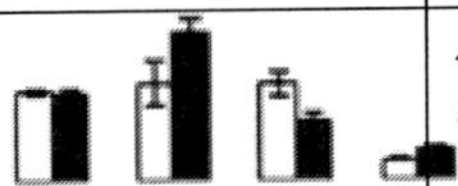	*Zea mays*	162458637	40.23/5.56	35.91/5.72	5	302
H8	NAD-malate dehydrogenase	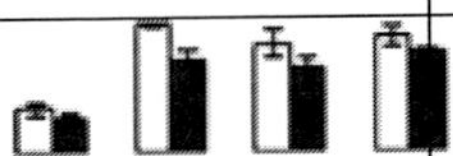	*Nicotiana tabacum*	5123836	43.67/8.03	34.10/5.38	11	436
H9	Putative2,3-bisphosphoglycerate-independent phosphoglycerate mutase	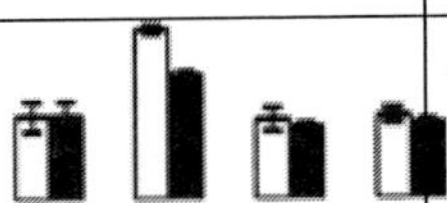	*Arabidopsis thaliana*	15982735	60.67/5.27	61.18/5.12	9	207

Redox regulation: 9

No.	Protein name	Protein accumulation □CK ■HT	Source organism	GI no.	Theo Mr/pI	Expt Mr/pI	PC	MS
H10	Isoflavone reductase related protein		*Pyrus communis*	3243234	33.80/6.02	33.12/4.53	5	92
H11	Oxidoreductase, zinc-binding dehydrog	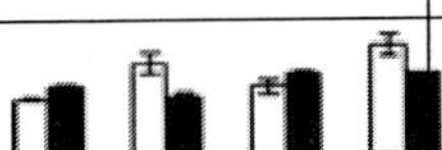	*Arabidopsis thaliana*	15220854	41.13/8.46	37.58/5.41	6	93

	enase family protein							
H 1 2	Putative NAD(P)H oxidoreductase, isoflavone reductase	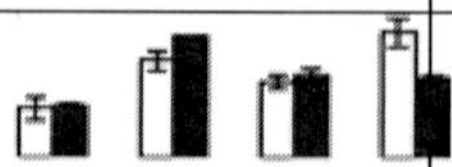	*Arabidopsis thaliana*	19310585	34.24 /6.61	34.43/4.71	4	67
H 1 3	Aldo/keto reductase family protein	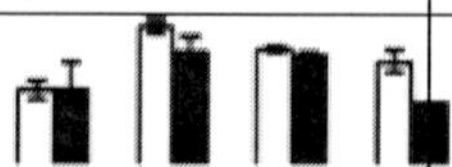	*Arabidopsis thaliana*	42571931	27.76 /6.33	38.36/5.16	6	164
H 1 4	2-oxoacid dehydrogenase family protein	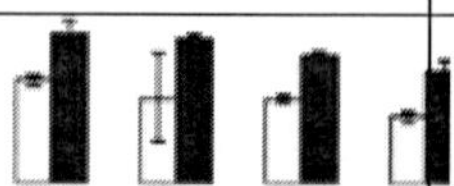	*Arabidopsis thaliana*	15240454	50.27 /9.19	47.44/4.92	7	168
H 1 5	Mitochondrial processing peptidase	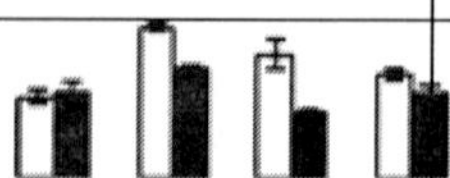	*Solanum tuberosum*	587562	54.62 /5.99	49.29/5.42	4	65
H 1 6	Copper/zinc superoxide dismutase		*Citrus limon*	33340236	15.20 /5.46	17.02/5.27	4	263
H 1 7	Putative NAD(P)H oxidoreductase, isoflavone reductase	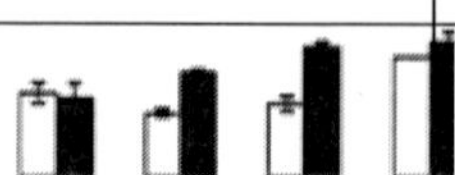	*Arabidopsis thaliana*	19310585	34.24 /6.61	34.47/4.92	3	85
H 1 8	Chloroplast stromal ascorbate peroxidase	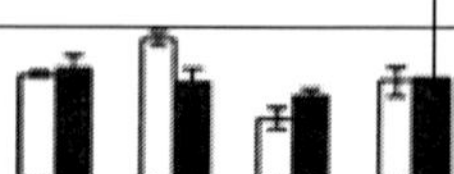	*Vigna unguiculata*	45268439	39.95 /7.06	31.00/5.06	9	191

Stress response: 7								
No.	**Protein name**	**Protein accumulation** 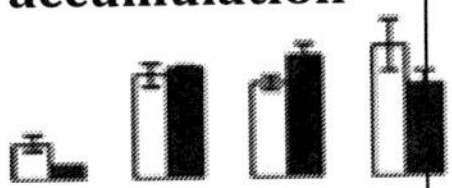	**Source organism**	**GI no.**	**Theo Mr/pI**	**Expt Mr/pI**	**PC**	**MS**
H19	Abscisic stress ripening-like protein	□CK ■HT	*Prunus persica*	16588758	20.74/5.68	29.89/5.25	4	316
H20	Class III chitinase	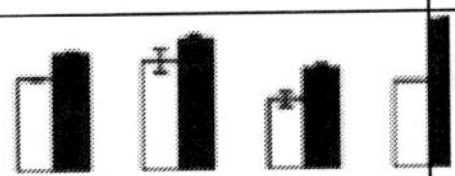	*Benincasa hispida*	5919201	33.03/9.8	26.84/6.32	3	65
H21	Heat shock protein 60	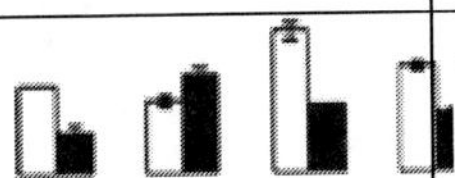	*Prunus dulcis*	24637539	58.06/5.26	88.79/5.66	10	205
H22	17.7 kDa heat shock protein	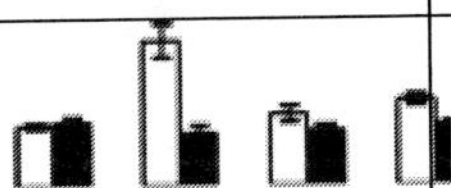	*Carica papaya*	37933812	17.77/6.39	19.12/5.95	2	98
H23	Putative heat shock 70 KD protein, mitochondrial precursor	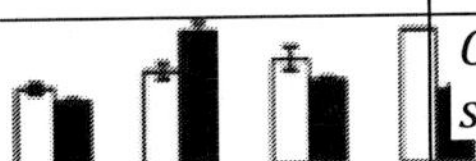	*Oryza sativa*	27476086	70.68/5.45	92.99/5.65	17	358
H24	Low molecular weight heat-shock protein		*Pseudotsuga menziesii*	1213116	18.18/5.82	17.78/6.39	4	122
H25	Beta-1, 3-glucanase	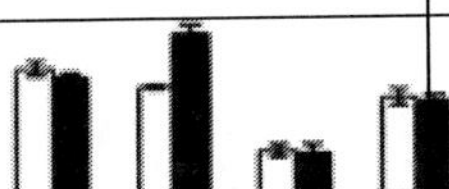	*Citrus sinensis*	2274915	37.32/9.19	34.38/6.18	7	98

Other metabolism: 6								
No.	**Protein name**	**Protein accumulation** 	**Source organism**	**GI no.**	**Theo Mr/pI**	**Expt Mr/pI**	**PC**	**MS**
H26	Fiber annexin	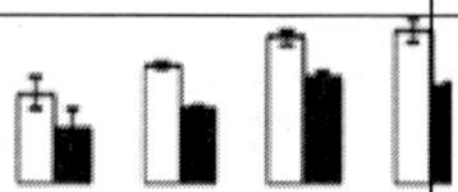	*Gossypium hirsutum*	3493172	36.20/6.34	36.23/5.53	5	70
H27	Alpha-amylase	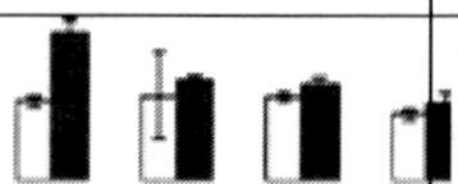	*Citrus reticulata*	20336385	17.05/4.91	44.16/5.64	8	177
H28	Glutamine synthetase	□CK ■HT	*Elaeagnus umbellata*	47933888	39.33/6.10	39.38/5.16	5	134
H29	Putative inorganic pyrophosphatase	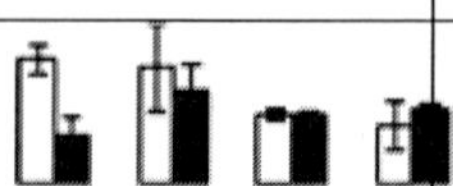	*Oryza sativa*	46805453	20.45/4.61	28.58/6.31	4	74
H30	Thiazole biosynthetic enzyme, chloroplast precursor	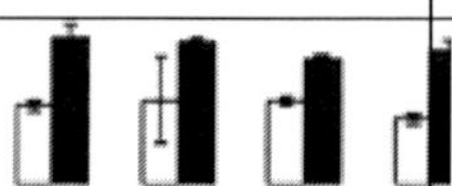	*Citrus sinensis*	6094476	37.74/5.40	31.35/6.04	12	447
H31	Lipocalin protein	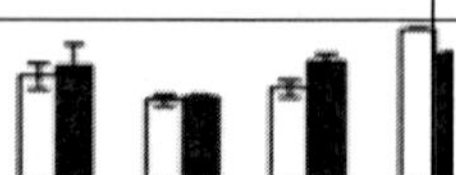	*Capsicum annuum*	50236424	21.41/7.66	19.0/4.33	3	143
Protein folding: 6								
No.	**Protein name**	**Protein accumulation** 	**Source organism**	**GI no.**	**Theo Mr/pI**	**Expt Mr/pI**	**PC**	**MS**

H32	Catalytic/ coenzyme binding	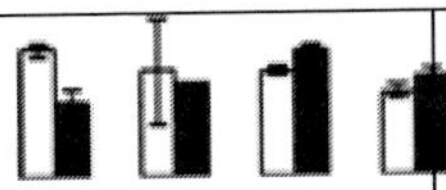	*Arabidopsis thaliana*	18404496	34.97/8.37	25.54/4.50	6	252
H33	Chaperonin-60 beta subunit	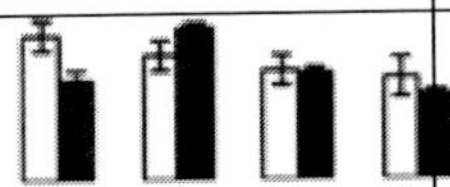	*Solanum tuberosum*	1762130	63.26/5.72	57.6/5.77	12	283
H34	Protein disulfide isomerase	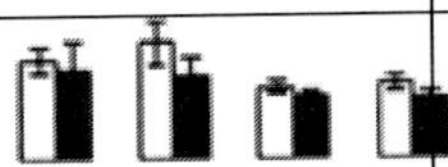	*Zea mays]*	16246192 5	40.43/6.29	37.7/5.26	9	165
H35	Transcription factor APFI	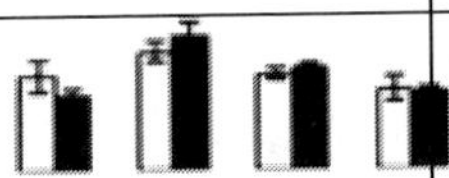	*Arabidopsis thaliana*	13507025	30.14/6.24	28.15/5.14	3	89
H36	Putative ATP synthase beta subunit		*Oryza sativa*	56784991	45.93/5.33	51.13/5.66	20	695
H37	Putative ATP synthase beta subunit	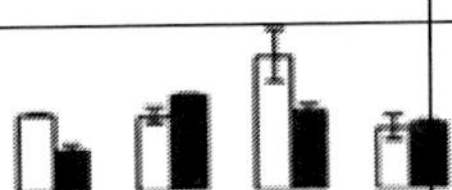	*Oryza sativa*	56784991	45.93/5.33	50.0/5.6	20	808

Samples at 1, 6, 12 and 32 DAT were used for differential analysis of proteins accumulation. After 2-DE, protein spot intensities were quantified using PDQuest 2-D software version 7.4. Comparative and statistical analysis shows there were only 50 differentially displayed protein spots whose abundance was altered at least 1.5-fold between the HT and control during the same time period. Of these, 34 proteins were successfully identified using Applied Biosystems 4800 matrix-assisted laser desorption/ionization-time-of-flight tandem mass spectrometry (MALDI-TOF MS) analysis. Protein relative accumulation is represented by the column configuration, and accumulation at 1, 6, 16 and 32 DAT is shown from left to right. Protein name and GI No. are from the National Center for Biotechnology Information (NCBI) database (http://www.ncbi.nlm.nih.gov/guide/). Experimental molecular mass and pI (Exp. M/pI) were calculated using PDQuest software (version 7.4), and mass is shown in kDa. PC: Peptide count. MS: Mascot score.

Differentially accumulated metabolites in HT citrus pericarp

Differentially accumulated metabolites were studied to investigate the changes in metabolic composition between HT and control pericarp. The biochemical changes were determined by GC-MS and HPLC-QTOF-MS. A parallel analysis of non-polar and polar extracts using methanol extraction enabled the detection of a large number of compounds of different classes.

A total of 62 identified metabolites were monitored in the same sample sets using GC-MS, including mainly alcohols, amino acids, sugars, organic acids, and fatty acids. After conducting an analysis of significance with ANOVA ($P < 0.05$), 45 detected metabolites, mostly organic acids, sugars and amino acids, were found to be significantly up/down-regulated in heat treated pericarp when compared to control pericarp with the same storage period (Table 2).

Table 2 Differential accumulated primary metabolites in heat treated pericarp compared to control pericarp

Sugars	**2 h**	**1 d**	**6 d**	**32 d**
Turanose	1.19 *	1.13	1.36 *	1.25 *
Galactose	1.44 *	1.59 *	1.09	1.17
Fructose	1.29 *	1.36 *	1.02	1.08
Glucose	1.4 *	1.51 *	1.06	1.14
Sucrose	1.21 *	1.09	1.07	1.19
4-Keto-glucose	1.59 *	1.09	1.43 *	0.84
Arabinose	1.36 *	0.71 *	1.23 *	0.18 *
Mannose	1.06	1.11	0.88	0.73 *
Xylose	0.82 *	1.37 *	0.82 *	0.98
Glucopyranose	0.81 *	2.09 *	0.84	1.41 *
Organic acids	**2 h**	**1 d**	**6 d**	**32 d**
Pentonic acid	1.77 *	0.78 *	1.04	1.46 *
Succinic acid	1.13	1.2 *	1.26 *	1.54 *

Hydroxypyruvic acid	1.16	1.07	1.28 *	0.71 *
2-Ketoglutaric acid	1.06	1.1	0.65 *	1.31 *
2-Keto-d-gluconic acid	1.05	1.56 *	1.13	2.52 *
Gluconic acid	1	1.81 *	0.92	1.04
Isocitric acid	0.85	2.1 *	1.11	1.15
2,3,4-Trihydroxybutyric acid	0.84	1.67 *	0.61 *	1.2 *
Ethanedioic acid	0.83 *	1.26 *	1.11	1.55 *
Citric acid	0.83 *	1.53 *	1.22 *	1.92 *
Malic acid	0.63 *	1	0.77 *	1.41 *
4-N-Trimethylsilylmethylaminobutyric acid	0.56 *	1.03	0.76 *	1.33 *
GABA	0.41 *	1.23 *	0.76 *	1.25 *
Amino acids	**2 h**	**1 d**	**6 d**	**32 d**
Ornithine	2.63 *	3.69 *	2.34 *	2.87 *
Valine	0.48 *	0.99	0.46 *	0.7 *
5-oxo-L-proline	0.45 *	1.08	0.73 *	1.14
Glycine	0.37 *	1.42 *	0.65 *	1.44 *
Alanine	0.31 *	2.22 *	0.69 *	1.25 *
Threonine	0.29 *	1.12	0.47 *	0.92
Aspartic acid	0.14 *	0.99	0.35 *	0.8 *
L-proline	0.13 *	0.99	0.19 *	0.65 *

Serine	0.13 *	1.27 *	0.2 *	0.64 *
Glutamine	0.11 *	0.86	0.3 *	0.73 *
Asparagine	0.03 *	0.34 *	0.25 *	0.89
Alcohols	**2 h**	**1 d**	**6 d**	**32 d**
Glycerol	1.61 *	1.2 *	0.91	2.23 *
Arabitol	1.05	1.04	0.39 *	0.92
Rhamnitol	0.98	0.97	0.76 *	1.73 *
Sorbitol	0.94	0.68 *	1.87 *	1.8 *
Fatty acids	**2 h**	**1 d**	**6 d**	**32 d**
Oleic acid	UP *	UP *	UP *	UP *
Tetradecanoic acid	UP *	UP *	UP *	UP *
Octadecanoic acid	0.65 *	1.11	0.94	1.66 *
9,12-Octadecadienoic acid	0.57 *	0.84	1.05	0.93
Hexadecanoic acid	0.62 *	1.06	0.99	1.38 *
Hexadecanoic acid,2,3-bisoxypropylester	0.68 *	0.39 *	0.81 *	2.18 *
Others	**2 h**	**1 d**	**6 d**	**32 d**
Phosphate	0.89	1.48 *	1.33 *	1.48 *

Samples at 2 h after treatment and 1, 6 and 32 d after harvest were used for the differential primary metabolic profiling analysis. A 300 mg sample was extracted in 2,700 µl of methanol. Extracts were determined with GC-MS used ribitol as internal standard. A total of 45 metabolites were differentially accumulated in heat treated pericarp. This table shows the ratio (HT/control during the same period of storage). *: significant difference ($P<0.05$). UP: metabolite was detected in the heat treated pericarp but not in the control pericarp.

GC-MS metabolite profiling (primary metabolites) was complemented using HPLC-QTOF-MS analysis of mostly secondary metabolites. Statistical filtering was applied to the mass signals data to estimate the number of up/down-regulated metabolites in heat treated pericarp. After MS/MS analysis, a total of 58 metabolites were found to be significantly regulated (2-fold; $P < 0.05$) in heat treated pericarp when compared to control pericarp during the same period of storage, and 27 of these metabolites were identified (Table 3). The metabolites detected by the HPLC-QTOF-MS technology contained mainly polyphenols and flavonoids (Table 3).

Table 3 Secondary metabolites specifically accumulated in heat treated pericarp

Secondary metabolites	**2 h**	**1 d**	**6 d**	**32 d**
Quercetin-dihexose-deoxyhexose	16.71	14.81	15.42	n.s.
Hesperetin	12.32	2.08	n.s.	n.s.
Naringenin chalcone-hexose	15.04	n.s.	4.77	n.s.
Vanillic acid	14.73	n.s.	n.s.	14.37
Hydroxybenzoic acid-hexose	14.99	n.s.	n.s.	n.s.
Diosmin	4.06	n.s.	n.s.	n.s.
Rutin	3.69	n.s.	n.s.	n.s.
Phloretin-C-diglycoside	14.97	n.s.	0.06	n.s.
Chlorogenic acid	15.96	0.06	0.12	n.s.
Isosakuranetin	n.s.	16.42	n.s.	n.s.
Qucercetin	n.s.	16.03	0.14	0.12

Neohesperidin	n.s.	1.31	0.14	n.s.
Protocatechuic acid	n.s.	0.09	n.s.	n.s.
Caffeic acid hexose	n.s.	0.06	0.20	15.90
Naringenin	n.s.	n.s.	15.31	n.s.
Caffeic acid	n.s.	n.s.	0.09	n.s.
Hydroxylated naringenin-hexose	n.s.	n.s.	0.10	n.s.
Sinensetin	n.s.	n.s.	0.08	n.s.
p-Coumaroyl quinic acid	n.s.	n.s.	0.07	n.s.
Feruloylquinic acid	n.s.	n.s.	0.06	n.s.
Pantothenic acid -hexose	n.s.	n.s.	0.06	n.s.
Quercetin-O-dihexoside	0.08	n.s.	1.13	0.46
Cinnamic acid	0.11	n.s.	n.s.	n.s.
Eriodictyol-O-dihexoside	0.06	n.s.	n.s.	0.06
Ferulic acid	0.42	n.s.	0.09	n.s.
Sinapic acid	0.05	n.s.	0.07	n.s.
Jasmonic acid	0.07	n.s.	0.06	n.s.

Samples at 2 h and 1, 6 and 32 DAT were used for the differential secondary metabolic profiling analysis using the HPLC-QTOF-MS instrument. Four replicates were compared and statistically analysed per sample using Student's t-test; $P < 0.05$. After MS/MS analysis, a total of 58 metabolites were found to be significantly regulated (2-fold) in heat treated pericarp when compared to control pericarp (during the same period of storage), and 27 of these metabolites were identified. This table shows the ratio (HT/control during the same period of storage). n.s.: not significant.

HT up-regulated stress response proteins in citrus pericarp

In disease response, HT induced Beta-1, 3-glucanase (H25) and Class III chitinase (H20) accumulated in pericarp (Table 1), where these proteins are involved in reducing stress resistance in plants. In the stress response, HT up-regulated 17.7 kDa heat shock protein (H22) and low molecular weight heat-shock protein (H24) accumulated in pericarp, especially at early storage stage (Table 1). Also, other proteins accumulated in heat treated pericarp in higher amounts when compared to control pericarp during the same period of storage, which might be indirectly involved in the fruit's response to stress, including the pyruvate dehydrogenase E1 beta subunit isoform 2 (H7), chloroplast stromal ascorbate peroxidase (H18), putative ATP synthase beta subunit (H36), putative inorganic pyrophosphatase, isoflavone reductase related protein (H10), and cytoplasmic malate dehydrogenase (H4, Table 1).

HT down-regulated reactive oxygen species metabolism-related proteins in citrus pericarp

HT down-regulated reactive oxygen species metabolism-related proteins, especially during later periods of storage (Table 1), including oxidoreductase, zinc-binding dehydrogenase family protein (H11), putative NAD(P)H oxidoreductase, isoflavone reductase (H12), aldo/keto reductase family protein (H13), copper/zinc superoxide dismutase (H16), and putative NAD(P)H oxidoreductase, isoflavone reductase (H17). Also, other proteins were down-regulated in heat treated pericarp (relative to control), such as triosephosphate isomerase, cytosolic (H5), abscisic stress ripening-like protein (H19), and thiazole biosynthetic enzyme, chloroplast precursor (H30). In addition, some proteins were up-regulated in heat treated pericarp (relative to control) only in the middle of the experiment, including, phosphoglycerate kinase, putative (H2), pyruvate dehydrogenase E1 beta subunit isoform 2 (H7), aldo/keto reductase family protein (H13), mitochondrial processing peptidase (H15), putative inorganic pyrophosphatase (H29). Putative ATP synthase beta subunit (H36) was up-regulated at the end but down-regulated at the beginning in heat treated pericarps (relative to control).

HT increased contents of sugars and flavonoids in citrus pericarp

Primary metabolite profiling of the HT and control pericarp revealed prominent changes in a group of metabolites during the entire storage phases (Table 2), mainly including sugars and fatty acids. Interestingly, the contents of many kinds of sugars were increased in heat treated pericarp (relative to control) at 2 h after HT, but decreased to the same level with control at later storage phases, such as galactose, fructose, glucose, sucrose and 4-keto-glucose. Some primary metabolites contents were increased in heat treated pericarp (relative to control) during whole storage, including oleic acid, tetradecanoic acid and ornithine (Table 2).

Secondary metabolite profiling of the HT and control pericarp revealed that a group of metabolites were increased in heat treated pericarp (relative to control) during some periods of storage, including quercetin-dihexose-deoxyhexose, hesperetin, hydroxybenzoic acid-hexose, naringenin, naringenin chalcone-hexose, vanillic acid, diosmin, rutin and isosakuranetin, which all might play a vital role in the response to stress.

HT decreased contents of organic acids, amino acids and phenylpropanoids

Primary metabolic profiling revealed organic acids, fatty acids and amino acids were down-regulated in heat treated pericarp when compared to control pericarp, especially at 2 h after heat treatment. Levels of five types of organic acids (citric acid, malic acid, 4-N-Trimethylsilylmethylaminobutyric acid, oxalic acid and GABA) had decreased in heat treated pericarp compared to the control at 2 h after HT (Table 2), but levels of ten types of organic acid increased in heat treated pericarp (relative to the control) at 32 DAT (Table 2). The content of ten types of amino acids were lower in heat treated pericarp than in the control at 2 h after HT, while only five of them were lower in heat treated pericarp than in the control at 32 DAT, although two of them were higher in heat treated pericarp than in the control at 32 DAT (Table 2). In a similar way, compared to organic acids, levels of four types of fatty acids decreased in heat treated pericarp compared to the control at 2 h after heat treatment; levels of three of them increased in heat treated pericarp compared to the control at 32 DAT (Table 2). There is common occurrence here; levels of many types of organic acids, amino acids and fatty acids decreased in heat treated pericarp compared to the control at 2 h after HT, but levels of many of them increased in heat treated pericarp compared to the control at 32 DAT. Compared to increased sugar concentrations, it appears there is a sudden transformation of organic, amino and fatty acids into sugars which is driven by HT; also, levels of many types of organic, amino and fatty acids increased in heat treated pericarp compared to the control during the later stages of storage.

Further metabolic characterization of chemicals which induced resistance in heat treated pericarp was performed using an HPLC-QTOF-MS. Contents of ferulic acid, sinapic acid, cinnamic acid and caffeic acid, the precursors of lignin synthesis, decreased in heat treated pericarp when compared to the control pericarp, especially at 2 h and 12 DAT. Moreover, the contents of other secondary metabolites decreased in heat treated pericarp, especially at 2 h or 12 DAT, such as hydroxylated naringenin-hexose, sinensetin, p-coumaroyl quinic acid, feruloylquinic acid, pantothenic acid-hexose, quercetin-O-dihexoside, eriodictyol-O-dihexoside, and jasmonic acid.

HT decreased H_2O_2 content in citrus pericarp

The content of H_2O_2 decreased in heat treated pericarp during the storage phase while the content of H_2O_2increased in the control pericarp (Figure 4A). This suggests HT decreased the content of H_2O_2 in pericarp, and the content of H_2O_2 was held at a lower level in heat treated pericarp when compared to control pericarp during storage.

Superoxide dismutase (SOD) and peroxidase (POD) play important roles in deoxidization of H_2O_2. In this study, the activity of POD showed no significant change between HT and the control (Figure 4C). The activity of SOD was higher in heat treated pericarp relative to the control at 6 DAT, but was lower in heat treated pericarp relative to the control at 12 and 32 DAT (Figure 4D). So, the activity of SOD fluctuated during the entire storage period in heat treated pericarp compared to the control (Figure 4D).

Concerning redox regulation, most of the proteins were down-regulated in heat treated pericarp when compared to control pericarp, especially in the later stages of storage (Table 1). This suggests the redox reaction was weaker in heat treated pericarp during storage.

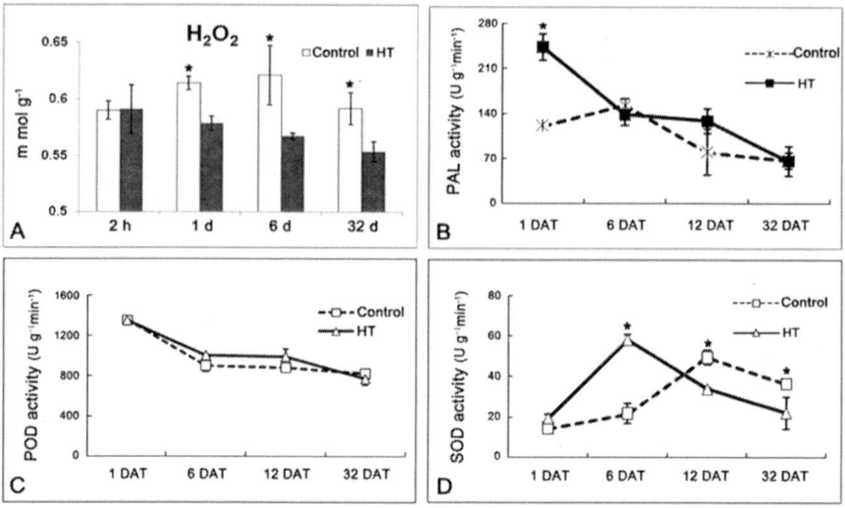

Figure 4 Changes of H_2O_2 content and enzyme activity in HT and control pericarp. A: H_2O_2 content. Samples at 2 h and 1, 6 and 32 DAT were used for H_2O_2 content determined in HT and control pericarp. **B-D**: Enzyme activity. Samples at 1, 6, 12 and 32 DAT were used for enzyme activity determination. PAL: phenylalanine ammonia-lyase; SOD: superoxide dismutase; POD: peroxidase. Four replicates were completed and statistically significant differences analysis was conducted between HT and control pericarp during the same period using Student's t-test. *: significant difference ($P < 0.05$). Mean values and SE bars are provided.

HT increased fruit firmness and the content of lignin in citrus pericarp

Fruit hardness gradually decreased during storage (Figure 5A); however, HT increased fruit hardness, especially at 1 DAT (Figure 5A), which led to the heat treated fruit remaining firmer than the control fruits during the entire storage period. Lignin is an important component responsible for fruit hardness. So, samples from 2 h after treatment and 1, 6 and 32 DAT were used to detect lignin content changes over time. Results show HT significantly induced lignin accumulation in pericarp (Figure 5B). Also, the activity of phenylalanine ammonia-lyase (PAL), a critical enzyme in lignin synthesis, initially increased in heat treated pericarp at 1 DAT but subsequently fell into the same level as in the control (Figure 4B). Furthermore, levels of ferulic acid, sinapic acid, cinnamic acid and caffeic acid, the precursors of lignin synthesis, declined in heat treated pericarp when compared to control pericarp (Table 3). This suggests a rapid transformation occurred from ferulic acid, sinapic acid, cinnamic acid and caffeic acid to lignin in heat treated pericarp after HT.

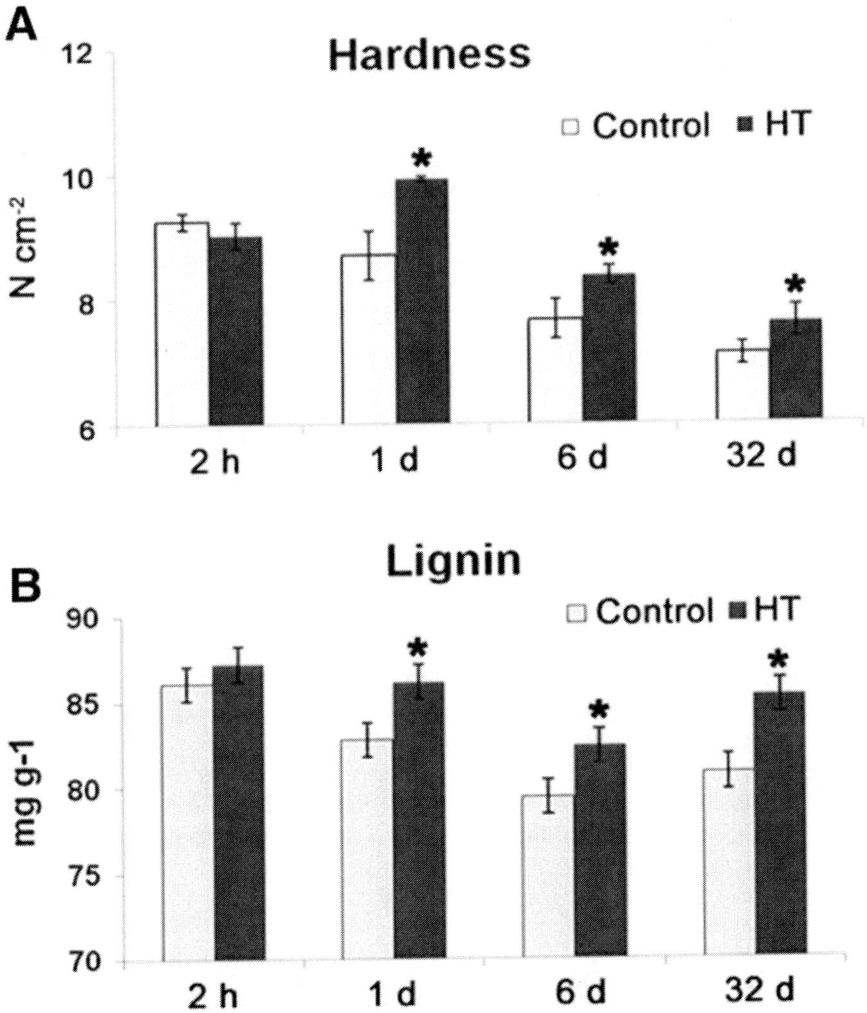

Figure 5 Changes of fruit hardness (A) and lignin content (B) in HT and control pericarp. Samples at 2 h and 1, 6 and 32 DAT were used for fruit hardness and lignin content determination. An analysis of statistically significant differences was conducted between HT and control pericarp during the same period using Student's t-test. *: significant difference ($P < 0.05$). Mean values and SE bars are provided.

DISCUSSION

HT improved fruit pathogen resistance without negatively impacting fruit quality

Physical treatments are used frequently to keep fruit fresh, including microporous membranes, pre-storage treatment (sweating), ultraviolet illumination, radiofrequency treatment, heat treatments (heat therapy), and other storage techniques. Postharvest heat treatments such as hot water treatment and hot air treatment provide a type of quarantine, inhibit the development of pathogens, enhance fruit resistance to chilling injury in cold-sensitive cultivars, and help preserve fruit quality during storage and extend its shelf life [7]. In the present study, HT effectively inhibited the development of pathogens. Also, HT (52°C, 2 min) caused a statistically significant reduction the incidence of chilling injury [9, 11].Also, postharvest HT had no obvious negative effects on fruit flesh quality during subsequent storage. This suggests HT increased the ability of fruit to tolerate storage stress without affecting their commercial qualities, and shows strong potential application in the citrus industry.

HT induced stress response proteins

HT can effectively inhibit fruit chilling injury during cold storage [12]. After HT, some proteins were up-regulated, while others were down-regulated. The up-regulated proteins might be responsible for heat induced fruit resistance. Interesting, some proteins were up-regulated at 1 d after HT, and decreased to the previous level of control during the storage period. Those kinds of proteins might be just involved in response to HT, but not play a vital role in fruit disease resistance. The proteins, which are up-regulated during whole storage period might be directly involved in fruit disease resistance.

In the present study, heat shock protein (HSP) 17.7, putative HSP 70 and low molecular weight HSP were up-regulated in heat treated pericarp compared to the control during entire storage period (Table 1). HSPs are encoded by a multigene family and are located in different subcellular compartments [13]. These molecular chaperones play a crucial role in protecting plants against stress by re-establishing normal protein configuration and thus helping to maintain cellular homeostasis [14]. Abiotic stresses usually causes protein dysfunction; so maintaining proteins in their functional configurations and preventing the aggregation of non-native proteins is particularly important for cell survival under stress [14]. HSPs are responsible for protein folding, assembly, translocation and degradation during cellular processes. HSPs also stabilize proteins and membranes, and can assist in protein refolding under stressful conditions [15]. Our results suggest HSPs play a crucial role in reducing the susceptibility of HT fruit to biotic and abiotic stress.

HT significantly inhibited pathogen development and fruit decay during storage [16]. Chitinase and β-1, 3-glucanase were up-regulated in heat treated fruit in this study; these enzymes have long been thought to be active in fruit antifungal defences. Pathogenesis-related (PR) proteins have been classified into five groups and two of these groups contain proteins with β-1, 3-glucanase and chitinase [17]. Chitinase catalyses the hydrolysis of chitin, a β-1, 4-linked polymer of N-acetyl D-glucosamine and is a major component of cell walls of most phytopathogenic fungi [18]. Crude protein extracts of chitinase and β-1, 3-glucanase were observed to effectively inhibit fungal activity [19]. Also, β-1, 3-glucanase and chitinase are thought to contribute to the level of resistance provided and are known to inhibit mycelial growth of a wide range of fungal pathogens [20]. This suggests β-1, 3-glucanase and chitinase were directly involved in HT induced pathogen resistance in fruits.

HT increased the content of stress response metabolites

Primary metabolites are directly involved in normal growth, development, and reproduction. In the present study, the content of ornithine, oleic acid and tetradecanoic acid increased in heat treated pericarp compared to the control during storage, and levels of seven types of sugars increased in heat treated pericarp compared to the control at 2 h after HT (Table 2). In plants, ornithine is required for the synthesis of polyamines and alkaloids, which contribute to oxidative stress tolerance in plants subjected to severe water stress [21].

Flavonoids are important defensive compounds which protect fruit against the development of pathogens, acting either as phytoanticipins or phytoalexins [22]. Flavonoids were observed to be up-regulated in heat treated pericarp (compared to the control), including quercetin-dihexose-deoxyhexose, hesperetin, naringenin, naringenin chalcone-hexose, hydroxybenzoic acid-hexose, isosakuranetin, diosmin, and rutin, which may play a vital role in response to stress [23]. Rutin is a bioflavonoid with strong antioxidant activity, which had an antimicrobial effect on pathogens, as evidenced by reduced conidial germination and appressorium formation of the pathogen [22]. Quercetin protected cells from oxidative stress-induced cell death by blocking both the cyclooxygenase and lipooxygenase pathways [24]. Quercetin has a protective function against oxidative stress by reducing oxidative injury to the cells during H_2O_2 treatment [25]. Also, plant extracts rich in quercetin have been applied as an interesting resource for functional food products and other foods used to reduce the risks of age-related macular degeneration [25]. Hesperidin acts as an antioxidant, based on in vitro studies, and hesperidin also provides strong cellular antioxidant protection against the damaging effects induced by stress [26]. Diosmin, hesperidin and naringin are flavonoid glycosides that occur naturally in citrus fruits; they exert a variety of properties such as antioxidant and free radical scavenging [27]. Interestingly, naringenin and rutin are not only involved in response to stress, but also were assumed to be signalling molecules [23].

HT induced fruit pathogen resistance and increased lignin content

The functions of phenylpropanoid compounds in plant defence range from preformed or inducible physical and chemical barriers against infection [10]. Defensive functions are not restricted to a particular class of phenylpropanoid compounds, but are found in the lignification of monolignols to lignin [28]. In the present study, HT decreased the level of precursors needed in lignin synthesis (ferulic acid, sinapic acid, cinnamic acid and caffeic acid), but increased lignin content (Table 3, Figure 5). Lignin is one of the major components of secondary cell walls, providing cells with mechanical support and isolation from the environment. In plants, cells can sense cell wall polymer structure and this activates trans-plasma membrane sensor proteins which might trigger a cascade of signal transduction events interwoven with other pathways that might modulate cellular functions by the activation or inhibition of specific transcription factors as well as by affecting posttranslational control of gene expression and protein function [29]. If a cell wall is damaged, cells activate a series of responses to the invading disease organisms, which might negatively impact fruit quality. The accumulation of lignin induced by HT creates a thickening of cell walls which forms an effective physical barrier to pathogens, delaying the invasion of disease organisms. Also, HT increased fruit firmness; this might be a result of HT induced lignin thickened fruit cell walls.

HT reduced the content of H_2O_2

HT is an effective method for improving citrus storability. In storage, fruits suffer a variety of biotic and abiotic environmental stresses. All these stresses could lead to the generation of reactive oxygen species (ROS). To avoid the accumulation of ROS, which may lead to cell death, plants have up-regulated enzymatic and non-enzymatic antioxidants to scavenge ROS and alleviate their negative effects [30]. In this study, HT down-regulated

H_2O_2 content in pericarp, which was consistent with the effect of HT on yeast cells. HT yeast cells accumulated less ROS than untreated cells in response to stresses [31].

Also, isoflavone reductase (IFR) might have an important role as an antioxidant [32]. IFR is an enzyme involved in the production of isoflavone phytoalexins and it accumulates in response to pathogenic attack, fungal elicitor and biotic stress [33]. Isoflavonoids are important secondary metabolites for essential physiological processes. Kim *et al.*[32] reported the isoflavone reductase-like gene may act as a down-regulator by producing antioxidant chemicals to prevent the elevation of ROS. In this study, isoflavone reductase, a related protein (H10), was up-regulated in heat treated pericarp when compared to control pericarp (Table 1), which might be another factor leading to lower H_2O_2 levels and stronger resistance to pathogens in heat treated pericarp when compared to control pericarp.

Ascorbate peroxidase scavenges H_2O_2 in specific organelles of the cell [34]. Also, Cu/Zn-SOD is the main cellular isoform of SODs which play a key role in the antioxidant defence system through the disproportionation of H_2O_2[34]. Both of these can prevent ROS caused damages to the cellular membrane and act as a primary defence mechanism when organisms were exposed to oxidative stress. These proteins are considered the main enzymatic and non-enzymatic systems for protecting cells against oxidative damage, and are responsible for disease resistance. In response to biotic stress, these proteins also act as the primary defence mechanism against oxidative stresses caused by pathogens, therefore preventing damage of ROS to cellular membranes. However, the activities of POD and SOD were not elevated in Heat treated pericarp when compared to control pericarp, which might be caused by the lower H_2O_2 content in heat treated pericarp. Also, other proteins involved in scavenging ROS were down-regulated in heat treated pericarp when compared to control pericarp, including oxidoreductase (H11), oxidoreductase (H12, 17), aldo/keto reductase (H13), 2-oxoacid dehydrogenase (H14), superoxide dismutase (H16). This suggests HT induced a level of resistance to pathogens attacking fruits by causing a decline in fruit H_2O_2 content and by down-regulated scavenging ROS enzymes.

Role of sugars, organic acids and amino acids in improving storability

In this study, the profiling of primary metabolics showed HT induced the accumulation of sugars at early stages of storage (Table 2). Soluble sugars, especially sucrose, glucose, and fructose, accumulated when fruits were subjected to stress, and accumulated soluble sugars play an obviously vital role in the response of fruit to a number of stressors [35]. Soluble sugars do not only act as nutrients and metabolites, but also act as signal substance in the modification of redox homeostasis [36]. However, less information was available on how other sugars are involved in the stress response, especially arabinose, mannose, galactose, and ribose.

Moreover, primary metabolic data shows HT reduced organic acids and amino acids content especially at 2 h after treatment in this study. Although organic acids and amino acids play an obvious central role in energy and metabolism at the cellular and whole-organism levels, there seems to be a rapid conversion from organic acids and amino acids to sugars in heat treated pericarp. However, little is known about the decreased levels of organic acids and amino acids seen during a stress response, and further study will be required to validate this response.

CONCLUSIONS

This study provides a broad picture of differential accumulation of proteins and metabolites, and gives a new insight into how HT increases fruit stress resistance during subsequent storage of citrus fruit. In metabolic substrates, the up-accumulation of secondary metabolites played vital roles in increasing the ability of fruit to deal with stress. Flavonoids are directly involved in the response to external stress. The increased lignin and decreased H_2O_2 contents involved in HT probably increased fruit resistibility in response to external stress.

METHODS

Sample collection

'Kamei' Satsuma mandarin (*Citrus unshiu* Marc.) fruits were harvested at commercial maturity (fruit colour has turned completely orange and the level of soluble solids approaches 9%) from an orchard in the city of Yichang, Hubei Province, China, in October 2009. Fruits of a uniform size and colour, free of visible injury or blemishes, were selected for the experiments. Three hundred kilograms of fruit were divided into two equal parts, 150 kg were used for the HT and the other 150 kg used for a control. After treatments, fruits were stored in a ventilated warehouse (storage temperature: 12–16°C; relative humidity: 90–95%). Samples were collected at 2 h, and 1, 2, 3, 4, 6, 9, 12, 16, 20, 24, 28, 32, 38, 44, 50 and 60 DAT. The pericarp was sampled along the equatorial plane of each fruit; 30 fruits were selected as one sample and ground to powder in liquid nitrogen, then stored at –80°C for further analysis.

Heat treatment (HT)

Heat treated fruits were dipped in a 52°C warm water bath for 2 min, and control fruits were dipped in a 25°C water bath for 2 min. After treatments, fruits were air dried and stored in a ventilated warehouse prior to sampling.

Fruit quality determination

Fruits were checked for weight loss (%), respiration rate (mg kg^{-1} h^{-1}) and total soluble solids (TSS,%) at 1, 2, 3, 4, 6, 9, 12, 16, 20, 24, 28, 32, 38, 44, 50 and 60 DAT. The weight was measured using an electronic hydrostatic balance (Model: MP31001, Shanghai Selon Scientific Instrument, Co., Ltd, Shanghai, China) with an accuracy of ± 0.01 g. The respiration rate of fruit was measured with an infrared gas analyser (Model: GXH-305H, Junfang Science & Technology Institute of Physical and Chemical Research, Beijing, China). TSS were determined with a refractometer (Model: Pocket PAL-1, Atago Inc., Toyko, Japan) according to the manufacturer's instructions.

Antifungal assay of HT in wounded citrus fruits

Blue mould (*Penicillium italicum*) was chosen as an indicator to evaluate the resistance of treated fruit to fungal infections. 150 fruits from each treatment were used for fungal inoculation. A uniform lesion (3 mm deep, 4 mm wide) was made at the equator of the fruit using a sterile nail. Aliquots of 10 µl suspension of *P. italicum* at 1×10^5 spore ml^{-1} were inoculated into each wound site. After fungal inoculation, fruits were stored in a storage chamber (relative humidity, 95%; temperature, 25°C), and disease incidence was detected at 1, 2, 3, 4, 5, 6 and 7 d after inoculation. Disease incidence rates and lesion diameters were recorded based on the following equations:

Disease incidence rate (%) = ∑ Number of diseased/ decayingcitrus/ totalnumber of fruit in the treatment × 100;

Lesion diameter (cm) = ∑ Lesion diameter in diseased/ total number of fruit in the treatment

Two-dimensional gel electrophoresis and MAILDI-TOF/TOF

Samples at 1, 6, 12 and 32 DAT were used for total protein extraction based on the method described by Isaacson *et al.*[37]. The precipitate was then nitrogen gas-dried and solubilized in the lysis buffer containing 8 M urea, 0.2% (w/v) Bio-Lyte, 4% CHAPS, 65 mM DTT. The protein concentration was determined with a Bio-Rad protein assay kit (Bio-Rad Laboratories Inc., Hercules, CA, USA) based on the Bradford method using BSA as the standard. Two-dimensional gel electrophoresis and gel staining was carried out based on the method of Yun *et al.*[38]. The stained gels were imaged with a UVP imager (Bio-Rad) in the 'trans white' mode using PDQuest 2-D analysis software version 7.4 (Bio-Rad). Protein spots matching and differential protein spots analysis were completed based on the method of Yun *et al.*[38]. Isoelectric points and molecular weights of the proteins were determined by comparison with the markers.

In-gel digestion and protein identification were performed using a 4800 Proteomics Analyzer MALDI TOF/TOF (Applied Biosystems, Foster City, CA) as described by Yun *et al.*[38]. MS/MS data were then submitted to the MASCOT program for protein identification against the green plants database. The following are the search parameters: trypsin enzyme, one missed cleavage, fixed modifications of carbamidomethyl (C), fragment mass tolerance of ± 0.5 Da, variable modifications of oxidation (Met), peptide tolerance of 100 ppm, and peptide charge of 1+. Only results of peptides with MS/MS $P < 0.05$ confidence were accepted.

The primary metabolic profiling

Samples collected at 2 h and 1, 6 and 32 DAT were used for the differential primary metabolic profiling analysis. A 300 mg sample was extracted in 2,700 µl of methanol as described by Roessner-Tunali *et al.*[39]. 300 µl of 0.2 mg ml^{-1} ribitol in water was added as a quantification internal standard. A derivatization reaction was performed based on the protocol of Zhang *et al.*[40] with little modification. Extracts were incubated in 50 µl of 20 mg ml^{-1} methoxyamine hydrochloride in pyridine for 30 min at 50°C, followed by a 40 min treatment at 60°C using 50 µl BSTFA (containing 1% TMCS). Each sample (1 µl) was injected into the gas chromatograph system through a fused-silica capillary column (30 m × 0.25 mm i.d., 0.25 µm) DB-5 MS stationary phase. The injector temperature was 250°C, with a carrier gas flow rate of 1.0 ml min^{-1}. The column temperature was held at 100°C for 1 min; increased to 184°C at a rate of 3°C min^{-1}, increased to 190°C at a rate of 0.5°C min^{-1}, increased to 280°C at 15°C min^{-1}. The flow rate of the carrier helium (99.999%) gas was 1 ml min^{-1}. The following were MS operating parameters: ionization voltage, 70 eV (electron impact ionization); ion source temperature, 200°C; interface temperature, 250°C. TIC (total ioncurrent) spectra were recorded in the mass range of 45–600 atomic mass units in scanning mode.

The secondary metabolic profiling

Samples at 2 h and 1, 6 and 32 DAT were used for differential secondary metabolic profiling analysis using HPLC-MS. 300 mg dried powder was extracted with methanol, the mixture was filtered through a 0.22-mm polytetrafluoroethylene membrane filter.

Metabolite profiling was performed using a QTOF 6520 mass spectrometer (Agilent Technologies, Palo Alto, CA, USA) coupled to a 1200 series Rapid Resolution HPLC system as described by Page *et al.*[41]. 2 µl of sample extract was loaded onto a Zorbax Eclipse Plus C18 1.8 µm, 2.1 × 100 mm reverse-phase analytical column (Agilent Technologies). Mobile phase A comprised 0.1% formic acid in water and mobile phase B was acetonitrile. The following gradient was used: 0 min–10% B; 20 min–95% B; 22 min–95% B; 22.1 min–10% B; 30 min–10% B. The flow rate was 0.3 ml min^{-1} and the column temperature was held at 35°C for the duration. The source conditions for electrospray ionization were as follows: gas temperature was 350°C with a drying gas flow rate of 10 l min^{-1} and a nebulizer pressure of 40 psig. The capillary voltage was 3.5 kV in both positive and negative ion mode. The fragmentor voltage was 135 V and the skimmer was 65 V.

Assay of phenylalanine ammonia-lyase (PAL), superoxide dismutase (SOD) and peroxidase (POD) activities

Samples collected at 1, 6, 12 and 32 DAT were used for enzyme activity determination. A 5 g sample was homogenized in 25 ml of 0.05 M sodium borate buffer (pH 8.8 for PAL, containing 5 mM β-mercaptoethanol) and 25 ml of ice-cold 100 mM sodium phosphate (pH 7.8 for POD and SOD, containing 0.5 g polyvinyl polypyrrolidone). PAL activity was determined by measuring the absorbance of cinnamic acid at 290 nm over a period of 30 min at 30°C [42]. POD activity was analysed using guaiacol as the substrate [43]. SOD activity was determined using the xanthine-xanthine oxidase method, and nitro blue tetrazolium was used as the indicator of superoxide radical production [44].

Determination the content of H_2O_2 in the peel

Samples collected at 2 h and 1, 6 and 32 DAT were used to measure hydrogen peroxide levels based on the methods of Patrick and Wagner [45] with some modifications. Briefly, peel tissues (0.5 g) were ground in liquid nitrogen and homogenized in 5 ml of physiological saline. Samples were then subjected to an ultrasonic bath for 15 min and centrifuged at 6,000 g for 20 min. The supernatants (50 μl) were subsequently added into 0.5 ml of solution I and warmed in a 37°C water bath for 10 min before 0.5 ml of solution II was added. The absorbance of the mixture was measured at 405 nm and the content of H_2O_2 in each sample was calculated by comparison with a standard calibration curve.

Determination of fruit firmness

Samples collected at 2 h and 1, 6 and 32 DAT were used for the determination of fruit firmness. Eighteen fruits were measured and every fruit was measured on three different spots around the equatorial plane. Firmness was measured with a fruit sclerometer (GY-B, Jilin, China) with 4 mm diameter and expressed as N cm^{-2} and the average value per fruit was used for further analysis, after removing three largest and three smallest measurements; a total twelve replicates were comparative statistically analysed per sample using Student's t-test ($P < 0.05$).

Determination of lignin in pericarp

Samples collected at 2 h and 1, 6 and 32 DAT were used to measure fruit lignin content. Lignin levels were determined using the method described by Lei *et al.*[46].

Statistical analysis

The experimental design was completely randomized with more than three replications. Four individual replicates were completed in 2-DE, GC-MS, HPLC-MS, H_2O_2 content, lignin content and enzyme activity determination. Twelve replicates were completed to measure weight loss, respiration rate, total soluble solids and fruit firmness. Data for each sample was statistically analysed using Student's t-test ($P < 0.05$).

NOTES

Ze Yun, Huijun Gao contributed equally to this work.

ABBREVIATIONS

2-DE: Two-dimensional gel electrophoresis **GC-MS:** Gas chromatography coupled to mass spectrometry **HSP:** Heat shock protein **HT:** Heat treatment **IFR:** Isoflavone reductase **LC-QTOF-MS:** Liquid chromatography quadrupole time-of-flight mass spectrometry **MALDI-TOF MS:** Matrix-assisted laser desorption/ionization-time-of-flight tandem mass spectrometry **NCBI:** National center for biotechnology information **PAL:** Phenylalanine ammonia-lyase **POD:** Peroxidise **PR:** Pathogenesis-related **ROS:** Reactive oxygen species **SOD:** Superoxide dismutase **TSS:** Total soluble solids.

ACKNOWLEDGEMENTS

We thank Prof. Hanhui Kuang of Huazhong Agricultural University for help in revising the manuscript. This work was supported by Huazhong Agricultural University Scientific & Technological Self-innovation Foundation, the Natural Science Foundation of China (NSFC No. 31271968), The Important Project of Ministry of Education (311029), and National Modern Agriculture (Citrus) Technology Systems of China (No. CARS-27).

REFERENCES

1. Gonzalez-Candelas L, Alamar S, Sanchez-Torres P, Zacarias L, Marcos JF: A transcriptomic approach highlights induction of secondary metabolism in citrus fruit in response to Penicillium digitatum infection. BMC Plant Biol. 2010, 10: 194. 10.1186/1471-2229-10-194.
2. Tu K, Shao XF, Tu SC, Su J, Zhao Y: Effects of heat treatment on wound healing in gala and Red fuji apple fruits. J Agr Food Chem. 2010, 58: 4303-4309. 10.1021/jf904273m.
3. Lara MV, Borsani J, Budde CO, Lauxmann MA, Lombardo VA, Murray R, Andreo CS, Drincovich MF: Biochemical and proteomic analysis of 'Dixiland' peach fruit (Prunus persica) upon heat treatment. J Exp Bot. 2009, 60: 4315-4333. 10.1093/jxb/erp267.
4. Gould GW: History of science - spores - lewis B perry memorial lecture 2005. J Appl Microbiol. 2006, 101: 507-513. 10.1111/j.1365-2672.2006.02888.x.

5. Hong SI, Lee HH, Kim D: Effects of hot water treatment on the storage stability of satsuma mandarin as a postharvest decay control. Postharvest Biol Tec. 2007, 43: 271-279. 10.1016/j.postharvbio.2006.09.008.
6. Zhang ZK, Bi Y, Ge YH, Wang JJ, Deng JJ, Xie DF, Wang Y: Multiple pre-harvest treatments with acibenzolar-S-methyl reduce latent infection and induce resistance in muskmelon fruit. Sci Hortic-Amsterdam. 2011, 130: 126-132. 10.1016/j.scienta.2011.06.024.
7. Schirra M, D'Aquino S, Cabras P, Angioni A: Control of postharvest diseases of fruit by heat and fungicides: efficacy, residue levels, and residue persistence. A Review. J Agr Food Chem. 2011, 59: 8531-8542. 10.1021/jf201899t.
8. Perotti VE, Del Vecchio HA, Sansevich A, Meier G, Bello F, Cocco M, Garrán SM, Anderson C, Vázquez D, Podestá FE: Proteomic, metabalomic, and biochemical analysis of heat treated Valencia oranges during storage. Postharvest Biol Tec. 2011, 62: 97-114. 10.1016/j.postharvbio.2011.05.015.
9. Sapitnitskaya M, Maul P, McCollum GT, Guy CL, Weiss B, Samach A, Porat R: Postharvest heat and conditioning treatments activate different molecular responses and reduce chilling injuries in grapefruit. J Exp Bot. 2006, 57: 2943-2953. 10.1093/jxb/erl055.
10. Dixon RA, Achnine L, Kota P, Liu CJ, Reddy MSS, Wang LJ: The phenylpropanoid pathway and plant defence - a genomics perspective. Mol Plant Pathol. 2002, 3: 371-390. 10.1046/j.1364-3703.2002.00131.x.
11. Ghasernnezhad M, Marsh K, Shilton R, Babalar M, Woolf A: Effect of hot water treatments on chilling injury and heat damage in 'satsuma' mandarins: Antioxidant enzymes and vacuolar ATPase, and pyrophosphatase. Postharvest Biol Tec. 2008, 48: 364-371. 10.1016/j.postharvbio.2007.09.014.
12. Vigneault C, Lu J, Charles MT, Goyette B, Raghavan GSV: Effect of heat treatment uniformity on tomato ripening and chilling injury. Postharvest Biol Tec. 2010, 56: 155-162. 10.1016/j.postharvbio.2010.01.005.
13. Vierling E: The roles of heat shock proteins in plants. Annu Rev Plant Biol. 1991, 42: 579-620. 10.1146/annurev.pp.42.060191.003051.
14. Wang WX, Vinocur B, Shoseyov O, Altman A: Role of plant heat-shock proteins and molecular chaperones in the abiotic stress response. Trends Plant Sci. 2004, 9: 244-252. 10.1016/j.tplants.2004.03.006.
15. Basha E, Jones C, Wysocki V, Vierling E: Mechanistic differences between two conserved classes of small heat shock proteins found in the plant cytosol. J Biol Chem. 2010, 285: 11489-11497. 10.1074/jbc.M109.074088
16. Hu WZ, Jiang AL, Jin LM, Liu CH, Tian MX, Wang YY: Effect of heat treatment on quality, thermal and pasting properties of sweet potato starch during yearlong storage. J Sci Food Agr. 2011, 91: 1499-1504. 10.1002/jsfa.4340.
17. Schraudner M, Ernst D, Langebartels C, Sandermann H: Biochemical plant responses to ozone: III. Activation of the defence-related proteins β-1, 3-glucanase and chitinase in tobacco leaves. Plant Physiol. 1992, 99: 1321. 10.1104/pp.99.4.1321.
18. Chalavi V, Tabaeizadeh Z, Thibodeau P: Enhanced resistance to Verticillium dahliae in transgenic strawberry plants expressing a Lycopersicon chilense chitinase gene. J Am Soc Hortic Sci. 2003, 128: 747-753.
19. Mauch F, Mauch-Mani B, Boller T: Antifungal hydrolases in pea tissue: II. Inhibition of fungal growth by combinations of chitinase and β-1, 3-glucanase. Plant Physiol. 1988, 88: 936-942. 10.1104/pp.88.3.936.
20. Magnin-Robert M, Trotel-Aziz P, Quantinet D, Biagianti S, Aziz A: Biological control of Botrytis cinerea by selected grapevine-associated bacteria and stimulation of chitinase and beta-1,3 glucanase activities under field conditions. Eur J Plant Pathol. 2007, 118: 43-57. 10.1007/s10658-007-9111-2.
21. Kalamaki MS, Alexandrou D, Lazari D, Merkouropoulos G, Fotopoulos V, Pateraki I, Aggelis A, Carrillo-Lopez A, Rubio-Cabetas MJ, Kanellis AK: Over-expression of a tomato N-acetyl-L-glutamate synthase gene (SlNAGS1) in Arabidopsis thaliana

results in high ornithine levels and increased tolerance in salt and drought stresses. J Exp Bot. 2009, 60: 1859-1871. 10.1093/jxb/erp072.

22. Radhakrishna Shetty R, Jensen JD: Silicon-induced changes in antifungal phenolic acids, flavonoids and Key phenylpropanoid pathway genes during the interaction between miniature roses and the biotrophic pathogen podosphaera pannosa. Plant Physiol. 2011, 157: 2194-2205. 10.1104/pp.111.185215.
23. Treutter D: Significance of flavonoids in plant resistance: a review. Environ Chem Lett. 2006, 4: 147-157. 10.1007/s10311-006-0068-8.
24. Nijveldt RJ, van Nood E, van Hoorn DEC, Boelens PG, van Norren K, van Leeuwen PAM: Flavonoids: a review of probable mechanisms of action and potential applications. Am J Clin Nutr. 2001, 74: 418-425.
25. Saviranta NMM, Veeroos L, Granlund LJ, Hassinen VH, Kaarniranta K, Karjalainen RO: Plant flavonol quercetin and isoflavone biochanin A differentially induce protection against oxidative stress and inflammation in ARPE-19 cells. Food Res Int. 2011, 44: 109-113. 10.1016/j.foodres.2010.10.056.
26. Wilmsen PK, Spada DS, Salvador M: Antioxidant activity of the flavonoid hesperidin in chemical and biological systems. J Agr Food Chem. 2005, 53: 4757-4761. 10.1021/jf0502000.
27. Kanaze FI, Gabrieli C, Kokkalou E, Georgarakis M, Niopas I: Simultaneous reversed-phase high-performance liquid chromatographic method for the determination of diosmin, hesperidin and naringin in different citrus fruit juices and pharmaceutical formulations. J Pharmaceut Biomed. 2003, 33: 243-249. 10.1016/S0731-7085(03)00289-9.
28. Denness L, McKenna JF, Segonzac C, Wormit A, Madhou P, Bennett M, Mansfield J, Zipfel C, Hamann T: Cell wall damage-induced lignin biosynthesis is regulated by a reactive oxygen species- and jasmonic acid-dependent process in arabidopsis. Plant Physiol. 2011, 156: 1364-1374. 10.1104/pp.111.175737.
29. Seifert GJ, Blaukopf C: Irritable walls: the plant extracellular matrix and signaling. Plant Physiol. 2010, 153: 467-478. 10.1104/pp.110.153940.
30. Apel K, Hirt H: Reactive oxygen species: metabolism, oxidative stress, and signal transduction. Annu Rev Plant Biol. 2004, 55: 373-399. 10.1146/annurev.arplant.55.031903.141701.
31. Liu J, Wisniewski M, Droby S, Tian SP, Hershkovitz V, Tworkoski T: Effect of heat shock treatment on stress tolerance and biocontrol efficacy of Metschnikowia fructicola. FEMS Microbiol Ecol. 2011, 76: 145-155. 10.1111/j.1574-6941.2010.01037.x.
32. Kim SG, Kim ST, Wang YM, Kim SK, Lee CH, Kim KK, Kim JK, Lee SY, Kang KY: Overexpression of rice isoflavone reductase-like gene (OsIRL) confers tolerance to reactive oxygen species. Physiol Plantarum. 2010, 138: 1-9. 10.1111/j.1399-3054.2009.01290.x
33. Kim ST, Cho KS, Yu S, Kim SG, Hong JC, Han CD, Bae DW, Myung AE, Kang KY: Proteomic analysis of differentially expressed proteins induced by rice blast fungus and elicitor in suspension-cultured rice cells. Proteomics. 2003, 3: 2368-2378. 10.1002/pmic.200300577.
34. Zhang HX, Xia Y, Wang GP, Shen ZG: Excess copper induces accumulation of hydrogen peroxide and increases lipid peroxidation and total activity of copper-zinc superoxide dismutase in roots of Elsholtzia haichowensis. Planta. 2008, 227: 465-475.
35. Rosa M, Prado C, Podazza G, Interdonato R, González JA, Hilal M, Prado FE: Soluble sugars—metabolism, sensing and abiotic stress: a complex network in the life of plants. Plant Signal Behav. 2009, 4: 388. 10.4161/psb.4.5.8294.
36. Rosenwasser S, Rot I, Sollner E, Meyer AJ, Smith Y, Leviatan N, Fluhr R, Friedman H: Organelles contribute differentially to reactive oxygen species-related events during extended darkness. Plant Physiol. 2011, 156: 185. 10.1104/pp.110.169797.

37. Isaacson T, Damasceno CMB, Saravanan RS, He Y, Catala C, Saladie M, Rose JKC: Sample extraction techniques for enhanced proteomic analysis of plant tissues. Nat Protoc. 2006, 1: 769-774. 10.1038/nprot.2006.102.
38. Yun Z, Li WY, Pan ZY, Xu J, Cheng YJ, Deng XX: Comparative proteomics analysis of differentially accumulated proteins in juice sacs of ponkan (Citrus reticulata) fruit during postharvest cold storage. Postharvest Biology and Technology. 2010, 56: 189-201. 10.1016/j.postharvbio.2010.01.002.
39. Roessner-Tunali U, Hegemann B, Lytovchenko A, Carrari F, Bruedigam C, Granot D, Fernie AR: Metabolic profiling of transgenic tomato plants overexpressing hexokinase reveals that the influence of hexose phosphorylation diminishes during fruit development. Plant Physiol. 2003, 133: 84-99. 10.1104/pp.103.023572.
40. Zhang JJ, Wang X, Yu O, Tang JJ, Gu XG, Wan XC, Fang CB: Metabolic profiling of strawberry (Fragariaxananassa Duch.) during fruit development and maturation. J Exp Bot. 2011, 62: 1103-1118. 10.1093/jxb/erq343.
41. Page M, Sultana N, Paszkiewicz K, Florance H, Smirnoff N: The influence of ascorbate on anthocyanin accumulation during high light acclimation in Arabidopsis thaliana: further evidence for redox control of anthocyanin synthesis. Cell & Environment: Plant; 2011.
42. Ballester AR, Lafuente MT, Gonzalez-Candelas L: Spatial study of antioxidant enzymes, peroxidase and phenylalanine ammonia-lyase in the citrus fruit-Penicillium digitatum interaction. Postharvest Biol Tec. 2006, 39: 115-124. 10.1016/j.postharvbio.2005.10.002.
43. Meng XH, Han J, Wang Q, Tian SP: Changes in physiology and quality of peach fruits treated by methyl jasmonate under low temperature stress. Food Chem. 2009, 114: 1028-1035. 10.1016/j.foodchem.2008.09.109.
44. Sala JM, Lafuente MT: Antioxidant enzymes activities and rindstaining in 'Navelina' oranges as affected by storage relative humidity and ethylene conditioning. Postharvest Biol Tec. 2004, 31: 277-285. 10.1016/j.postharvbio.2003.10.002.
45. Patrick WA, Wagner HB: Determination of hydrogen peroxide in small concentrations. Anal Chem. 1949, 21: 1279-1280. 10.1021/ac60034a038.
46. Lei Y, Liu Y-Z, Gu Q-Q, Yang X-Y, Deng X-X, Chen J-Y: Comparison of cell wall metabolism in the pulp of three cultivars of 'Nanfeng' tangerine differing in mastication trait. J Sci Food Agr. 2011, 92: 496-502.

CHAPTER 8

Postharvest Processing of Large Cardamom in the Eastern Himalaya

Angom Ingocha Singh and Anand Kumar Pothula*

Department of Agricultural Engineering, North Eastern Regional Institute of Science and Technology, Nirjuli, Arunachal Pradesh 791109, India

ABSTRACT

The dried fruits of large cardamom (*Amomum subulatum* Roxb.), a high-value, low-volume spice crop grown only in the three eastern Himalayan countries, are widely used in foods, beverages, perfumes, and medicines. Production is currently declining, and improved postharvest management would be one way to help ensure the sustainability of this niche crop. The value chain for large cardamom consists largely of traditional practices; scientific improvements are needed in a number of postharvest steps, including marketing. This article reviews the crop's postharvest processing (with emphasis on curing, calyx cutting, packaging, and storage), quality issues, and trade patterns, and identifies research topics that could contribute to increasing its quality and value and thereby to protecting and promoting the livelihoods of several thousands of people in the value chain.

Keywords: Large cardamom, postharvest processing, quality, curing, grading, calyx cutting, trade

INTRODUCTION

Large cardamom (*Amomum subulatum* Roxb.), a member of Zingiberaceae family under the order Scitaminae is a perennial soft-stemmed plant (Figure 1). The large cardamom plant is 1.5 to 2.5 m tall (Bisht et al 2011). It grows in the vicinity of mountain streams in swampy, cool, and humid areas in the shade of forest trees, of which nitrogen-fixing trees are the more suitable shade trees. The Himalayan alder (*Alnus nepalensis*) is often used as a shade tree (Chempakam and Sindhu 2008; Sharma et al 2008). The plant can grow at altitudes ranging from 600 to 2000 m above mean sea level and in areas with an annual rainfall varying from 2800 to 3500 mm (Sharma et al 2000; Bisht et al 2011). The useful portion of large cardamom is the dried capsule, which has 40 to 50 small seeds and is grayish brown to dark red brown. The capsules are held together inside the spike with viscous sugary pulp and are 20–25 mm long and oval to globule in shape (Thomas et al 2009).

FIGURE 1 (A) Large cardamom plantation at Lyako village, Old Zero, Arunachal Pradesh; (B) single bush of large cardamom; (C and D) close views of mature large cardamom spikes. (Photos by A. Ingocha Singh)

A. subulatum is grown only in the eastern Himalayan countries—Nepal, northeast India, and Bhutan—and originated in Sikkim, India (Sharma et al 2009). Hence, the worldwide production of this crop is estimated as the sum of production in these three countries. Total world production is about

12,278.20 metric tons. Nepal leads in the production of large cardamom (52%), followed by India (37%), and Bhutan (11%). Production data for Nepal, India, and Bhutan for 2008–2011 are presented in Table 1. Taplejung, Panchthar, Ilam, Dhankura, Bhojpur, Terhathum, and Sankhuwasabha are the major large-cardamom–growing districts of Nepal (ECCOS 2010). In India it is cultivated in Sikkim, the Darjeeling district of West Bengal, Arunachal Pradesh, Nagaland, and Uttarakhand (Sharma et al 2009; Bisht et al 2010). According to the Spices Board (2012a), the major share of India's production is from Sikkim.

TABLE 1. Worldwide production of large cardamom (in metric tons), 2005–2011.

Country	2005–2006	2006–2007	2007–2008	2008–2009	2009–2010	2010–2011	6-year average
Nepal	6647.00	7150.00	7087.00	7037.00	5233.00	5517.00	6445.17
India	5185.00	4303.00	4920.00	4300.00	4180.00	3918.00	4467.67
Bhutan	1190.00	3477.00	986.90	942.28	433.00	1163.00	1365.36
Total	13,022.00	14,930.00	12,993.90	12,279.28	9846.00	10,598.00	12,278.20

Sources: Nepal: MoAC (2008, 2011), Chapagain (2011); India: Spices Board (2012a); Bhutan: Dema (2011).

Large cardamom has been a major agricultural cash crop and export commodity of Southeast Asian countries in recent years (Sharma et al 2009). Large cardamom farming is one of the important livelihood sources for mountain people in the Himalayan region. It is a low-volume, high-value crop (Avasthe et al 2011). In India, most of the crop is consumed domestically, and the rest is exported. The main market for Nepal's crop is Siliguri, a town in North Bengal, India (Sharma et al 2009).

In Sikkim, large cardamom is called *alainchii* and is believed to be one of the oldest known spices. Dried large cardamom capsules are used as a spice in various dish preparations, food essences, perfumes, and medicines. It has been important for many centuries in Ayurvedic preparations and the Unani system of medicine (Jafri et al 2001; Madhusoodanan and Rao 2001). Large cardamom can be used to treat several ailments (NIIR 2006). The volatile oil of large cardamom seed contains 1,8-cineole, α-terpineol, α-pinene, β-pinene, and allo-aromadendrene (Gurudutt et al 1996). The essential oil of large cardamom is reported to have antimicrobial properties (Agnihotri and Wakode 2010), and many studies have been conducted on it (Mishra and Dubey 1990; Adegoke et al 1998; Naik et al 2004; Gilani et al 2006; Kaskoos et al 2008; Singh et al 2008).

Large cardamom cultivation, harvesting, and processing are carried out in traditional ways with very little scientific influence (Sharma et al 2009). Dried capsules produced in the traditional way fail to meet some contemporary market requirements, resulting in lower prices to farmers. Mature capsules are manually removed from the spikes after they are harvested. Curing of fresh capsules, which have about 80% moisture content, is accomplished in a traditional *bhatti*(furnace-based curing structure), which often results in poor quality (Rao et al 2001). A number of improved drying systems have been developed; these have yielded better quality, but farmers are not yet convinced of their value as each system has some problems (Sharma et al 2000). The capsule tail (calyx) is removed manually using scissors, a laborious and time-consuming process. Capsules are graded according to size, but no reports of use of mechanical graders have been found.

This article reviews the harvesting and postharvest processing methods (curing, calyx cutting, packaging, and storage), quality issues and their impacts on the value chain, and trade patterns; it also suggests approaches for future research and development that could make the crop more sustainable. In cost discussions, the following currency conversion values (as of 1 March 2012) are used: US$ 1 = 78.81 Nepalese rupees, 49.33 Indian rupees, and 49.08 Bhutanese ngultrum.

Harvesting

Large cardamom is harvested when the capsules are fully matured. This is usually done during September to November, depending on the altitudes and cultivars. Capsule maturity is commonly assessed by opening the topmost capsule of a spike. Ripening of large cardamom capsules on a spike is not uniform, however; the topmost capsules ripen first and those on the bottom last. The full maturity of a capsule is indicated by the brown color of its seeds. Harvesting at the correct stage of maturity is essential to produce high-quality capsules. When the topmost capsule is fully matured, the spike-bearing shoots are cut at 45 cm height and left for another 10–15 days to ensure maturity of all the capsules (Spices Board 2001). The spikes are harvested using a special knife known as an *elaichichhuri* (Spices Board 2001). The harvested spikes are stored for 2–3 days after harvesting, which makes it easier to separate the capsules (Spices Board 2001). Separation is done by hand, and no device is available for this operation so far. The separated capsules are also manually cleaned from other plant materials before curing. Figure 2 presents several views of harvested and cured capsules.

FIGURE 2 (A) Heaped harvested spikes of large cardamom; (B) close view of a spike; (C) mature fresh capsules manually separated from the spike; (D) cured capsules with calyx. (Photos by A. Ingocha Singh)

Curing

Fresh capsules have a moisture content of 70–80% (Mande et al 1999). In curing, the moisture content is reduced to a level that is safe for storage, about 10–12%. Curing substantially reduces the weight of the capsules; the final weight depends on several parameters, including initial size, final moisture content, method of curing, and curing temperature. Depending on capsule size and curing methods, the weight ratio from fresh (Figure 2C) to cured capsule (Figure 2D) is 4:1 to 5:1 (Madhusoodanan and Rao 2001).

Curing is the most crucial step in large cardamom processing, as capsule quality largely depends on curing conditions and methods. It is obvious that a certain amount of easily evaporable substances that are part of the essential oil content of large cardamom are lost during curing (Rout et al 2003). Curing at too high a temperature results in loss of volatile oil and charring of capsules, and curing at too low a temperature results in longer drying times and increased chances of mold growth. Optimum curing temperature is 45–55°C (Deka et al 2003). Curing is usually done in traditional *bhattis*. However, as mentioned earlier, a few types of improved curing systems have also been introduced (Deka et al 2003; Anonymous 2006; TERI 2012). The effect of various curing methods on quality is discussed in the sections that follow.

Traditional *bhatti*

The traditional *bhatti* (Figure 3A) is a drying kiln developed by farmers and used for curing fresh large cardamom capsules (Sharma et al 2000); it is based on a direct heating system, and drying time is 25–40 hours (Mande et al 1999). The fuel efficiency of this system is very poor (Rao et al 2001); it requires 2.5 kg of fuelwood to produce 1 kg of dried capsules (Sharma et al 2009). The quality of large cardamom capsules cured in traditional *bhattis* is poor. They are dark brown and have a smoky flavor; the quantity of charred and cracked capsules is high, as is the loss of volatile oil. However, the cost of building and maintaining a *bhatti* is low. A traditional *bhatti* can easily be made using locally available materials and requires no scientific knowledge.

Gasifier-based system

Gasification is a thermochemical conversion process that converts organic matter into high-value fuel gas. The Tata Energy Research Institute developed and introduced a gasifier-based curing system consisting of an updraft-type biomass gasifier connected to a traditional *bhatti*. Instead of direct burning of fuelwood, it burns producer gas from the gasifier, functioning as a kind of smoke-free kiln. The institute found the quality of these dried capsules to be better than that of capsules dried in traditional *bhatti* (TERI 2012). The gasifier system has several advantages over traditional *bhatti*, including better conversion efficiency (above 70%), controlled combustion, production of clean flue gases, a better controlled flame, and fuelwood savings of up to 65% (Rao et al 2001). It also produces dried capsules that have a more attractive color and greater volatile oil content. Nevertheless, presently no such curing system is being used by farmers in Sikkim. Chipping of fuelwoods and the additional cost of installing the gasifier might be the reason. The cost of this curing system is about US$ 243 (Deka et al 2003).

FIGURE 3 (A) Traditional *bhatti*; (B) improved *bhatti.* (Photos by A. Ingocha Singh)

Improved *bhatti*

The Indian Cardamom Research Institute developed an improved *bhatti*, an indirect heating system that uses heated air and a flue gas pipe arrangement to dry the capsules (Figure 3B). The capacity of this *bhatti* varies from 200 kg to 400 kg of fresh capsules. Drying time is reported as 17–24 hours, and it gives excellent product quality with maroon color and volatile oil content of 2–2.4%. One such unit costs about US$ 102 (Deka et al 2003). Improved *bhattis* are used by a few farmers in Arunachal Pradesh state in India. They were also introduced in the state of

Sikkim by the Spices Board of India, but farmers are reluctant to adopt them (Anonymous 2007). Use of a similar curing system in Nepal has also been reported (Stoep 2010).

Mechanical-trolley system

The Indian Council of Agricultural Research has developed another indirect-heating curing system, which can be operated by diesel or electricity. This curing system consists of a blower, a heating unit, and a multitray curing chamber similar to a mechanical cabinet tray dryer. This system works effectively and produces high-quality capsules. Its capacity is 600 kg, and curing time is 12 hours (Anonymous 2007). The cost of a diesel-fired system is about US$ 1317, and that of the electric system varies from about US$ 2027 to US$ 3041 (Anonymous 2007). However, because of the difficulty of transporting it in the hilly terrain and its high initial cost, there is no report of farmers using this system.

Other systems

In India, there are also reports of curing systems developed by other research organizations, such as the G.B. Pant Institute of Himalayan Environment and the Development and Central Food Technological Research Institute in Mysore, but none is in use by farmers so far (Madhusoodanan and Rao 2001; Deka et al 2003). In Nepal, another type of indirect-heating curing system, called the rocket stove dryer, has recently been introduced as part of a Netherlands Development Organization initiative (Stoep 2010). This flue-gas-based system produces less smoke and is reported to have produced high-quality capsules. However, very few niche markets have been identified in which higher-quality capsules will yield a higher price (Stoep 2010). In India also, except for a few growers who participated in the large cardamom auction center at Rangpo, Sikkim, farmers have been unable to negotiate with local dealers for higher prices for the higher-quality products of improved curing systems.

Thus, none of the improved curing systems have been successfully adopted by farmers. Additional research on the reasons for this will be helpful for the further development of new curing systems. Nevertheless, additional modification of the new systems is still recommended to bring down the costs of construction (Chhetri et al 2008).

Calyx cutting

The calyx or tail of large cardamom capsules (Figure 4A) is partially detached when it is rubbed against a wire mesh just after curing. In this way, the bristly outer layer of capsules can also be removed (Mande et al 1999). The tails are usually manually cut with scissors by the local traders. Capsules with the tail removed (Figure 4B) are graded as *kainchi-cut* and those with the tail intact as *non-kainchi-cut.*

FIGURE 4 (A) Large cardamom capsules with calyx; (B) large cardamom capsules with calyx removed. (Photos by A. Ingocha Singh)

No machine for removing capsule tails has been devised. As revealed by local dealers in Sikkim, extra labor needs to be employed at the cost of about US$ 0.41 per kg of capsules for this labor-intensive step. However, the College of Agricultural Engineering and Post Harvest Technology in Ranipool, Sikkim, and the Indian Cardamom Research Institute (Spices Board) in Tadong, Sikkim, jointly evaluated a cardamom polisher for possible use as a tail-cutting machine (Yurembam 2010). The Indian Cardamom Research Institute is also in search of alternative capsule-tail-cutting devices. Such a device, if cost effective, will have a strong potential application in the large cardamom value chain.

Packaging and storage

Dried large cardamom capsules are usually packed in polythene-lined jute bags. Polypropylene and ethylene terepthalate/polyethylene have been reported to considerably reduce moisture and volatile oil exchange under normal storage conditions (Sulochanamma et al 2008). The hot capsules taken out of the curing chamber are allowed to cool and then are placed into the bags, which are sealed and stored on wooden platforms to avoid moisture absorption, which could lead to mold growth. Storage stability has been maintained for large cardamom capsules with up to 11% moisture content (Naik et al 2000).

Loss of capsule weight and insect damage can also occur during storage. Moisture content of 13–15% is conducive to insect breeding; therefore, the Central Food Technological Research Institute in Mysore has recommended use of fumigants like methyl bromide (0.016 kg/m^3), phosphine (0.0015 kg/m^3), and ethyl formate (0.30 kg/m^3) to control insect infestation without affecting quality (Naik et al 2005). No report addresses the need for consumer packs (smaller packages) suitable for local markets, which may also have benefits in this regard.

Grading and quality standards

Finished large cardamom capsules are commercially graded in local markets as *badadana* (big capsules) or *chotadana*(small capsules) and, as discussed previously, as *kainchi-cut* (capsule tail removed) or *non-kainchi-cut* (capsule tail intact) (Sharma et al 2009). The difference in capsule size may be due to cultivar difference or preharvest conditions. For example, capsules of the Golsey cultivar are generally bigger. Size grading can be done using manual screens. Use of mechanical grading machines is so far not reported, except for manually operated sieves in Nepal. Quality grading is only done by local dealers and wholesalers, who employ large numbers of laborers for this purpose.

The Bureau of Indian Standards (BIS 1999) has established quality standards for large cardamom capsules (Table 2) based on the Prevention of Food Adulteration Act of 1954. Some importing countries allow only the product that conforms to these standards. However, awareness about the standards is minimal among growers and traders.

TABLE 2 Indian government quality standards for large cardamom.

Quality parameters	Requirements
Odor and taste	Free from foreign odor and taste, including rancidity and mustiness
Insects, molds, and other infestations	Not more than 10% on visual observation
Extraneous matter	Not more than 5% (by weight)
Empty and malformed capsules	Not more than 5% (by count)
Immature and shriveled capsules	Not more than 7% (by weight)
Light seeds	Not more than 5% (by weight)
Moisture	Not more than 12% (by weight)
Volatile oil	Not less than 1% (mL/100 g) on dry basis

Source: BIS (1999).

Value chains

The large cardamom postharvest value chain consists of growers, collectors, traders, and exporters. The primary processing steps required by the present market are curing, tail cutting, and grading. Curing is carried out by the farmers, and the remaining steps are done by wholesalers. The cost per kilogram of processing cardamom capsules has been estimated at US$ 0.07 for curing (including fuelwood), US$ 0.04 for tail cutting, and US $0.02 for grading (cleaning and sorting); the volume losses per kilogram in the value chain have been estimated at US$ 0.25 (MoAC 2008).

In Nepal, large cardamom is marketed to village traders, road head collectors (local collectors), regional traders, national traders or exporters, and finally buyers and consumers abroad. The price of 1 kg of cured capsules at the farmer, local trader, wholesaler, and exporter levels are US$ 2.47, US$ 2.66, US$ 2.85, and US$ 3.17, respectively. The margins at each level are US$ 0.15, US$ 0.11, US$ 0.02, and US$ 0.19, respectively (MoAC 2008).

In India, large cardamom moves through 2 market channels. In the first, farmers sell cured capsules through aggregators, and in the second, farmers sell them through contractors or bidders in an auction center (SFAC 2012). In both channels, the capsules then move on to wholesalers, then retailers, and finally consumers. The prices for 1 kg at the farmer, aggregator, wholesaler, and retailer levels are US$ 12.16, US$ 13.38, US$ 16.26, and US$ 20.33, respectively. The margins enjoyed by aggregators, wholesalers, and retailers are reported as US$ 1.22, US$ 2.71, and US$ 4.07, respectively (SFAC 2012).

Similar information for Bhutan is not available. Value-added products of large cardamom, such as essential oil and oleoresin, have high market values and export potentials (EXIM Bank 2009). However, no equipment for extracting such products is present in the production region.

Trade

In India, a major portion of the product is consumed in the domestic market. The rest is exported to other countries, including Australia, Canada, Pakistan, South Africa, the United Arab Emirates, the United Kingdom, and the United States (Table 3). Pakistan has the largest share of the market, as the product is exported to other countries is through Pakistan. Delhi, Kolkata, and Guwahati are the major Indian domestic markets. About 90% of large cardamom produced in Nepal is exported to India (ECCOS 2010). Less than 9% of the product is exported directly to Afghanistan, Pakistan, the United Arab Emirates, and other Gulf countries.

TABLE 3 Exports of large cardamom from India, 2006–2011. (Table extended on next page.)

Country	2006–2007		2007–2008	
	Quantity (metric tons)	Value (million US$)	Quantity (metric tons)	Value (million US$)
Pakistan	1199	2.54	1082	2.24
United Arab Emirates	146	0.39	97	0.27
United Kingdom	81	0.25	67	0.23
South Africa	12	0.03	11	0.03
United States	18	0.08	13	0.06
Canada	17	0.05	10	0.04
Australia	3	0.02	3	0.01

Source: Spices Board (2012b). Conversion at a fixed rate of US$ 1 = INR 49.325.

TABLE 3 Extended. (First part of Table 3 on previous page.)

Country	2008–2009		2009–2010		2010–2011	
	Quantity (metric tons)	Value (million US$)	Quantity (metric tons)	Value (million US$)	Quantity (metric tons)	Value (million US$)
Pakistan	1482	3.33	758	2.42	581	6.55
United Arab Emirates	159	0.46	58	0.26	46	0.77
United Kingdom	92	0.33	79	0.39	39	0.50
South Africa	19	0.06	16	0.08	21	0.38
United States	22	0.11	18	0.11	33	0.24
Canada	7	0.03	16	0.07	7	0.09
Australia	2	0.01	8	0.04	9	0.05

In 2010–2011, Nepal exported about 4820.85 metric tons (worth about US$ 25.92 million) to India and Singapore (TEPC 2012) and Bhutan exported 292.5 metric tons (worth US$ 2.15 million) to Bangladesh and 183.7 metric tons (US$ 2.37 million) to India (RNR 2011). However, the total volume exported from India in the same year was about 736 metric tons (worth about US$ 8.58 million). For all three countries, export volumes and values for the years 2008–2009 to 2010–2011 are presented in Figure 5. The export volume decreased for all three countries from 2008–2009 to 2010–2011, whereas the export value increased (except for Nepal in 2009–2010). This pattern might be due to global increases in large cardamom prices in response to market demands. Dema (2011) reported that farmers were paid half as much in 2010 as in 2011, when exporters paid them US$ 733.48 to US$ 814.97 per *mon* (a unit equivalent to 40 kg).

The large cardamom trade involves a number of groups in between producers (farmers) and end users. In Nepal, local dealers (or wholesalers) collect the product from farmers and sell it to exporters based in Birtamod and Biratnagar. Export price fluctuation is the major problem for farmers and traders. Marginal farmers used to take advances from local traders before the harvest season and repay the amount with interest by selling their products to them. Another system is the *dahadani*, the selling of crops in the field to local merchants, in which farmers display the harvested produce after drying (MoAC 2008; Stoep 2010). These systems of marketing are common among large cardamom growers of Nepal who are in need of cash in advance, but the price they get is lower because of it (MoAC 2008; Stoep 2010).

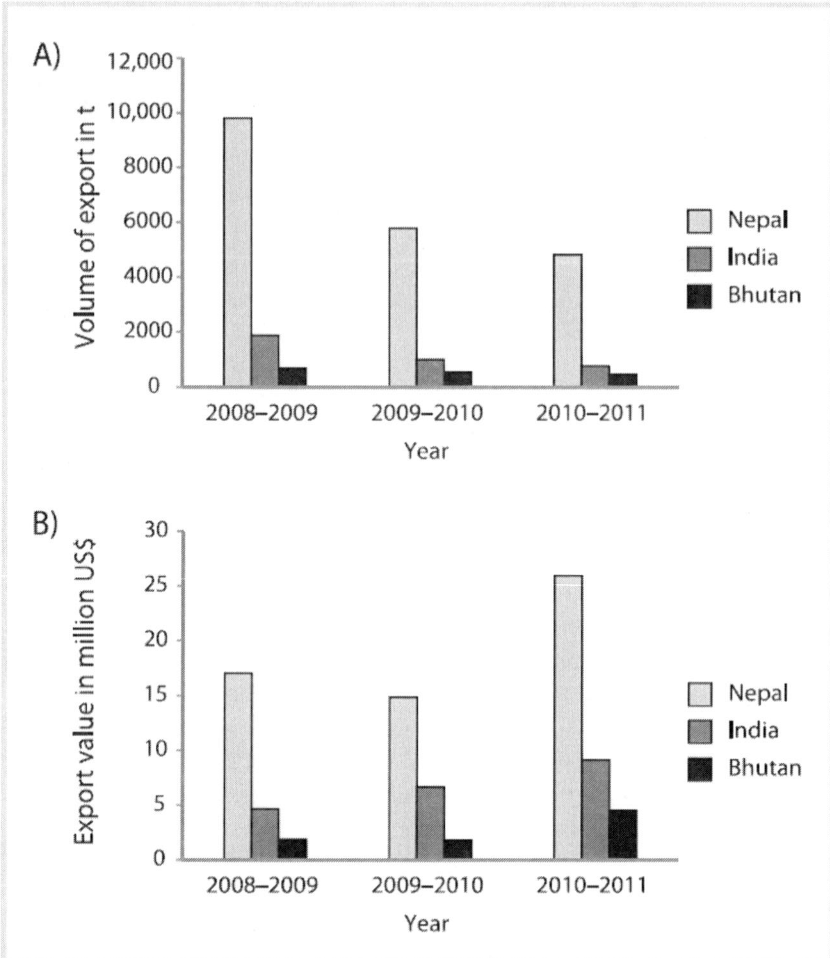

FIGURE 5 Exports of large cardamom capsules, 2008–2011: (A) by volume; (B) by value.

The Indian large cardamom market is a complex structure, as product inflow and outflow take place simultaneously. However, it is apparent that a large quantity is consumed in the domestic market, as the Indian export volume is comparatively low. The Spices Board of India controls and monitors the spice trade in this country. Local dealers or wholesalers collect dried large cardamom capsules from the farmers, perform minimal quality grading, and sell in bulk to exporters. In this system, the price of the commodity is fixed between the farmer and the local dealer, and farmers are usually paid less than the market average. Singtam, Gangtok, Jorethang, Gyalshing, Naya Bazar, Rongli, and Mangan are the major local large cardamom markets in Sikkim (Anonymous 2006). Siliguri in North Bengal is the main trade junction for Indian large cardamom; from there, it goes to other collection centers, such as Guwahati, Kolkata, Delhi, and Amritsar (ECCOS 2010; Stoep 2010; SFAC 2012).

Price protection for farmers will be an important aspect in large cardamom trading, as this plays a major role in the value chain. The Spices Board recently introduced auctions with the help of the North Eastern Regional Agricultural Marketing Corporation, a government enterprise. Thus, the Indian large cardamom trade is becoming more organized, except for products from some remote places. In Nepal and Bhutan, growers are compelled to sell their products at prices set by local wholesalers.

Processing and marketing policies

Large-cardamom–growing countries are implementing various policies to improve the crops. Most such policies seem to emphasize replanting, rejuvenation, and crop maintenance. However, positive interventions are also being undertaken in the area of postharvest processing and marketing.

The government of Nepal set up the National Cardamom Development Centre in Ilam to provide various supports for large cardamom development. Technical know-how in production, processing, and marketing is also offered through District Agricultural Development Offices, the Trade and Export Promotion Centre, and the Agro Enterprise Centre (MoAC 2008).

Similarly, in India, the Spices Board (2013) set up the Indian Cardamom Research Institute in Sikkim. The board is offering assistance worth US$ 246–444 to farmers for constructing improved *bhatti* under the Export and Development Fund, administered by the Department of Commerce. The Indian Cardamom Research Institute, in collaboration with the Indian Council of Agricultural Research, is also implementing the National Agricultural Innovation Project on Improvement of Large Cardamom, including popularization of improved *bhatti* in Sikkim (Bhattarai et al 2013). The Ministry of Commerce approved the establishment of a unit for processing and extracting essential oil and oleoresin from large cardamom in 2009–2010 (DoC 2010). In the market sector, the North Eastern Regional Agricultural Marketing Corporation is helping large cardamom growers to sell their products by auction so that they can fetch a better price (NERAMAC 2012).

In spite of these improvement policies, farmers continue to face difficulties in all aspects of improvement of the crop, including postharvest aspects. This may be because of farmers' lack of awareness about the policies, as many of them work in remote places. Better extension services through nongovernmental organizations (NGOs) and state government agencies and other ways of creating awareness among farmers may increase the utilization of existing support.

Constraints and supports

Large cardamom production has declined in recent years (Sharma et al 2009). One of the main reasons for this is viral diseases like *chirkey* and *phurkey* (MoAC 2008; ECCOS 2010; Stoep 2010). Improved curing methods known to produce higher-quality capsules are difficult for farmers to adopt because of their cost. Moreover, these *bhattis* are designed to be used only for one harvesting season in a year, and hence, it is difficult for farmers to pay back the costs invested. Lack of effective grading mechanisms also limits farmers' ability to charge higher prices for their products. Market channelization for quality capsules is also a major challenge (MoAC 2008; Sharma et al 2009).

However, many governmental organizations and NGOs are being engaged to encourage large cardamom farming in the sub-Himalayan region, such as the National Cardamom Development Center and the Netherlands Development Organization in Nepal (MoAC 2008). In India, the Indian Cardamom Research Institute, Department of Horticulture and Cash Crop Development, Government of Sikkim, and North Eastern Regional Agricultural Marketing Corporation are providing supports in crop development, postharvest technology, and marketing.

Conclusions and future prospects

The worldwide production of large cardamom, a high-value, low-volume crop, has fallen in recent years due to several factors, including disease. Adoption of proper postharvest processing techniques can help to compensate for decreased production by reducing postharvest losses and adding value. Improved *bhattis* yield better quality than traditional *bhattis*, but these devices have not been well accepted by farmers. Farmers need a low-cost curing system that can produce good-quality capsules. Several other labor-intensive postharvest processing operations—such as separating capsules from spikes, cleaning, tail cutting, and grading—have not received the attention they need from researchers. Mechanical systems that reduce human drudgery and make postharvest processing more efficient will decrease losses and increase the value of the capsules produced.

Apart from primary processing, it is also necessary to explore other ways to add value to this crop, such as production of essential oil and oleoresin. Compared with 2008–2009, the quantity of the cardamom traded in 2010–2011 decreased. However, the export value was greater in 2010–2011, which shows the increased value and demand for the crop. Therefore, an increase in production of this important crop will improve the livelihoods of many mountain people in the sub-Himalayan region.

In spite of several policies implemented to improve the crop, large cardamom farmers still face difficulties in postharvest management. Better extension services by NGOs or government agencies or both will improve awareness about the policies among farmers and help ensure their successful implementation. Organizations like the North Eastern Regional Agricultural Marketing Corporation can also help farmers get better prices by acting as a link between farmers and traders. Finally, proper pricing for high-quality product can also help to make large cardamom farming an attractive and profitable livelihood.

ACKNOWLEDGMENTS

The authors wish to acknowledge the Indian Cardamom Research Institute, Regional Research Station, Gangtok, Sikkim, for technical discussions on large cardamom issues and field visits.

REFERENCES

1. Adegoke GO, Rao LJM, Shankaracharya NB, Jagmohan RL. 1998. A comparison of the essential oils of Afromum daniellii and Amomum subulatum. *Flavour and Fragrance Journal* 13:345–352.
2. Agnihotri S, Wakode S. 2010. Antimicrobial activity of essential oil and various extracts of fruits of greater cardamom. *Indian Journal of Pharmaceutical Sciences* 72(5):657–659.,
3. [Anonymous]. 2006. *Training Manual of Horticultural Crops.* Gangtok, India: Horticulture and Cash Crop Development Department-Sikkim.

4. [Anonymous]. 2007. *Proceedings of the Group Meeting Held on 06-11-2007*. Tadong, Sikkim, India: Indian Cardamom Research Institute, Regional Research Station.
5. Avasthe RK, Singh KK, Tomar JMS. 2011. Large cardamom (*Amomum subulatum* Roxb.) based agroforestry systems for production, resource conservation and livelihood security in the Sikkim Himalaya. *Indian Journal of Soil Conservation* 39(2):155–160.
6. Bhattarai NK, Deka TN, Chhetri P, Harsha KN, Gupta U. 2013. Livelihood improvement through sustainable large cardamom cultivation in North Sikkim. *International Journal of Scientific and Research Publications*3(5):1–4.
7. BIS [Bureau of Indian Standards]. 1999. *Spices and Condiments—Large Cardamom—Capsules and Seeds—Specification*. New Delhi, India: Bureau of Indian Standards IS 13446:1999.
8. Bisht VK, Negi AK, Bhandari AK, Sundriyal RC. 2011. *Amomum subulatum* Roxb.: Traditional, phytochemical and biological activities—An overview. *African Journal of Agricultural Research* 6(24):5386–5390.
9. Bisht VK, Purohit V, Negi JS, Bhandari AK. 2010. Introduction and advancement of large cardamom (*Amomum subulatum* Roxb.) in Uttarakhand, India. *Research Journal of Agricultural Sciences* 1(3):205–208.
10. Chapagain D. 2011. *Assessment of Climate Change Impact on Large Cardamom and Proposed Adaptation Measures in Eastern Hill of Nepal*. Kathmandu, Nepal: Central Department of Environmental Science, Tribhuvan University. www.climatenepal.org.np/main/downloadFile.php?fn=l5smn4oiz1e.pdf&ft=application/pdf&d=publication; accessed on 6 March 2012.
11. Chempakam B, Sindhu S. 2008. Large cardamom. In: Parthasarathy VA, Chempakam B, Zachariah TJ, editors. *Chemistry of Spices*. Wallingford, United Kingdom: CAB International, pp 60–69.
12. Chhetri PB, Khatiwada E, Lloret A, Jabegu K. 2008. Benefits of value addition: A success story from the hills of Nepal. *Enterprise Development and Microfinance* 19(1):69–83.
13. Deka TN, Biswas AK, Gopakumar B, Potty SN. 2003. Large cardamom curing through ICRI improved bhati. *Journal of Hill Research* 16(1):57–60.
14. Dema K. 2011. Happy days are here again! Blog post. www.kuenselonline.com/2011/?p=19883; accessed on 3 February 2012.
15. DoC [Department of Commerce, Government of India]. 2010. Trade promotion initiatives in the North East Region. In: Annual Report 2009–2010. www.commerce.nic.in/publications/anualreport_chapter13-2009-10.asp; accessed on 20 August 2013.
16. ECCOS [Economics Consultancy Services]. 2010. *Market Survey Medicinal and Aromatic Products (MAP) Nepal. Report of Economics Consultancy Services, Germany*. www.includenepal.org/alluploads/MAPs%20Market%20Survey%2C%20ECCOS%202010.pdf; accessed on 3 February 2012.
17. EXIM Bank [Export Import Bank]. 2009. *Sikkim: Export Potential and Prospects. Occasional paper No. 123*. Mumbai, India: EXIM Bank and Quest Publications. www.eximbankindia.org.in/op/op134.pdf; accessed on 8 August 2013.
18. Gilani SR, Shahid I, Javed M, Mehmud S, Ahmed R. 2006. Antimicrobial activities and physico-chemical properties of the essential oil from *Amomum subulatum*. *International Journal of Applied Chemistry* 2(2):81–86.
19. Gurudutt KN, Naik JP, Srinivas P, Ravindranath B. 1996. Volatile constituents of large cardamom (*Amomum subulatum* Roxb.). *Flavour and Fragrance Journal* 11(1):7–9.
20. Jafri MA, Javed KF, Singh S. 2001. Evaluation of the gastric antiulcerogenic effect of large cardamom (fruits of *Amomum subulatum* Roxb.). *Journal of Ethno pharmacology* 75:89–94.,
21. Kaskoos RA, Mir SR, Kapoor R, Ali M. 2008. Essential oil composition of fruits of *Amomum subulatum* Roxb. *Journal of Essential Oil Bearing Plants* 11:184–187.
22. Madhusoodanan KJ, Rao YS. 2001. Cardamom (large). In: Peter KV, editor. *Handbook of Herbs and Spices, vol. 1*. Cambridge, United Kingdom: Woodhead, pp. 139.

23. Mande S, Kumar A, Kishore VVN. 1999. A study of large-cardamom curing chambers in Sikkim. *Biomass and Bioenergy* 16:463–473.
24. Mishra AK, Dubey NK. 1990. Fungitoxicity of essential oil of *Amomum subulatum* against *Aspergillus flavus*. *Economic Botany* 44(4):530–533.
25. MoAC [Ministry of Agriculture and Cooperatives]. 2008. *Final Report: Product Chain Study Cardamom*. Biratnagar, Nepal: MoAC. www.cadp.gov.np /Document_Uploads/Reports/dd1d5812506492836355.pdf; accessed on 4 February 2012.
26. MoAC [Ministry of Agriculture and Cooperatives]. 2011. *Area, Production and Yield of Major Spice Crops 2010/2011. Statistical Information on Nepalese Agriculture: 2010/2011. Year Book 2011*. Singha Durbar, Kathmandu, Nepal: Government of Nepal, Ministry of Agriculture and Co-operatives. www.moad. gov.np/ downloadfile/cover_1331268679.docx; accessed on 19 September 2013.
27. Naik HP, Balasubrahmanyam N, Dhanaraj S, Gurudutt KN. 2000. Packaging and storage studies in flue-cured large cardamom (*Amomum subulatum* Roxb.). *Journal of Food Science and Technology (Mysore)* 37(6):577–581.
28. Naik JP, Ramesh BS, Gurudutt KN. 2005. Fumigation studies on cured large cardamom (*Amomum subulatum*Roxb.) capsules. *Journal of Food Science and Technology (India)* 42(6):531–533.
29. Naik JP, Rao LJM, Kumar TMM, Sampathu SR. 2004. Chemical composition of the volatile oil from the pericarp (husk) of large cardamom (*Amomum subulatum* Roxb.). *Flavour and Fragrance Journal* 19(5):441–444.
30. NERAMAC [North Eastern Regional Agricultural Marketing Corporation Limited]. 2012. *30th Annual Report 2011–2012*. www.neramac.com/ Uploads /File/AR_11-12_eng.pdf; accessed on 20 August 20013.
31. NIIR [National Institute of Industrial Research]. 2006. *The Complete Book on Spices & Condiments (with Cultivation, Processing & Uses)*. Delhi, India: Asia Pacific Business Press.
32. Rao VG, Mande S, Kishore VVN. 2001. Study of drying characteristics of large cardamom. *Biomass and Bioenergy* 20:37–43.
33. RNR [Renewable Natural Resources]. 2011. *Statistics on Volume and Monetary Value of Major Exporting Commodities by Countries of Destination*. Thimphu, Bhutan. www.moaf.gov.bt/moaf/?wpfb_dl=846; accessed on 3 June 2012.
34. Rout PK, Sahoo D, Jena KS, Rao YR. 2003. Analysis of the oil of large cardamom (*Amomum subulatum* Roxb.) growing in Sikkim. *Journal of Essential Oil Research* 15(4):265–266.
35. SFAC [Small Farmers' Agribusiness Consortium]. 2012. *Value Chain Analysis of Selected Crops in North Eastern States*. New Delhi, India: SFAC. www.sfacindia.com/PDF/SFAC_Value-Chain-Analysis.pdf; accessed on 8 August 2013.
36. Sharma G, Sharma R, Sharma E. 2008. Influence of stand age on nutrient and energy release through decomposition in alder-cardamom agroforestry systems of eastern Himalayas. *Ecological Research* 23(1):99–106.
37. Sharma G, Sharma R, Sharma E. 2009. Traditional knowledge systems in large cardamom farming: Biophysical and management diversity in Indian mountainous regions. *Indian Journal of Traditional Knowledge* 8(1):17–22.
38. Sharma E, Sharma R, Singh KK, Sharma G. 2000. A boon for mountain populations: Large cardamom farming in the Sikkim Himalaya. *Mountain Research and Development* 20(2):108–111.
39. Singh P, Srivastava B, Kumar A, Dubey NK, Gupta R, Tanu. 2008. Efficacy of essential oil of *Amomum subulatum* as a novel aflatoxin B_1 suppressor. *Journal of Herbs, Spices & Medicinal Plants* 14(3–4):208–218.
40. Spices Board. 2001. *Guidelines for Production of Organic Spices in India*. Cochin, India: Spices Board.
41. Spices Board. 2012a. *Spices Board Annual Report 2010-11*. Cochin, India: Spices Board, Ministry of Commerce & Industries, Government of India. http://www.indianspices.com/pdf/Spice_Annual_Report_2010_11.pdf; accessed on 12 June 2012.

42. Spices Board 2012b. *Major Item/Country-Wise Export of Spices from India*. Cochin, India: Spices Board. www.indianspices.com/html/maj_impcon.htm; accessed on 12 June 2012.
43. Spices Board. 2013. *Export Oriented Production & Post Harvest Improvement of Spices*. Cochin, India: Spices Board. www.indianspices.com/html/prodev_sc_01.htm; accessed on 20 August 2013.
44. Stoep GAV. 2010. *Enhancing Competitiveness of Nepal's Large Cardamom Value Chain*. Kathmandu, Nepal: SNV Netherlands Development Organization.
45. Sulochanamma G, Ramalakshmi K, Kumar TMM, Indiramma AR, Ramesh BS, Sampathu SR, Naik JP. 2008. Storage characteristics of large cardamom (*Amomum subulatum* Roxb.) and seeds in different packages. *Journal of Food Science and Technology (Mysore)* 45(2):183–186.
46. TEPC [Trade and Export Promotion Centre]. 2012. *Trade Statistics*. Kathmandu, Nepal: TEPC. www.tepc.gov.np/tradestatistics/gl-16-major-country-exports-sel-comm.php; accessed on 12 June 2012.
47. TERI [Tata Energy Research Institute]. 2012. *A New Curing Chamber for Large Cardamom*. New Delhi, India: TERI. www.teriin.org/index.php?option= com_casestudy&task=details&sid=2; accessed on 12 June 2012.
48. Thomas VP, Sabu M, Gupta U. 2009. Taxonomic studies on cultivars of *Amomum subulatum*. *Rheedea* 19(1&2):25–36.
49. Yurembam GS. 2010. *Performance Evaluation of Cardamom Polisher as Large Cardamom Detailing Machine [B.Tech thesis]*. Sikkim, India: College of Agricultural Engineering and Post Harvest Technology, Ranipool.

CHAPTER 9

Energy Status of Ripening and Postharvest Senescent Fruit of Litchi (*Litchi Chinensis* Sonn.)

Hui Wang[1,2], Zhengjiang Qian[1,2], Sanmei Ma[3], Yuchuan Zhou[4], John W Patrick[5], Xuewu Duan[1], Yueming Jiang[1] and Hongxia Qu[1*]

[1]*Key Laboratory of Plant Resources Conservation and Sustainable Utilization, South China Botanical Garden, Chinese Academy of Sciences, Guangzhou 510650, P R China.*

[2]*University of Chinese Academy of Sciences, Beijing 100049, P R China.*

[3]*Department of Biotechnology, Jinan University, Guangzhou 510632, P R China.*

[4]*Australian Institute for Bioengineering and Nanotechnology, the University of Queensland, Brisbane St Lucia, QLD 4072, Australia.*

[5]*School of Environmental & Life Sciences, the University of Newcastle, Callaghan, NSW 2308, Australia.*

ABSTRACT

Background: Recent studies have demonstrated that cellular energy is a key factor switching on ripening and senescence of fruit. However, the factors that influence fruit energy status remain largely unknown.

Results: HPLC profiling showed that ATP abundance increased significantly in developing preharvest litchi fruit and was strongly correlated with fruit fresh weight. In contrast, ATP levels declined significantly during postharvest fruit senescence and were correlated with the decrease in the proportion of edible fruit. The five gene transcripts isolated from the litchi fruit pericarp were highly expressed in vegetative tissues and peaked at 70 days after flowering (DAF) consistent with fruit ADP concentrations, except for uncoupling mitochondrial protein 1 (*UCP1*), which was predominantly expressed in the root, and ATP synthase beta subunit (*AtpB*), which was up-regulated significantly before harvest and peaked 2 days after storage. These results indicated that the color-breaker stage at 70 DAF and 2 days after storage may be key turning points in fruit energy metabolism. Transcript abundance of alternative oxidase 1 (*AOX1*) increased after 2 days of storage to significantly higher levels than those of *LcAtpB*, and was down-regulated significantly by exogenous ATP. ATP supplementation had no significant effect on transcript abundance of ADP/ATP carrier 1 (*AAC1*) and slowed the changes in sucrose non-fermenting-1-related kinase 2 (*SnRK2*) expression, but maintained ATP and energy charge levels, which were correlated with delayed senescence.

Conclusions: Our results suggest that senescence of litchi fruit is closely related with energy. A surge of *LcAtpB* expression marked the beginning of fruit senescence. The findings may provide a new strategy to extend fruit shelf life by regulating its energy level.

BACKGROUND

Fruit ripening and senescence are active processes initiated by internal and environmental factors. Considerable evidence suggests that cellular energy supply is a key factor controlling ripening and senescence events, and that aging and browning of postharvest horticultural crops may be related to inadequate supplies and reduced efficiency of cellular energy generation [1, 2, 3]. A salient feature initiating the aging process is a decline in adenosine triphosphate (ATP) levels. For instance, increased membrane permeability and enhanced reactive oxygen species (ROS) production by harvested fruit are related to low ATP and energy charge (EC) levels [4, 5]. Preharvest application of boron (B) and calcium (Ca) [6], and postharvest treatments, including ethylene inhibitors [7], high oxygen [4], anaerobic conditions [8], controlled atmosphere (CA) storage [2, 3, 5, 9], an exogenous carbon source [1] and ATP [10, 11], can maintain tissue levels of ATP and EC, thereby delaying pericarp browning of litchi and longan and internal flesh browning in 'Conference' pears. In addition, exogenous ATP elevates ATP levels, inhibits ROS accumulation, and maintains unsaturated fatty acid levels and membrane integrity, thus delaying senescence and deterioration of horticultural products [5, 7, 8, 9, 10, 12, 13, 14]. Therefore, browning and senescence are closely related with cellular energy status, and measures to maintain energy levels can delay aging and inhibit occurrence of browning and deterioration to a certain extent. However, the mechanism contributing to the energy deficit remains unclear owing to the diversity of energy regulatory elements and complexity of energy production, transfer and control. Moreover, a growing body of evidence indicates that cellular energy levels are closely related to the activity, transcription and translation of certain regulatory elements [15, 16, 17, 18]. Among these regulatory elements, proteins that synthesize (ATP synthase), dissipate (alternative oxidase, AOX; and mitochondrial inner membrane uncoupling proteins, UCP), transport [adenosine diphosphate (ADP)/ATP carrier, AAC] and regulate (sucrose non-fermenting-1-related kinase, SnRK) ATP have attracted considerable attention.

F1F0-ATP synthase is a multimeric enzyme that catalyzes the final step of oxidative phosphorylation and photophosphorylation, the synthesis of ATP from ADP and inorganic phosphate (P_i) [19, 20]. The general structure of the core subunits of the enzyme are highly conserved in both prokaryotic and eukaryotic organisms, and are designated as αβγδε [16]. As a novel cell death regulator, subunit β (AtpB) plays a pivotal role in stress responses. Up-regulation of *AtpB* transcription marks initiation of the aging process [21, 22]. Complex V cannot be assembled successfully and respiration rates are reduced resulting in inhibition of ATP synthesis in the absence of the β subunit [21]. However, it is unknown whether *AtpB* acts as a senescence marker gene in fruit as in photosynthetic tissues.

AOX and UCP are ubiquitous in plants and have various physiological roles. These include heat production and protection against oxygen free radicals mediating the non-phosphorylating bypasses that affect ATP synthesis [16, 23, 24, 25, 26]. In plants UCP and AOX belong to multigene families, of which members exhibit specific organ and temporal expression patterns that respond differentially to stress conditions [27]. AOX expression and activity are generally induced under stress conditions, and consequently AOX is used as a marker for cell reprogramming under stress [28, 29, 30, 31]. The alternate oxidation pathway may slow down the aging process and quality deterioration of postharvest horticultural crops [29]. Furthermore, changes in AOX and UCP accumulation in vine-ripened fruit differ significantly from those of postharvest-ripened fruit [32, 33]. This implies that a complementary relationship exists between the two energy dissipation pathways. Several reports have described the complex regulation of AOX and UCP at the translational and posttranslational levels. However, most research on these two genes has focused on climacteric fruit, such as tomato, apple, banana, and mango [34, 35, 36, 37], and their roles in non-climacteric fruit remains unclear.

ADP/ATP carrier (AAC) belongs to the high-abundance mitochondrial transporter protein family [38]. It catalyzes a counter-exchange of cytosolic ADP with matrix ATP, enabling mitochondria to supply the cytosol, and subsequently other organelles, with energy [39]. In higher organisms AAC is the core protein of the mitochondrial adenine nucleotide translocator and the main regulator of mitochondrial ATP concentration [15, 40, 41]. As 'the energy regulator' in the cell, sucrose non-fermenting-1-related protein kinase (SnRK) can sense the internal energy status by measuring cytosolic ATP/ [adenosine monophosphate (AMP)] ratios and controlling the expression and phosphorylation of key metabolic enzymes [15]. However, to the best of our knowledge, no report has documented these two genes in the development and senescence of fruit.

Overall, in contrast to the relatively well-known physiological roles of these genes in stress responses, little information is available on how these genes contribute to the energy status of developing and senescing fruit. Studying the transcript level of these genes in relation to cellular energy status is crucial to identify causal agents contributing to energy deficits and their role in fruit ripening and senescence. This knowledge may provide opportunities to develop new strategies to control fruit ripening and the aging process and ultimately extend the storage life of harvested fruit.

Litchi is a non-climacteric, tropical and subtropical fruit with a high commercial value on the international market. Litchi fruit deteriorate rapidly after harvest because of water loss, pericarp browning and rot development [42]. Our previous studies reported that an exogenous ATP supply enhanced antioxidant systems and maintained membrane integrity to delay browning and senescence of litchi fruit [11, 13, 14]. However, the molecular mechanisms underlying these phenomena remain unclear. In the current study, full-length sequences of *AtpB*, *AOX1*, *UCP1*, *AAC1* and *SnRK2* were cloned from litchi fruit. Transcript abundance of these energy-related genes, respiration intensity and fruit energy status were determined for developing and postharvest-senescent litchi fruit.

RESULTS

Preharvest

Growth curve and pericarp color of litchi fruit

Figure 1A shows the appearance of fruit at five developmental stages ranging from 50 days after flowering (DAF) (young) to 90 DAF (fully ripe) at 10-day intervals. Fruit growth was divided into two stages. Stage I (50-70 DAF) represented a rapid growth phase, whereas stage II (70-90 DAF) was a slow growth phase as fruit approached maturity (Figure 1B). Lightness values significantly increased during stage I (50-70 DAF), but chroma values increased more markedly at 50-60 DAF, which indicated that color intensity rose before it decreased (Table 1). Hue angle decreased slowly at

50-60 DAF, but more substantially from 60 to 80 DAF before attaining a plateau (Table 1), which was consistent with the change in color of the pericarp (Figure 1A).

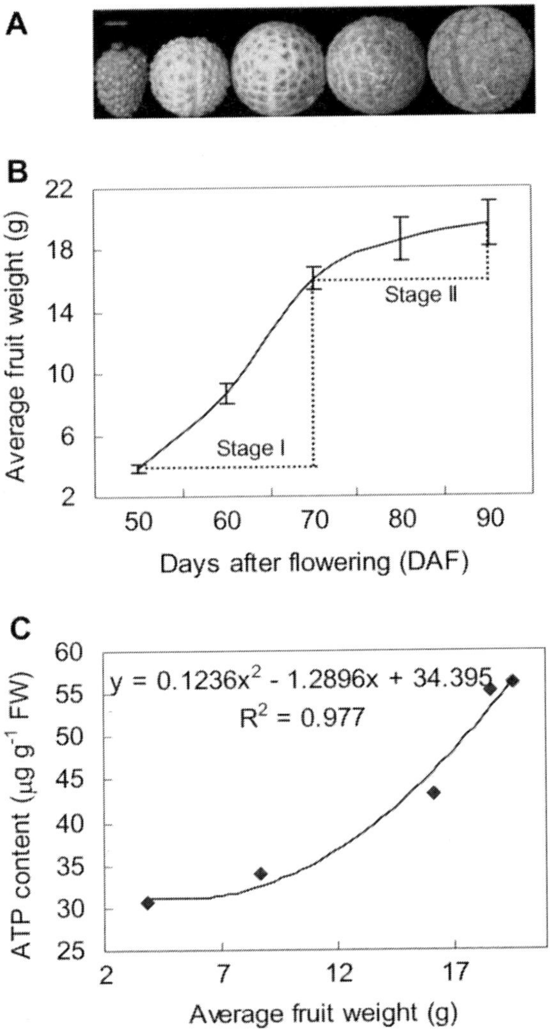

Figure 1 Appearance (A), average fruit weight (B), and the correlation between average fruit weight and ATP level (C) during development and ripening of litchi fruit. Left to right: fruit at 50 DAF, 60 DAF, color-breaker stage at 70 DAF, red-ripe stage at 80 DAF, and over-ripe stage at 90 DAF. Bar = 1 cm; DAF, days after flowering; FW, fresh weight. Data are the mean ± standard deviation (n = 20 in B; n = 3–5 in C).

Table 1 Color parameters (lightness, chroma, and hue angle) of litchi fruit at different development and ripening stages

Color parameters	Days after flowering (DAF)				
	50	60	70	80	90
Lightness	45.60 ± 1.36	47.93 ± 1.39	51.99 ± 2.77	43.14 ± 2.23	36.48 ± 2.72
Chroma	32.87 ± 1.09	38.34 ± 0.95	36.03 ± 1.15	34.63 ± 0.94	28.21 ± 1.52
Hue angle	115.19 ± 0.79	109.74 ± 1.37	75.94 ± 6.60	36.67 ± 2.91	34.62 ± 5.50

Data are the mean ± standard deviation (n = 15–20).

Respiration intensity, oxygen consumption rate and cyanide-resistant respiration

Whole-fruit respiration intensity decreased significantly during stage I, but increased slightly during stage II (Figure 2A). The decrease in respiration intensity during stage I may be feedback to enhanced ADP accumulation (Figure 3B). Oxygen consumption rate (V_t) declined significantly after 60 DAF and thereafter did not change (Figure 2B). Cyanide-resistant respiration capacity (the proportion of alternative oxidative respiration to total respiration, $\rho V_{alt}/V_t$) increased significantly during fruit development and ripening at 50–90 DAF (Figure 2C), which implies it has a pivotal role in the balance between respiration and production of ROS at these stages.

Energy status

ATP concentrations were correlated strongly with fruit fresh weight (Figure 1C), and increased rapidly between 50 and 80 DAF before slowing at the final stage of maturation (80–90 DAF) (Figure 3A). ADP concentrations increased during stage I but decreased progressively during stage II (Figure 3B). AMP concentrations decreased gradually throughout fruit development, but increased slightly in the last 10 days before harvest (Figure 3C). Total adenylate levels increased steadily between 50 and 80 DAF, and did not change thereafter (Figure 3D). The EC level of the fruit remained relatively constant in young fruit, and increased markedly as the fruit ripened between 80 and 90 DAF (Figure 3E). The ATP/AMP ratio peaked at 80 DAF, and thereafter declined until maturity (Figure 3F).

Isolation and sequence analysis of energy-related genes

Partial sequence fragments of energy-related genes were isolated by reverse transcription- PCR (RT-PCR) using degenerate primers (Table 2). Sequencing of several of these fragments revealed the existence of one isoform for each gene. These genes were *LcAtpB* (1188 bp), *LcAOX1* (676

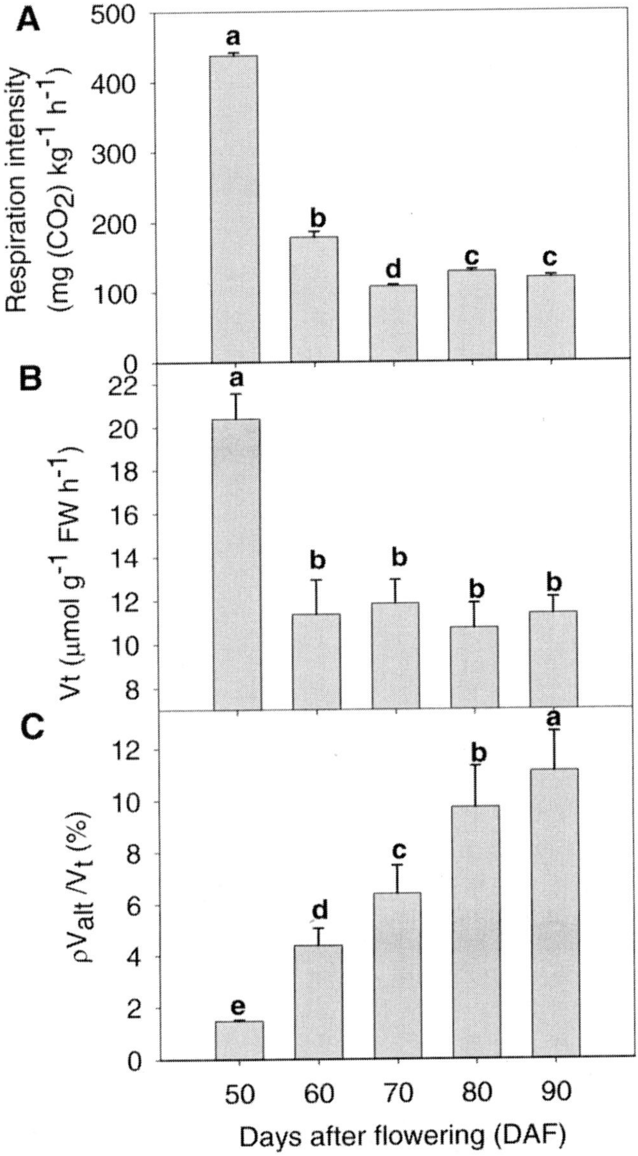

Figure 2 Respiration intensity (A), V_t(B) and ρ V_{alt}/ V_t(C) during development and ripening of litchi fruit. V_t, oxygen consumption rate; $\rho V_{alt}/V_t$, contributions of alternative oxidative respiration to total respiration. Data are the mean ± standard deviation (n = 3–6). Means with the same letter are not significantly different ($p < 0.05$) among the different developmental stages.

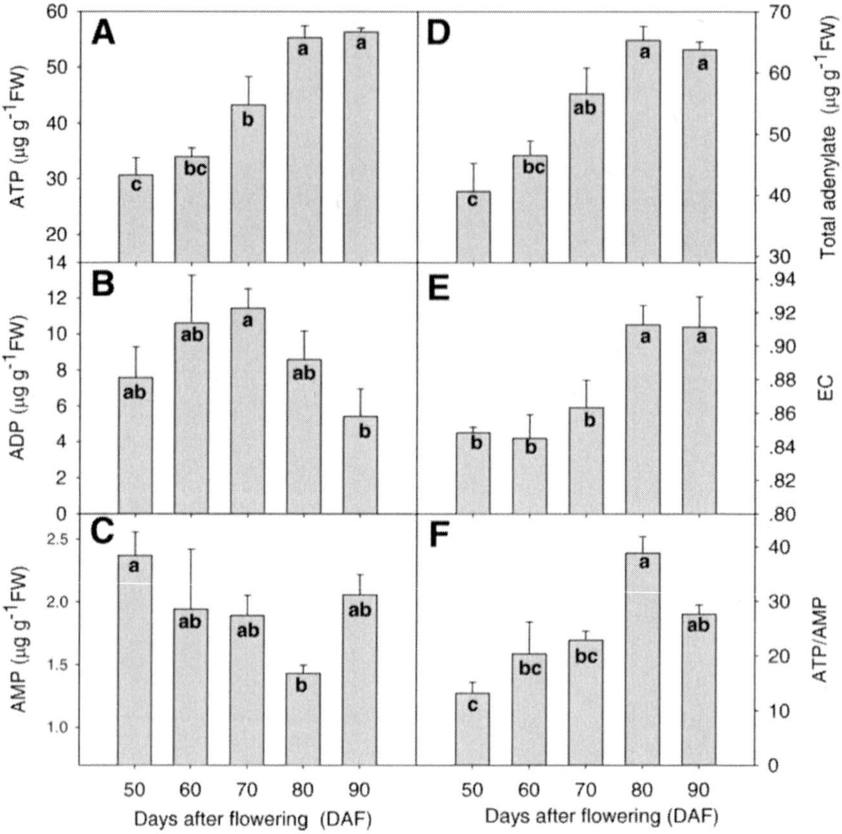

Figure 3 Energy status during development and ripening of litchi fruit. (**A**) Adenosine triphosphate (ATP); (**B**) adenosine diphosphate (ADP); (**C**) adenosine monophosphate (AMP); (**D**) total adenylate; (**E**) energy charge (EC); and (**F**) ratio of ATP/AMP. Contents of ATP, ADP, AMP and total adenylate are expressed as μg g^{-1} FW. FW, fresh weight. Data are the mean ± standard deviation ($n = 3$). Means with same letter are not significantly different ($p < 0.05$) among the different developmental stages.

bp), *LcUCP1* (652 bp), *LcAAC1* (850 bp) and *LcSnRK2* (773 bp). Full-length sequences of these genes were obtained after 3′ and 5′-RACE. These sequences were compared with known sequences from other species using the NCBI BLAST server. The GenBank accession codes of the sequences are listed in Table 3.

Table 2 Degenerate primers for cloning of energy-related genes in litchi pericarp

Gene	Forward primer (5′ to 3′)	Reverse primer (5′ to 3′)	Product (bp)
AtpB	GGGCCGTGGCCatgwsngcnac	TCGGGCAGGCCGtcnarytcncc	1188
AOX1	GGAGGACGGCACCGAntggmmntg	GCGTCCTTGGGCAGCCkccartartcda	676
UCP1	ACCATCGCCCGGGARGARGGNNT	AGGTGCCCAGCCGGSYRAARTTNGG	652
AAC1	GCCGCCGCCccnathgarmg	CGCCGCCGGAGCcrtayttyttnc	850
SnRK2	GGCGCCGGCAACttyggngtngc	AGCTCCCTGGGCAGGttyttnarraac	773

AtpB, ATP synthase β subunit; *AOX1*, alternative oxidase 1; *UCP1*, uncoupling mitochondrial protein 1; *AAC1*, ADP/ATP carrier 1; *SnRK2*, sucrose non-fermenting-1-related kinase 2.

Table 3 Homologies based on nucleotide sequences for energy-related genes isolated from litchi cv. Huaizhi

Gene	GenBank number	Top *Arabidopsis* BLAST match	Top BLAST match excluding *Arabidopsis*	Homology (%)
LcAtpB	JQ349005	AJ271468.1 *AtAtpB*	AJ235513.2 *AtpB Koelreuteria paniculata*	60^a, 92^b
LcAOX1	JQ349006	NM_113135.3 *AtAOX1A*	EF523518.1 *AOX1a Nicotiana glutinosa*	69^a, 71^b
LcUCP1	JQ349009	NM_125287.4 *AtUCP2*	XM_002520396.1 *UCP Ricinus communis*	66^a, 65^b
LcAAC1	JQ349007	NM_121352.3 *AtAAC2*	XM_002531865.1 *AAC Ricinus communis*	71^a, 75^b
LcSnRK2	JQ349008	NM_001203118.1 *AtSnRK2.2*	XM_002513909.1 *SAPK1 Ricinus communis*	68^a, 73^b

See legend for Table 2 for names of genes.

The coding region of *LcAtpB* was 1377 bp in length, encoding a deduced 459-amino acid (aa) sequence, with a predicted MW of 49.90 kDa and a calculated isoelectric point (pI) of 5.12. LcAtpB was 92% homologous with the AtpB from *Koelreuteria paniculata,* a close relative of litchi. The C-terminal domain of the β subunit contained the highly conserved 'DELSEED' motif (Additional files 1 and 2) involved in mechanochemical coupling of ATP synthase [43]. *AOX* belongs to a multigene family in many plants, including mango, tomato and other horticultural crops. Two types of *AOX* genes are present in angiosperms [44], of which isoforms may be differentially expressed in fruit and other tissues, such as in tomato [45]. While in the present study, transcript evidence showed support for a single predicted isoform (*LcAOX1)* for all candidates in litchi fruit pericarp and consequently only this isoform was detected during experimental validation. *LcAOX1* (1032 bp), encoding a 344 aa sequence (predicted MW 39.13 kDa; pI 8.76), showed 71% identity with that of *Nicotiana glutinosa*. Site-directed mutagenesis has shown that several residues are required for AOX activity [44]. The highly conserved residues were identified in litchi fruit using multiple sequence alignments of AOX proteins from diverse organisms (Additional files 3 and 4). UCP family from diverse organisms are grouped into five subfamilies, plant UCP1 and UCP2 belong to subfamilies II. Moreover, investigations have identified 1 to 6 genes that encode UCPs in different plants [27]. The major feature of UCPs from plants is the presence of three energy transfer protein signatures (ETPS) [23] conserved in all UCP isoforms (Additional file 5). In the present study, three copies of the ETPS signature were identified in *LcUCP1*, which contained an open reading frame (ORF) of 915 bp encoding a 305 aa protein, with a predicted MW of 33.23 kDa and a pI of 9.01. A BLAST search of GenBank revealed that LcUCP1 shared 66% identity with AtUCP2 (NM_125287.4) from *Arabidopsis thaliana* and 65% identity with RcUCP (XM_002520396.1) from *Ricinus communis* at the protein level. A neighbor-joining cladogram further showed that *LcUCP1* was most similar to *UCP* from *R. communis*(Additional files 5 and 6). There are four groups of mitochondrial adenine nucleotide carriers presented in the five eukaryotic clades, among which the ADP/ATP carriers are the best known. Furthermore, three mitochondrial AAC genes: *AtAAC1*, *AtAAC2*, and *AtAAC3* were presented in the genome of *A. thaliana*[46]. In the present study, *LcAAC1* gene isolated from litchi fruit pericarp contained an ORF of 1266 bp encoding a 422 aa protein (predicted MW 46.00 kDa; pI 9.79), and exhibited 71% identity with AAC2 from *A. thaliana* and 75% identity to AAC from *R. communis*. LcAAC1 contained a highly conserved RRRMMM signature motif (Additional files 7 and 8) [17]. On the basis of sequence similarity, domain structure and cellular function, the plant SnRK family can be divided into three subfamilies: SnRK1, SnRK2 and SnRK3. Among these subfamilies, SnRK2 is unique to plants and is involved in responses to environmental stresses [47]. In the present study, *SnRK* cDNA isolated from litchi pericarp was 1394 bp in length with a 1014 bp ORF encoding a predicted polypeptide of 338 aa (MW 38.00 kDa; pI 5.36). The deduced amino acid sequence showed homology with counterpart SnRK2 family members, namely *R. communis*, *Glycine max*, *Zea mays*, and *A. thaliana*. Similar to other *SnRK2* genes, the N-terminal catalytic domain was highly conserved, containing an ATP-binding site (DIGSGNFGVAKLVRDTWTDELLAVK) and protein kinase activating signature (QICHRDLKLENTLL), whereas the C-terminal regulatory domain contained an acidic 'patch', which is essential for kinase activity (Additional file 9). In a neighbor-joining cladogram *LcSnRK2* and its counterparts *OsSAPK3*, *ZmSnRK2.2*, *GmSARK2*, and *RcSAPK1*were clustered in subclass 2 (Additional file 10) as reported by Zhang et al. [47].

Spatial and temporal expression of energy-related gene transcripts

Expression analysis of energy-related genes indicated ubiquitous expression in all tissues tested (Figure 4). The most notable features of the tissue expression profiles were that *LcUCP1* (Figure 4C) exhibited strongest expression in the root in which transcript levels were some 100-fold greater than those detected in the other tissues. In contrast, the remaining genes exhibited high expression in vegetative tissues (root, stem, and leaf). *LcAtpB* and *LcAAC1* were expressed most strongly in the leaf (Figure 4A and D). *LcAtpB* and *LcAOX1* showed the highest expression in the root and leaf, with lower levels of expression detected in the stem, and

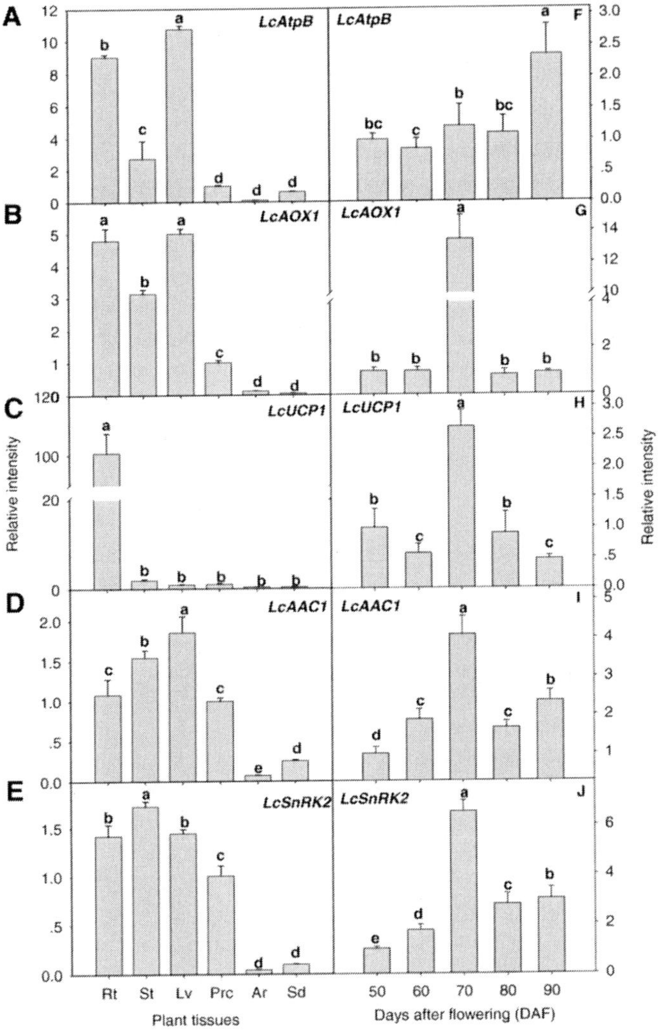

Figure 4 Spatial and temporal expression of gene transcripts. Total RNA was isolated from litchi plants 80 days after flowering for tissue-specific expression analysis. Transcript abundance was determined using qRT-PCR and was normalized using *LcACTIN*. Values are the mean ± standard deviation ($n = 3$) from three separate RNA extractions. Means with the same letter are not significantly different ($p < 0.05$) among the different tissues (**A, B, C, D,** and **E**) and among developmental stages (**F, G, H, I,** and **J**). *AtpB*, ATP synthase β subunit; *AOX1*, alternative oxidase 1; *UCP1*, uncoupling mitochondrial protein 1; *AAC1*, ADP/ATP carrier 1; *SnRK2*, sucrose non-fermenting-1-related kinase 2. Rt, root; St, stem; Lv, leaf; Prc, pericarp; Ar, aril; Sd, seed.

weakest expression was observed in the fruit pericarp, aril and seed (Figure 4A and B). Thus, *LcSnRK2* exhibited strongest expression in the stem, root, leaf, and pericarp, and weakest expression in the aril and seed (Figure 4D and E).

Expression of *LcAtpB* was up-regulated by 114% at 80–90 DAF (Figure 4F), which was consistent with changes in ATP concentration (Figure 3A). Increased transcript abundance of *LcAOX1* peaked at 14-fold at 70 DAF but significant differences in transcript abundance at other stages of fruit development were not observed (Figure 4G). Abundance of *LcUCP1* declined by 43% between 50 and 60 DAF before increasing by 363% at 70 DAF and thereafter declining rapidly by 48% at 90 DAF (Figure 4H). *LcAAC1* and *LcSnRK2* exhibited similar changes in transcript levels (Figure 4I and J). Transcription of these two genes increased significantly between 50 and 70 DAF before decreasing to the lowest levels between 70 and 80 DAF, but recovered partly in the final stage of maturation (80–90 DAF).

Postharvest

Disease incidence, pericarp browning, proportion of edible fruit, and membrane permeability

The fruit disease and pericarp browning indices increased after harvest, whereas the proportion of edible fruit decreased. Exogenous ATP treatment reduced disease incidence and delayed tissue browning of the fruit significantly (Figure 5A and B). The disease and browning indices were 19% and 80% in the non-ATP-treated (control) fruit, but only 3% and 49% in ATP-treated fruit, by the end of storage. The proportion of edible fruit significantly increased after ATP treatment, by 21% and 28% compared with control fruit, after 4 and 6 days of storage, respectively (Figure 5C). Membrane permeability reflects senescence and deterioration of plant tissue and is usually measured as relative conductivity. This parameter increased in harvested fruit with no significant difference between ATP-treated and control fruit observed during 0–4 days of storage, but was significantly lower in ATP-treated fruit at the end of storage. These results suggested that exogenous ATP treatment protected the membrane system and, thus, delayed senescence and deterioration of litchi fruit in storage (Figure 5D).

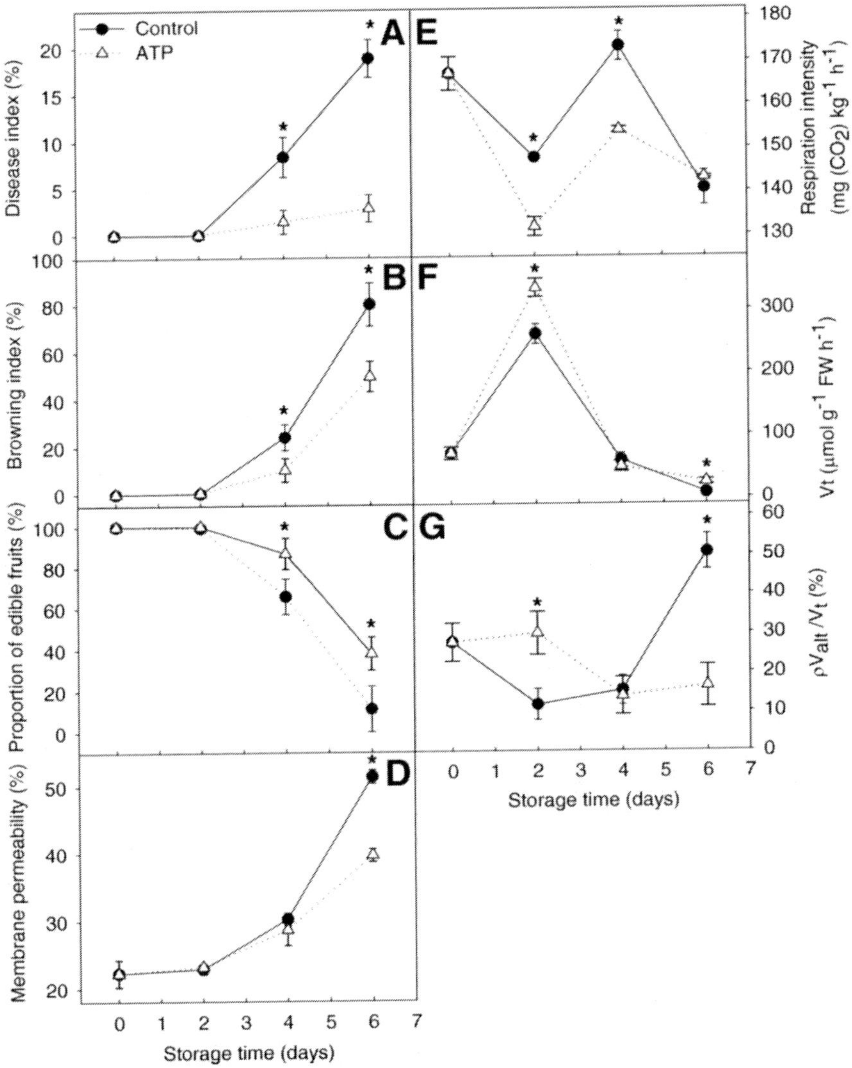

Figure 5 Senescence parameters and respiration activities of harvested litchi fruit. **(A)** Disease index; **(B)** browning index; **(C)** proportion of edible fruits; **(D)** membrane permeability; **(E)** respiration intensity; **(F)** V_t; and **(G)** $\rho V_{alt}/V_t$. Data are the mean ± standard deviation (n = 3). An asterisk represents a significant difference ($p < 0.05$) between ATP-treated fruit and the untreated control.

Respiration intensity, oxygen consumption rate, and cyanide-resistant respiration

Respiration intensity fluctuated and was significantly blocked by ATP treatment (Figure 5E). The oxygen consumption rate peaked after 2 days of storage and was enhanced significantly by ATP application (Figure 5F). In the first 2 days of storage $\rho V_{alt}/V_t$ decreased significantly, then rose slowly during the medial period of storage and increased sharply by the end of storage; exogenous ATP treatment significantly slowed down these effects (Figure 5G). In the ATP-treated fruit $\rho Valt/Vt$ was relatively stable throughout the storage process, which indicated that stability in $\rho Valt/Vt$ played an important role in maintenance of fruit storability.

Energy status

The ATP concentration significantly decreased by 69% during 6 days of storage, and the severity of the decline was reduced in ATP-treated fruit (Figure 6A). No significant difference in ADP concentration was observed during the 6 days of storage, whereas ATP treatment reduced ADP levels by 37% compared to that of control fruit after 4 days of storage (Figure 6B). The AMP concentration significantly increased during the first 2 days of storage before decreasing as storage progressed, and ATP treatment significantly depressed (by 75%) AMP levels (Figure 6C). The total adenylate concentration decreased significantly during postharvest storage and was 57% lower by the end of storage, but was up-regulated by ATP treatment (Figure 6D). Energy charge showed the same pattern but no significant difference in EC level between control and ATP-treated fruit was observed by the end of storage (Figure 6E). The ATP/AMP ratio decreased significantly after harvest. ATP treatment significantly attenuated the decrease in the ATP/AMP ratio until 6 days of storage (Figure 6F).

Abundance of energy-related gene transcripts

LcAtpB expression significantly increased and peaked 2 days after harvest before declining, whereas exogenous ATP treatment blocked the expression peak at 2 days postharvest (Figure 7A). *LcAOX1* transcript abundance remained stable until 2 days of storage, and thereafter increased constantly to 17-fold higher than the initial level at 6 days of storage. Exogenous ATP significantly inhibited the rise in *LcAOX1* transcript level after 4 days of storage (Figure 7B). The *LcUCP1* transcript level declined during the first 4 days of storage, with a slight increase observed thereafter, and ATP slowed both of these changes (Figure 7C). The *LcAAC1*transcript level rose slowly after harvest and no marked effect was imposed by exogenous ATP application (Figure 7D). The transcript abundance of *LcSnRK2* decreased within the first 2 days of storage and subsequently increased, and ATP slowed both of these changes (Figure 7E).

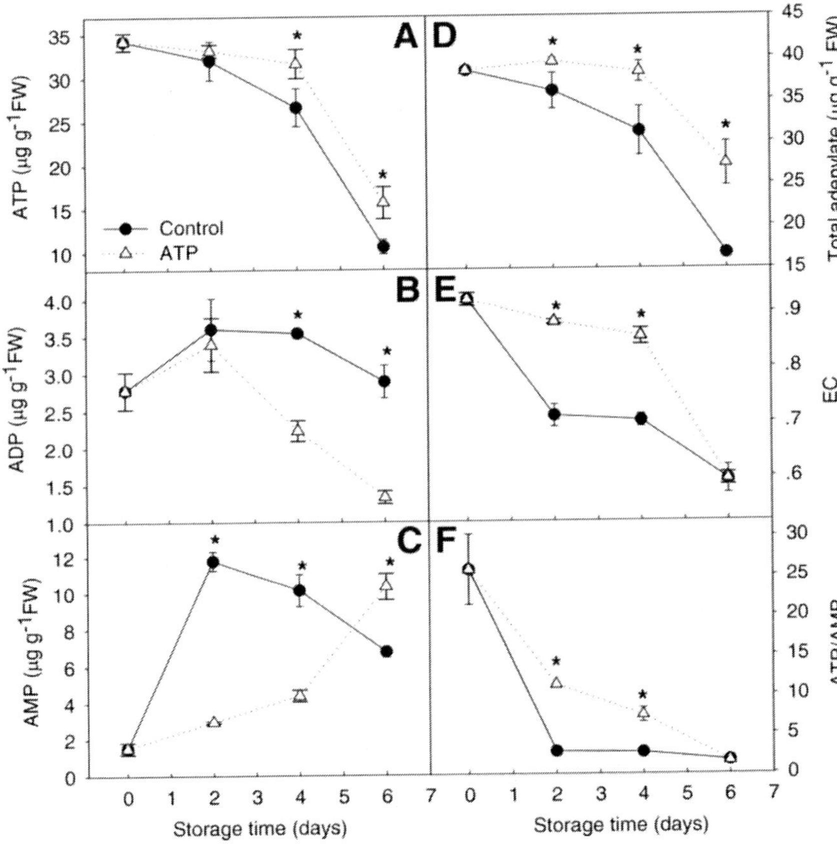

Figure 6 Energy status in harvested litchi fruit. (**A**) Adenosine triphosphate (ATP); (**B**) adenosine diphosphate (ADP); (**C**) adenosine monophosphate (AMP); (**D**) total adenylate; (**E**) energy charge (EC); and (**F**) ratio of ATP/AMP. Contents of ATP, ADP, AMP and total adenylate are expressed as μg g^{-1} FW. FW, fresh weight. Data are the mean ± standard deviation ($n = 3$). An asterisk represents a significant difference ($p < 0.05$) between ATP-treated fruit and untreated control.

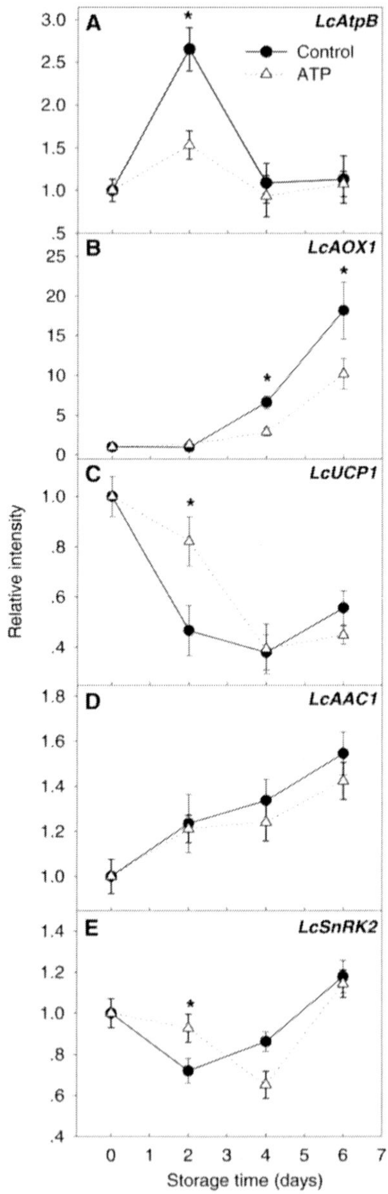

Figure 7 Relative intensity of gene transcripts in harvested litchi fruit. Transcript abundance was determined using qRT-PCR and was normalized using *LcACTIN*. Data are the mean ± standard deviation ($n = 3$). An asterisk represents a significant difference ($p < 0.05$) between ATP-treated fruit and untreated control. See Figure 4 legend for names of genes.

DISCUSSION

Preharvest

ATP level increased and energy charge remained relatively constant during fruit development and ripening

Changes in energy levels may be a key factor in switching on fruit ripening and senescence. Initiation of senescence is characterized by a reduced ATP level [48]. In the present study, ATP, EC, and total adenylate levels and the ATP/AMP ratio increased rapidly, whereas the AMP concentration decreased gradually, during fruit development (50–80 DAF) (Figure 3). ATP is produced mainly by oxidation of carbohydrates and is regulated by transcription of key proteins catalyzing nucleotide synthesis [15]. The respiratory rate of pericarp tissue of litchi fruit declined during fruit development, whilst the cyanide-resistant respiration to total respiration ratio ($\rho V_{alt}/V_t$) and energy production were enhanced (Figures 2 and 3). A declining respiratory rate and enhanced alternative respiration is an overflow protection mechanism to prevent ROS production [49].

Color-breaker stage at 70 DAF may be a key turning point in fruit energy metabolism

We isolated total RNAs from the root, stem, leaf, pericarp, aril and seed from fruit 80 DAF for real-time RT-PCR analysis. Five gene transcripts were predominantly expressed in the root, stem, and leaf at this specific developmental stage and increased significantly in developmental stage I except for *LcUCP1*, which was predominantly expressed only in the root, and *LcAtpB*, which was up-regulated by the end of developmental stage II. The *LcAtpB* expression level rose continuously during fruit development and maturation consistent with the increased ATP concentration (Figure 3A). Transcript abundance of *LcAOX1*, *LcUCP1*, *LcAAC1*, and *LcSnRK2* reached peaks at 70 DAF in the rank order *LcAOX1* > *LcSnRK2* > *LcAAC1* > *LcUCP1* (Figure 4F–J). Their transcription peaks coincided with the peak in ADP concentration (Figure 3D). Transcription peaks of *AOX1b*and *UCP* were also observed in mango fruit at the turning stage [35], which indicated that the color-breaker stage may be a key turning point in fruit energy metabolism. *LcAOX1*, *LcUCP1*, *LcAAC1* and *LcSnRK2* transcript levels significantly decreased after 70 DAF, allowing a surge in ATP synthesis. Under the combined effects of *AtpB*, *AOX1*, *UCP1*, *AAC1*, and *SnRK2*, the EC level remained relatively constant during fruit development and ripening. *LcAOX1* expression spiked at 70 DAF, which probably reduced levels of ROS generated by high rates of fruit photosynthesis. Recent studies indicate that the SnRK2 family plays key roles in stress responses, such as hyperosmotic stress and abscisic acid (ABA) signaling. ABA is involved in the maturation of many climacteric and non-climacteric fruit. Commonly, endogenous ABA concentrations increase during fruit maturation, whereas exogenous application of ABA accelerates fruit ripening or maturation [50]. Sugar signaling also regulates senescence in a complex network with other signals [47] resulting from biotic or abiotic stress. SnRK2 has been implicated in stress and ABA-mediated signaling pathways [51]. Enhanced transcript abundance of *LcSnRK2* at 70 DAF might be a consequence of increased ABA and sugar accumulation during maturation of litchi fruit [50].

Postharvest

ATP and energy charge levels decreased significantly in postharvest litchi fruit

The levels of ATP, total adenylate, and EC, and the ATP/AMP ratio significantly decreased, but the AMP concentration significantly increased, in stored harvested fruit (Figure 6). Similar results were reported for apples, pears, cut flowers, and other horticultural products [1, 2, 3]. ATP content decreased significantly in 'Jonagold' apples during controlled atmosphere storage. Brown heart disease of 'Conference' pears during storage was partly attributed to energy depletion [9]. As the petals of carnation cut flowers wilt (a senescence symptom), ATP synthesis decreases [1, 2]. ATP concentrations were negatively correlated with the browning index and proportion of edible fruit, which was consistent with our previous work [7]. ATP and EC levels in horticultural crops can be maintained by appropriate handling, thus effectively postponing senescence. Application of ethylene inhibitors, high oxygen, anaerobic conditions, exogenous sucrose, or ATP can maintain the ATP content and EC levels in litchi, longan, tulip and carnation flowers, thereby delaying pericarp browning and extending flowering longevity [1, 3, 4, 7, 10]. In the present study, the decrease in ATP, total adenylate, and EC levels and the ATP/AMP ratio were significantly slowed by exogenous ATP. At the same time, the disease and browning indices were lower and commodity rates were higher in ATP-treated fruit (Figure 5, Additional file 11). These data support the hypothesis that intracellular energy depletion or ATP deficiency induce senescence of horticultural crops. Respiratory intensity decreased in harvested litchi fruit but increased rapidly during 2–4 days in storage. Respiration rates might be activated by ADP accumulated during the first 2 days of fruit storage (Figure 6B). Energy levels, oxygen consumption rate, and $\rho V_{alt}/V_t$ were increased by exogenous ATP treatment. A reduced respiration rate is apparently favorable to extend the storage life of litchi fruit. The highly flexible nature of respiratory metabolism in plants allows survival under varying and often stressful environmental and nutritional conditions [52]. In the present study, exogenous ATP treatment extended the storage life of litchi fruit by reducing respiration intensity, which differed from the results of Klotz et al. [53] that suggested adenylates did not limit respiration in stored sugarbeet roots.

Decrease in ATP and energy charge levels may have resulted from imbalanced accumulation of LcAOX1 and LcAtpB transcripts in postharvest litchi fruit

LcAtpB transcription was up-regulated in harvested litchi fruit, which might restrain the rapid decline in ATP level during 0–4 days of storage. Exogenous ATP treatment inhibited *LcAtpB* expression, possibly by feedback inhibition by ATP. Chivasa [21] reported a new role for *AtpB* as a pro-cell death protein. In the current study, in preharvest (80–90 DAF) or postharvest (0–2 days in storage) litchi fruit, the rapid increase of *LcAtpB*expression followed by rapid senescence and decline in fruit quality suggested that the increase in *LcAtpB*transcript level marked initiation of the aging process. During abiotic stresses, such as heat shock and hypoxia, AMPKs (SnRK analogues) switch off ATP-consuming processes and activate ATP-generating catabolic pathways through direct enzyme regulation and transcriptional regulation [54]. SnRK may sense ATP deficiency and trigger downstream transcription of *AtpB*[55]. In the present study, the transcript abundance of *LcAtpB* was significantly increased in postharvest litchi fruit, which might be regulated by *LcSnRK2*.

UCP and AOX energy-dissipating systems play similar roles in mitochondrial energy-linked processes in plants, either through tissue-specific thermogenesis or by protecting plant cells against oxidative stress. The initial steps of cellular death are accompanied by an oxidative burst, depletion of ATP, and a strong stimulation of AOX expression [56]. In the present study, the transcript abundance of *LcAOX1* in litchi fruit was significantly up-regulated during storage. The lower ATP level in control fruit might have resulted from enhanced transcription and activity of AOX. Mitochondrial inner membrane UCPs catalyze a proton conductance that dissipates the proton electrochemical gradient established by the respiratory chain, thus affecting ATP synthesis. *LcUCP1* continued to decline in the first 4 days of storage, but thereafter rebounded slightly. Differential expression and complementary function of AOX and UCP is reported in a variety of fruits, vegetables, and plant tissues, and may be regulated by ethylene, low temperature, and other factors [27, 31, 33, 34, 35, 57, 58] through different catalytic efficiencies under distinct physiological conditions [26]. ADP/ATP carrier (AAC) is unable to modify the net content of adenine nucleotides (ATP + ADP + AMP) because it always exchanges one

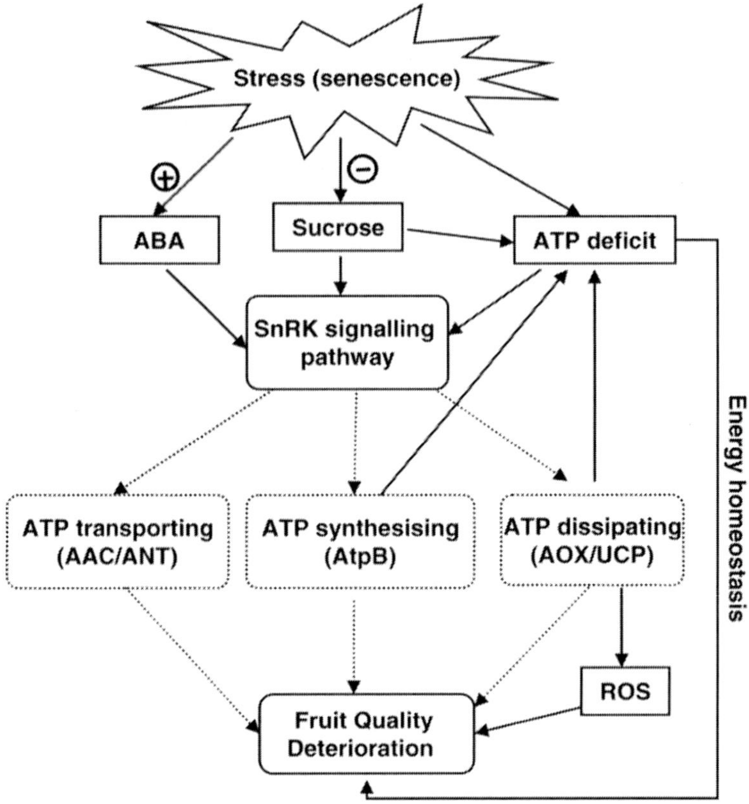

Figure 8 Possible mechanism to account for energy regulation in senescent litchi fruit. An energy deficit in intensive respiration during the senescence process may be sensed by sucrose non-fermenting-1-related kinase (SnRK), which controls the expression and phosphorylation of key metabolic enzymes, and this might involve ATP synthase, ADP/ATP carrier (AAC), alternative oxidase (AOX) and uncoupling mitochondrial protein (UCP). The energy homoeostatic condition can be maintained to a certain extent, but the equilibrium is ultimately disrupted, which is correlated with fruit deterioration.

nucleotide for another, whereas mutation of the AAC signature sequence (Arg triplet) inhibits ATP synthesis in cells and isolated mitochondria. In addition, loss of endogenous nucleotides is responsible for impeding ADP/ATP transport [17, 39]. In the present study, *LcAAC1* was up-regulated, and its transcript abundance negatively correlated with a decreased adenylate pool, in litchi fruit during storage. Furthermore, *LcAAC1* was not significantly affected by exogenous ATP treatment (Figures 6D and 7D), which indicated that *LcAAC1* was not sensitive to adenylate content. Transcript abundance of *LcAtpB*, *LcAOX1*, and *LcAAC1* were significantly up-regulated, whereas *LcUCP1* and *LcSnRK2* expression decreased before being enhanced, in harvested fruit and ATP supplementation slowed all these changes (Figure 7). Moreover, *LcAOX1* transcript abundance was significantly higher than that of *LcAtpB*. Under the joint action of *LcAtpB*, *LcAOX1*, *LcUCP1*, *LcAAC1*, and *LcSnRK2*, the EC remained relatively balanced between 2 and 4 days of storage. However, the dynamic energy equilibrium was eventually broken, probably because of dramatic increases in the transcript level of *LcAOX1* after 4 days of storage, causing a significant decrease in ATP and EC levels by the end of storage, which was correlated with rapid fruit deterioration (Figure 8).

CONCLUSIONS

This study investigated energy characteristics and the level of five transcripts that are involved in the modulation of ATP levels during preharvest development and postharvest senescence of litchi fruit. Measurements were also recorded after ATP treatment of harvested fruits. Senescence and deterioration of litchi fruit during storage was closely related to ATP content and EC levels. Dramatic increase in *LcAtpB*expression either in planta or postharvest suggested that *LcAtpB* expression may mark the onset of senescence of litchi fruit. The energy status in litchi fruit seems to be controlled by the combined effects of genes responsible for energy production, dissipation, transfer and regulation, in which *LcAtpB* and *LcAOX1*play a decisive role. Given their importance for energy regulation, these genes could be used to generate markers for the breeding of new litchi cultivars that show increased ATP production or fruit with a longer postharvest storage life.

METHODS

Plant materials and treatments

Seven 10-year-old litchi (*Litchi chinensis* Sonn. cv. Huaizhi) trees were chosen for these experiments from a commercial orchard in Guangzhou, China, in 2010 and 2011. Ten fruit on each tree were used to measure fruit diameter every 10 days over a period of 40 days beginning at 50 DAF (31 May, 2010; 2 June, 2011) and ending at 90 DAF (11 July, 2010; 12 July, 2011). At each stage, fruit were randomly sampled from different parts of the canopy between 9 and 11 a.m. Fresh weights of individual fruit were recorded, and respiration rate was measured using fresh fruit. The pericarp was collected, frozen in liquid nitrogen and stored at -80°C for RNA extraction. All trees used in the above studies were maintained in accordance with commercial litchi production practices.

Litchi fruit at about 80% maturity (80–82 DAF) were collected from the same orchard. In order to examine the expression pattern of energy-related gene transcripts, a variety of tissues (root, stem, leaf, pericarp, aril and seed) were collected at this maturation stage in 2012. Fruit were selected for uniformity of shape, color, and absence of blemishes or disease. The fruits were surface sterilized in 0.5% sodium hypochlorite solution for 5 s before rinsing twice in sterile distilled water, and then placed in 4.2 L glass desiccators containing sterile distilled water or 1.0 mmol L^{-1} ATP solution (200 fruit L^{-1}). A partial vacuum at 75 kPa was gently applied for 3 min. The vacuum-infiltrated fruits were air-dried, then packed in 0.015-mm-thick polyethylene bags (15 pieces/bag, 12 bags/treatment), and stored at 25°C and 85–90% relative humidity. Thirty fruit from different treatments were randomly sampled every 2 days after storage. Pericarp browning and disease incidence were recorded, and respiration rate and alternative oxidative respiratory activity were tested using fresh fruit.

Color analyses

The pericarp color was measured with a Konica Minolta CR-400 Chroma Meter (Minolta, Japan) using 15 fruits immediately after picking. The meter was calibrated using the manufacturer's standard white plate. Color was recorded in the L*, C* and h color space, and these values were measured randomly at the site opposite to the fruit suture (Table 1). L* indicated the lightness ranging from black (L* = 0) to white (L* = 100), chroma values (C*) represented the color saturation that varied from dull to vivid, lower and higher values, respectively, and hue angle (h*) referred to a color wheel, with red at an angle of 0°, yellow at 90°, green at 180°, and blue at 270°, in accordance with the method of McGuire [59].

HPLC analysis of ATP, ADP and AMP

Extraction of ATP, ADP and AMP were conducted with the method of Ozogul et al. [60] with a minor modification. Litchi fruit pericarp (2 g) was ground in liquid nitrogen and extracted with 6 mL of 0.6 mol L^{-1}perchloric acid. The homogenate was centrifuged at 4°C and 16,000 × *g* for 10 min. The supernatant (3 mL) was quickly neutralized to pH 6.5–6.8 using 1 mol L^{-1} potassium hydroxide. The solution was diluted to 4 mL and passed through a 0.45 μm filter. The filtrate was then stored at -30°C until analyzed. ATP, ADP and AMP measurements were performed in accordance with the method of Liu et al. [61] with a Waters 2695 HPLC (Waters, Inc., Milford, MA, USA) using a Pinnacle ll-C_{18} column (4.6 × 250 mm) and an ultraviolet (UV) detector at 254 nm. Mobile phase A consisted of 0.06 mol L^{-1} dipotassium hydrogen phosphate and 0.04 mol L^{-1}potassium dihydrogen phosphate dissolved in deionized water and adjusted to pH 7.0 with 0.1 mol L^{-1}potassium hydroxide. Mobile phase B was 100% acetonitrile. Elution was carried out by a linear gradient program with 75–100% A and 0–25% B for 7 min. The flow rate was 1.2 mL min^{-1}. Sample aliquots of 5 μL were injected into the HPLC. ATP, ADP and AMP concentrations were calculated according to the external standard curve and expressed on a fresh weight (FW) basis. Energy charge was calculated as: ([ATP] + 0.5 × [ADP])/([ATP] + [ADP] + [AMP]).

Pericarp browning and disease incidence

Pericarp browning was assessed after 0, 2, 4 and 6 days of storage using the following scale by an experienced technician [11]: 0, no browning (excellent quality); I, slight browning; II, <1/4 browning; III, 1/4 to 1/2 browning; and IV, >1/2 browning (poor quality). The browning index was calculated as: Σ (browning scale × percentage of corresponding fruit within each class).

Disease development was monitored using 30 randomly selected fruits and recorded as the ratio of mean proportion (%) of the fruit surface with fungal growth to the total number of fruit. The rating scale described above for browning was also applied for disease severity [11]. The proportion of edible fruit was calculated as the percentage of the total number of fruit exhibiting a browning scale value ≤1.

Membrane permeability

Membrane permeability was expressed as the relative electrolyte leakage. Fifteen discs (10 mm in diameter) from 30 fruits were washed three times in deionized water, dried with filter paper and then incubated for 30 min in 20 mL of 0.3 mol L^{-1} mannitol solution at 25°C. Initial electrolyte leakage rates were determined using a conductivity meter (model DDS-11A; Shanghai Scientific Instruments, Shanghai, China). Total electrolyte leakage was then determined after boiling for 20 min and cooling rapidly to 25°C. The relative leakage was expressed as the percentage of the initial electrolyte to the total electrolyte.

Whole-fruit respiration intensity

The respiratory activities of intact litchi fruit were measured by infrared gas analysis. A sample of 10 litchi fruit was weighed before being sealed in an easy-lock food container (2.2 L in volume) at 25°C, through which CO_2-free air was pumped. Increases in CO_2 concentration in the container were monitored for 5 min by passing the air stream through an infrared gas analyzer (Li-6262 CO_2/H_2O analyzer, LI-COR, Inc., Lincoln, NE, USA). Respiration rate was expressed in terms of carbon dioxide output per hour per kilogram of fresh weight (mg CO_2 kg^{-1} h^{-1}). The analysis was carried out using five replicate fruit samples.

Oxygen consumption rate and the contributions of alternative oxidative respiratory to total respiration

The oxygen consumption rate of fruit pericarp was measured using a Clark-type Oxygraph oxygen electrode (Hansatech Instruments, King's Lynn, UK) at 25°C by the method of Bingham and Farrar [62] with some modifications. Ten pericarp discs (5 mm in diameter) were dipped in distilled water before surface drying with filter paper. Potassium phosphate buffer (65 mmol L^{-1}, pH 6.8) was used as the respiratory medium. The total respiration rate (V_t) was determined in the absence of any respiratory inhibitors. The alternative respiratory pathway (ARP) rate (ρV_{alt}) was calculated from V_t minus the respiration rate in the presence of 25 mmol L^{-1}benzohydroxamate. The contribution of ARP to total respiration was expressed as $\rho V_{alt}/V_t$. The oxygen consumption rate was expressed as nmol min^{-1} cm^{-2}.

RNA extraction

Total RNA was extracted from litchi fruit pericarp using the hot borate method of Won and Wilkins [63], with some modifications as described by Kuang et al. [64]. Extracted RNA was digested with DNase I (TaKaRa Bio, Inc., Otsu, Shiga, Japan) and then confirmed by PCR to be free of detectable amounts of DNA. RNA concentration was determined by measuring UV absorbance at 260 nm with a SmartSpec Plus spectrophotometer (Bio-Rad, Hercules, CA, USA). The integrity and quality of the RNA were verified by agarose gel electrophoresis and calculation of the A_{260}/A_{280} ratio.

Gene cloning and sequencing

To isolate fragments of energy-related genes, degenerated oligonucleotides were constructed based on conserved sequence elements of *AtpB, AOX, UCP, AAC,* and *SnRK* from other species (Table 2). Using these oligonucleotides, amplification of fragments was carried out on a mixture of cDNA obtained from pericarp of litchi fruit stored for 0 and 2 days, because the genes displayed high expression at these stages in a preliminary analysis. The product of the first-strand cDNA was subjected to PCR amplification. The PCR product of the predicted size was purified, cloned into the pMD20-T vector (TaKaRa), and then transformed into *E. coli* DH5a cells (TaKaRa) in accordance with the manufacturer's protocol. Plasmid DNA isolated from positive *E. coli* cells was digested with *Eco*RI and *Hin*dIII. The inserted DNA was sequenced by Sangon Biotech Co., Shanghai, China. The DNA fragments were extended by 3′- and 5′-RACE using the 3′-Full RACE Core Set version 2.0 and the 5′ RACE Kit (TaKaRa) with gene-specific primers. PCR products from 3′- or 5′-RACE were sequenced, and their sequence information was used to design gene-specific primers for full-length amplification. Full-length clones from each gene were sequenced in double strands. The resulting sequences were designated as *LcAtpB, LcAOX1, LcUCP1, LcAAC1,* and *LcSnRK2,* and were deposited in GenBank (http://www.ncbi.nlm.nih.gov) with the accession numbers JQ349005 (*LcAtpB*), JQ349006 (*LcAOX1*), JQ349009 (*LcUCP1*), JQ349007 (*LcAAC1*), and JQ349008 (*LcSnRK2*) (Table 3).

Determination of gene transcript levels

Total RNA, extracted from the different plant tissues, was reverse-transcribed in 20 μL reaction mixture containing oligo $d(T)_{18}$ primer using the Prime-Script™ RT-PCR Kit (TaKaRa, Japan) in accordance with the manufacturer's protocol. Relative transcript abundance was analyzed by quantitative real-time PCR (qPCR) using the ABI 7500 Real-Time PCR System (Applied Biosystems, Carlsbad, CA, USA) and the LightCycler 480 SYBR Green I Master Mix (Roche Applied Science, http://www.roche-applied-science.com), in accordance with the manufacturer's instructions under the following conditions: 30 s at 95°C, 40 cycles of 5 s at 95°C, and 34 s at 58°C. The primer pairs are listed in Table 4. The *LcACTIN* gene was used for quantitative normalization [65, 66].

Table 4 Primers used for real-time PCR analysis

Gene	Forward primer (5′ to 3′)	Reverse primer (5′ to 3′)	Product (bp)
LcAtpB	GAGAGTTGGTTTGACTGCCCTAA	GAAGGCATTCTACCCAATAAGGC	141
LcAOX1	TCGGCTATCACTAAGATGTTGGTCA	ACAAGCACACACTCAACGGATTTAC	149
LcUCP1	AGTTGGCGGTGATTTTATTGGAG	GCATTCAAGGTCCCATAATAACG	177
LcAAC1	CTGAAGAATGAGGGTGCCAAGTC	CTAAAACAGCGAAAGGAATACCG	171
LcSnRK2	AGTTGAAGGACATCTGCTTGGAA	ATGCGAAAACCATACCTGTGTCA	153

See legend for Table 2 for names of genes.

Bioinformatics and statistical analysis

Identification of nucleotide sequences from RT-PCR clones was carried out with the NCBI BLAST program (http://www.ncbi.nlm.nih.gov/BLAST). Alignments were made using Clustal X and Jalview software, and cladograms were constructed by the neighbor-joining method using the MEGA program and visualized with TreeView software. The theoretical isoelectric point (pI) and mass value for mature peptides were calculated using the PeptideMass program (http://us.expasy.org/ tools/ peptidemass.html).

The experiment was repeated twice during two consecutive growing seasons. Experiments for respiration intensity were repeated six times, whereas other experiments were repeated three times. The results are presented as the mean values ± standard deviation. Significant differences among different developmental stages, different storage times, and those between ATP-treated fruit and the untreated control were determined by one-way analysis of variance (ANOVA) using SPSS® version 13.0 (SPSS, Inc., Chicago, IL, USA). Statistical differences were assessed with a significance level of 5%. Graphs were drawn with SigmaPlot 9.0 and Excel 2003.

Additional files

Additional file 1: Sequence alignment of LcAtpB and AtpBs from other plant species. The alignment was made using CLUSTAL X software. Identical and similar amino acids are indicated by black and grey shading, respectively. The highly conserved 'DELSEED' motif is underlined. Abbreviations on the left of each sequence: At, Arabidopsis thaliana; St, Solanum tuberosum; Si, Sesamum indicum; Ma, Melia azedarach; and Zm, Zea mays.

Additional file 2: Cladogram of LcAtpB and AtpBs from other plant species. The cladogram was constructed with the neighbor-joining method using MEGA software with default settings. Numbers at the nodes indicate bootstrap values. A scale of distance was shown at the bottom. The deduced amino acid sequences of plant AtpBs were obtained from the following sources, with their GenBank accession numbers enclosed in parentheses: Koelreuteria paniculata, KpAtpB (CAB89921.1); Cneorum tricoccon, CtAtpB (ACZ73599.1); Citrus sinensis, CsAtpB (YP_740482.1); Melia azedarach, MaAtpB (ABU75138.1); Picrasma excelsa, PeAtpB (ABU75177.1); Sesamum indicum, SiAtpB (CAB65433.1); Solanum tuberosum, StAtpB (ABB90048.1); Arabidopsis thaliana, AtAtpB (BAA84392.1); Oryza sativa, OsAtpB (NP_039390.1); and Zea mays, ZmAtpB (CAA60293.1).

Additional file 3: Sequence alignment of LcAOX1 and AOXs from other plant species. The alignment was made using CLUSTAL X software. Identical and similar amino acids are indicated by black and grey shading, respectively. The key residues required for catalysis are numbered based on the Litchi chinensis AOX1 protein. Abbreviations on the left of each sequence: St, Solanum tuberosum; Ng, Nicotiana glutinosa; At, Arabidopsis thaliana; Pt, Populus tremula x Populus tremuloides; and Os, Oryza sativa.

Additional file 4: Cladogram of LcAOX1 and AOXs from other plant species. The cladogram was constructed with the neighbor-joining method using the MEGA software with default settings. Numbers at the nodes indicate bootstrap values. A scale of distance was shown at the bottom. The deduced amino acid sequences of plant AOXs were obtained from the following sources, with their GenBank accession numbers enclosed in parentheses: Nicotiana glutinosa, NgAOX1a (ABU24346.1); Solanum tuberosum, StAOX (BAE92716.1); Arabidopsis thaliana, AtAOX1A (NP_188876.1), and AtAOX1B (NP_188875.1); Citrus sinensis, CsAOX (ACE95101.1); Vigna unguiculata, VuAOX1 (AAZ09196.1); Gossypium hirsutum, GhAOX (ABJ98721.1); Populus tremula x Populus tremuloides, PtAOX (CAB64356.1); Daucus carota, DcAOX1a (ABZ81227.2); Nelumbo nucifera, NnAOX1a (BAH56640.1); Glycine max, GmAOX1 (NP_001236166.1); Oryza sativa, OsAOX (BAA28774.1); Triticum aestivum, TaAOX (BAB88645.1); and Zea mays, ZmAOX (NP_001105180.1).

Additional file 5: Sequence alignment of LcUCP1 and UCPs from other plant species. The alignment was made using CLUSTAL X software. Identical and similar amino acids are indicated by black and grey shading, respectively. Energy transfer protein signatures (ETPS) are underlined. Abbreviations on the left of each sequence: At, Arabidopsis thaliana; Rc, Ricinus communis; Zm, Zea mays; and Gm, Glycine max.

Additional file 6: Cladogram of LcUCP1 and UCPs from other plant species. The cladogram was constructed with the neighbor-joining method using MEGA software with default settings. Numbers at the nodes indicate bootstrap values. A scale of distance was shown at the bottom. The deduced amino acid sequences of plant UCPs were obtained from the following sources, with their GenBank accession numbers enclosed in parentheses: Arabidopsis thaliana, AtUCP1 (NP_190979.1), AtUCP2 (NM_125287.4), and

AtUCP5 (NP_179836.1); Solanum tuberosum, StUCP (CAA72107.1); Glycine max, GmUCP3 (XP_003516932.1); Medicago truncatula, MtUCP (AES86982.1); Symplocarpus renifolius, SrUCP (BAI49702.1); Zea mays, ZmUCP3 (NP_001182792.1); Oryza sativa, OsUCp (BAB40658.1); Triticum aestivum, TaUCP (BAB16385.1); and Ricinus communis, RcUCP (XM_002520396.1).

Additional file 7: Sequence alignment of LcAAC1 and AACs from other plant species. The alignment was made using CLUSTAL X software. Identical and similar amino acids are indicated by black and grey shading, respectively. Asterisk indicates the mitochondria AAC signature motif (RRRMMM). Abbreviations on the left of each sequence: At, Arabidopsis thaliana; Rc, Ricinus communis; and Zm, Zea mays.

Additional file 8: Cladogram of LcAAC1 and AACs from other plant species. The cladogram was constructed with the neighbor-joining method using MEGA software with default settings. Numbers at the nodes indicate bootstrap values. A scale of distance was shown at the bottom. The deduced amino acid sequences of plant AACs were obtained from the following sources, with their GenBank accession numbers enclosed in parentheses: Ricinus communis, RcAAC (XM_002531865.1); Glycine max, GmAAC1 (XP_003546882.1); Zea mays, ZmANT (CAA33742.1); Arabidopsis thaliana, AtAAC1 (NP_187470.1), and AtAAC2 (NM_121352.3); Solanum lycopersicum, SlANT (NP_001234018.1); Medicago truncatula, MtAAC (XP_003627715.1); and Chlamydomonas incerta, CiAAC (ABA01103.1).

Additional file 9: Sequence alignment of LcSnRK2 and SnRKs from other plant species. The alignment was made using CLUSTAL X software. Identical and similar amino acids are indicated by black and grey shading, respectively. Regions 1–3 represent the ATP binding site, protein kinase activating signature, and C-terminal regulatory domain, respectively. Abbreviations on the left of each sequence: Rc, Ricinus communis; Gm, Glycine max; Zm, Zea mays; and At, Arabidopsis thaliana.

Additional file 10: Cladogram of LcSnRK2 and SnRKs from other plant species. The cladogram was constructed with the neighborjoining method using MEGA software with default settings. Numbers at the nodes indicate bootstrap values. A scale of distance was shown at the bottom. The deduced amino acid sequences of plant SnRKs were obtained from the following sources, with their GenBank accession numbers enclosed in parentheses: Glycine max, GmSAPK2 (XP_003519175.1); Ricinus communis, RcSAPK1 (XM_002513909.1), and RcSAPK3 (XP_002517501.1); Zea mays, ZmSnRK2.1 (ACG50005.1) and ZmSnRK2.2 (ACG50006.1); Solanum lycopersicum, SlSNF1 (NP_001234353.1); Oryza sativ, OsSAPK3 (BAD17999.1); Arabidopsis thaliana, AtSnRK1.3 (NM_123306.1), and AtSnRK2.2 (NM_001203118.1).

Additional file 11: Appearance of litchi fruit with or without exogenous ATP supply before storage at 25°C for 6 days. (A) Untreated control; (B) treated with exogenous ATP. Numbers 0 to IV represent the pericarp browning scale: 0, no browning (excellent quality); I, slight browning; II, <1/4 browning; III, 1/4 to 1/2 browning, and IV, >1/2 browning (poor quality), respectively.

ACKNOWLEDGEMENTS

This study was supported by the National Basic Research Program of China (973 program; No. 2013CB127100), National Key Technologies R&D Program (Grant No. 2012BAD38B03), National Natural Science Foundation of China (Grant Nos. 30972039 and 31272216), Guangdong Natural Science Foundation (Grant No. S2011020001156), and the Fund of Key Laboratory of Plant Resources Conservation and Sustainable Utilization, South China Botanical Garden, Chinese Academy of Sciences.

REFERENCES

1. Azad AK, Ishikawa T, Sawa Y, Shibata H: Intracellular energy depletion triggers programmed cell death during petal senescence in tulip. J Exp Bot. 2008, 59 (8): 2085-2095. 10.1093/jxb/ern066.
2. Saquet AA, Streif J, Bangerth F: Changes in ATP, ADP and pyridine nucleotide levels related to the incidence of physiological disorders in 'Conference' pears and 'Jonagold' apples during controlled atmosphere storage. J Hortic Sci Biotech. 2000, 75 (2): 243-249.
3. Trippi VS, Paulin A, Pradet A: Effect of oxygen concentration on the senescence and energy-metabolism of cut carnation flowers. Physiol Plantarum. 1988, 73 (3): 374-379. 10.1111/j.1399-3054.1988.tb00613.x.
4. Duan XW, Jiang YM, Su XG, Liu H, Li YB, Zhang ZQ, Zheng YH, Jiang WB: Role of pure oxygen treatment in browning of litchi fruit after harvest. Plant Sci. 2004, 167 (3): 665-668. 10.1016/j.plantsci.2004.05.009.
5. Saquet AA, Streif J, Bangerth F: Energy metabolism and membrane lipid alterations in relation to brown heart development in 'Conference' pears during delayed controlled atmosphere storage. Postharvest Biol Tec. 2003, 30 (2): 123-132. 10.1016/S0925-5214(03)00099-1.
6. Xuan H, Streif J, Saquet A, Romheld V, Bangerth F: Application of boron with calcium affects respiration and ATP/ADP ratio in 'Conference' pears during controlled atmosphere storage. J Hortic Sci Biotech. 2005, 80 (5): 633-637.
7. Qu H, Duan X, Su X, Liu H, Jiang Y: Effects of anti-ethylene treatments on browning and energy metabolism of harvested litchi fruit. Aust J Exp Agr. 2006, 46 (8): 1085-1090. 10.1071/EA04238.
8. Jiang YM, Jiang YL, Qu HX, Duan XW, Luo YB, Jiang WB: Energy aspects in ripening and senescence of harvested horticultural crops. Stewart Postharvest Rev. 2007, 2 (5): 1-5.

9. Veltman RH, Lentheric I, Van der Plas LHW, Peppelenbos HW: Internal browning in pear fruit (*Pyrus communis* L. cv Conference) may be a result of a limited availability of energy and antioxidants. Postharvest Biol Tec. 2003, 28 (2): 295-302. 10.1016/S0925-5214(02)00198-9.
10. Song L, Liu H, Su X, You Y, Jiang Y: Effects of adenosine triphosphate on the vase life of cut carnation flowers. Aust J Exp Agr. 2006, 46 (1): 137-139. 10.1071/EA05011.
11. Yi C, Qu HX, Jiang YM, Shi J, Duan XW, Joyce DC, Li YB: ATP-induced changes in energy status and membrane integrity of harvested litchi fruit and its relation to pathogen resistance. J Phytopathol. 2008, 156 (6): 365-371. 10.1111/j.1439-0434.2007.01371.x.
12. Lapaille M, Thiry M, Perez E, Gonzalez-Halphen D, Remacle C, Cardol P: Loss of mitochondrial ATP synthase subunit beta (Atp2) alters mitochondrial and chloroplastic function and morphology in *Chlamydomonas*. Bba-Bioenergetics. 2010, 1797 (8): 1533-1539. 10.1016/j.bbabio.2010.04.013.
13. Yi C, Jiang YM, Shi J, Qu HX, Duan XW, Yang B, Prasad NK, Liu T: Effect of adenosine triphosphate on changes of fatty acids in harvested litchi fruit infected by *Peronophythora litchii*. Postharvest Biol Tec. 2009, 54 (3): 159-164. 10.1016/j.postharvbio.2009.06.008.
14. Yi C, Jiang YM, Shi J, Qu HX, Xue S, Duan XW, Shi JY, Prasad NK: ATP-regulation of antioxidant properties and phenolics in litchi fruit during browning and pathogen infection process. Food Chem. 2010, 118 (1): 42-47. 10.1016/j. foodchem. 2009.04.074.
15. Geigenberger P, Riewe D, Fernie AR: The regulation of plant physiology by adenylates. Trends Plant Sci. 2010, 15 (2): 98-105. 10.1016/j.tplants.2009.11.004.
16. Millar AH, Whelan J, Soole KL, Day DA: Organization and regulation of mitochondrial respiration in plants. Annu Rev Plant Biol. 2011, 62: 79-104. 10.1146/annurev-arplant-042110-103857.
17. Traba J, Satrustegui J, del Arco A: Adenine nucleotide transporters in organelles: novel genes and functions. Cell Mol Life Sci. 2011, 68 (7): 1183-1206. 10.1007/s00018-010-0612-3.
18. Usukura E, Suzuki T, Furuike S, Soga N, Saita E, Hisabori T, Kinosita K, Yoshida M: Torque generation and utilization in motor enzyme F0F1-ATP synthase half-torque-1 with short-sized pushrod helix and reduced ATP synthesis by half-torque F0F1. J Biol Chem. 2012, 287 (3): 1884-1891. 10.1074/jbc.M111.305938.
19. Fillingame RH: Protein structure - Molecular rotary motors. Science. 1999, 286 (5445): 1687-1688. 10.1126/science.286.5445.1687.
20. Pedersen PL, Ko YH, Hong SJ: ATP syntheses in the year 2000: Evolving views about the structures of these remarkable enzyme complexes. J Bioenerg Biomembr. 2000, 32 (4): 325-332. 10.1023/A:1005594800983.
21. Chivasa S, Tome DFA, Hamilton JM, Slabas AR: Proteomic analysis of extracellular ATP-regulated proteins identifies ATP synthase beta-subunit as a novel plant cell death regulator. Mol Cell Proteomics. 2011, 10 (3): M110.003905. 10.1074/mcp.M110.003905.
22. Martinez-Cruz O, Garcia-Carreno F, Robles-Romo A, Varela-Romero A, Muhlia-Almazan A: Catalytic subunits atp alpha and atp beta from the Pacific white shrimp *Litopenaeus vannamei* F(O)F(1) ATP-synthase complex: cDNA sequences, phylogenies, and mRNA quantification during hypoxia. J Bioenerg Biomembr. 2011, 43 (2): 119-133. 10.1007/s10863-011-9340-0.
23. Borecky J, Nogueira FTS, de Oliveira KAP, Maia IG, Vercesi AE, Arruda P: The plant energy-dissipating mitochondrial systems: depicting the genomic structure and the expression profiles of the gene families of uncoupling protein and alternative oxidase in monocots and dicots. J Exp Bot. 2006, 57 (4): 849-864. 10.1093/jxb/erj070.
24. Considine MJ, Goodman M, Echtay KS, Laloi M, Whelan J, Brand MD, Sweetlove LJ: Superoxide stimulates a proton leak in potato mitochondria that is related to the

activity of uncoupling protein. J Biol Chem. 2003, 278 (25): 22298-22302. 10.1074/jbc.M301075200.
25. Ito-Inaba Y, Hida Y, Mori H, Inaba T: Molecular Identity of Uncoupling Proteins in Thermogenic Skunk Cabbage. Plant Cell Physiol. 2008, 49 (12): 1911-1916. 10.1093/pcp/pcn161.
26. Vercesi AE, Borecky J, Godoy Maia ID, Arruda P, Cuccovia IM, Chaimovich H: Plant uncoupling mitochondrial proteins. Annu Rev Plant Biol. 2006, 57: 383-404. 10.1146/annurev.arplant.57.032905.105335.
27. Figueira TRS, Arruda P: Differential expression of uncoupling mitochondrial protein and alternative oxidase in the plant response to stress. J Bioenerg Biomembr. 2011, 43 (1): 67-70. 10.1007/s10863-011-9333-z.
28. Arnholdt-Schmitt B, Costa JH, de Melo DF: AOX - a functional marker for efficient cell reprogramming under stress?. Trends Plant Sci. 2006, 11 (6): 281-287. 10.1016/j.tplants.2006.05.001.
29. Fung RWM, Wang CY, Smith DL, Gross KC, Tian MS: MeSA and MeJA increase steady-state transcript levels of alternative oxidase and resistance against chilling injury in sweet peppers (*Capsicum annuum* L.). Plant Sci. 2004, 166 (3): 711-719. 10.1016/j.plantsci.2003.11.009.
30. Matos AR, Mendes AT, Scotti-Campos P, Arrabaca JD: Study of the effects of salicylic acid on soybean mitochondrial lipids and respiratory properties using the alternative oxidase as a stress-reporter protein. Physiol Plantarum. 2009, 137 (4): 485-497. 10.1111/j.1399-3054.2009.01250.x.
31. Tian M, Gupta D, Lei XY, Prakash S, Xu C, Fung RWM: Effects of low temperature and ethylene on alternative oxidase in green pepper (*Capsicum annuum* L.). J Hortic Sci Biotech. 2004, 79 (3): 493-499.
32. Almeida AM, Jarmuszkiewicz W, Khomsi H, Arruda P, Vercesi AE, Sluse FE: Cyanide-resistant, ATP-synthesis-sustained, and uncoupling-protein-sustained respiration during postharvest ripening of tomato fruit. Plant Physiol. 1999, 119 (4): 1323-1329. 10.1104/pp.119.4.1323.
33. Holtzapffel RC, Finnegan PM, Millar AH, Badger MR, Day DA: Mitochondrial protein expression in tomato fruit during on-vine ripening and cold storage. Funct Plant Biol. 2002, 29 (7): 827-834. 10.1071/PP01245.
34. Almeida AM, Navet R, Jarmuszkiewicz W, Vercesi AE, Sluse-Goffart CM, Sluse FE: The energy-conserving and energy-dissipating processes in mitochondria isolated from wild type and nonripening tomato fruits during development on the plant. J Bioenerg Biomembr. 2002, 34 (6): 487-498. 10.1023/A:1022574327117.
35. Considine MJ, Daley DO, Whelan J: The expression of alternative oxidase and uncoupling protein during fruit ripening in mango. Plant Physiol. 2001, 126 (4): 1619-1629. 10.1104/pp.126.4.1619.
36. Duque P, Arrabaca JD: Respiratory metabolism during cold storage of apple fruit. II. Alternative oxidase is induced at the climacteric. Physiol Plantarum. 1999, 107 (1): 24-31. 10.1034/j.1399-3054.1999.100104.x.
37. Fung RWM, Wang CY, Smith DL, Gross KC, Tao Y, Tian MS: Characterization of alternative oxidase (AOX) gene expression in response to methyl salicylate and methyl jasmonate pre-treatment and low temperature in tomatoes. J Plant Physiol. 2006, 163 (10): 1049-1060. 10.1016/j.jplph.2005.11.003.
38. Haferkamp I: The diverse members of the mitochondrial carrier family in plants. FEBS Lett. 2007, 581 (12): 2375-2379. 10.1016/j.febslet.2007.02.020.
39. Klingenberg M: The ADP and ATP transport in mitochondria and its carrier. Bba-Biomembranes. 2008, 1778 (10): 1978-2021. 10.1016/j.bbamem.2008.04.011.
40. Haferkamp I, Fernie AR, Neuhaus HE: Adenine nucleotide transport in plants: much more than a mitochondrial issue. Trends Plant Sci. 2011, 16 (9): 507-515. 10.1016/j.tplants.2011.04.001.
41. Palmieri F, Pierri CL, De Grassi A, Nunes-Nesi A, Fernie AR: Evolution, structure and function of mitochondrial carriers: a review with new insights. Plant J. 2011, 66 (1): 161-181. 10.1111/j.1365-313X.2011.04516.x.

42. Jiang YM, Wang Y, Song L, Liu H, Lichter A, Kerdchoechuen O, Joyce DC, Shi J: Postharvest characteristics and handling of litchi fruit - an overview. Aust J Exp Agr. 2006, 46 (12): 1541-1556. 10.1071/EA05108.
43. Mnatsakanyan N, Kemboi SK, Salas J, Weber J: The beta subunit loop that couples catalysis and rotation in ATP synthase has a critical length. J Biol Chem. 2011, 286 (34): 29788-29796. 10.1074/jbc.M111.254730.
44. McDonald AE: Alternative oxidase: an inter-kingdom perspective on the function and regulation of this broadly distributed 'cyanide-resistant' terminal oxidase. Funct Plant Biol. 2008, 35 (7): 535-552. 10.1071/FP08025.
45. Holtzapffel RC, Castelli J, Finnegan PM, Millar AH, Whelan J, Day DA: A tomato alternative oxidase protein with altered regulatory properties. Bba-Bioenergetics. 2003, 1606 (1–3): 153-162.
46. Millar AH, Heazlewood JL: Genomic and proteomic analysis of mitochondrial carrier proteins in Arabidopsis. Plant Physiol. 2003, 131 (2): 443-453. 10.1104/pp.009985.
47. Zhang HY, Mao XG, Jing RL, Chang XP, Xie HM: Characterization of a common wheat (*Triticum aestivum* L.) TaSnRK2.7 gene involved in abiotic stress responses. J Exp Bot. 2011, 62 (3): 975-988. 10.1093/jxb/erq328.
48. She KC, Kusano H, Yaeshima M, Sasaki T, Satoh H, Shimada H: Reduced rice grain production under high-temperature stress closely correlates with ATP shortage during seed development. Plant Biotechnol J. 2010, 27 (1): 67-73. 10.5511/plantbiotechnology.27.67.
49. Peters K, Niessen M, Peterhansel C, Spath B, Holzle A, Binder S, Marchfelder A, Braun HP: Complex I-complex II ratio strongly differs in various organs of *Arabidopsis thaliana*. Plant Mol Biol. 2012, 79 (3): 273-284. 10.1007/s11103-012-9911-4.
50. Wang HC, Huang HB, Huang XM: Differential effects of abscisic acid and ethylene on the fruit maturation of *Litchi chinensis* Sonn. Plant Growth Regul. 2007, 52 (3): 189-198. 10.1007/s10725-007-9189-8.
51. Coello P, Hirano E, Hey SJ, Muttucumaru N, Martinez-Barajas E, Parry MA, Halford NG: Evidence that abscisic acid promotes degradation of SNF1-related protein kinase (SnRK) 1 in wheat and activation of a putative calcium-dependent SnRK2. J Exp Bot. 2012, 63 (2): 913-924. 10.1093/jxb/err320.
52. Plaxton WC, Podesta FE: The functional organization and control of plant respiration. Crit Rev Plant Sci. 2006, 25 (2): 159-198. 10.1080/07352680600563876.
53. Klotz KL, Finger FL, Anderson MD: Respiration in postharvest sugarbeet roots is not limited by respiratory capacity or adenylates. J Plant Physiol. 2008, 165 (14): 1500-1510. 10.1016/j.jplph.2007.12.001.
54. Hardie DG: AMP-activated/SNF1 protein kinases: conserved guardians of cellular energy. Nat Rev Mol Cell Bio. 2007, 8 (10): 774-785. 10.1038/nrm2249.
55. Baena-Gonzalez E: Energy signaling in the regulation of gene expression during stress. Mol Plant. 2010, 3 (2): 300-313. 10.1093/mp/ssp113.
56. Krause M, Durner J: Harpin inactivates mitochondria in Arabidopsis suspension cells. Mol Plant Microbe In. 2004, 17 (2): 131-139. 10.1094/MPMI.2004.17.2.131.
57. Costa ADT, Nantes IL, Jezek P, Leite A, Arruda P, Vercesi AE: Plant uncoupling mitochondrial protein activity in mitochondria isolated from tomatoes at different stages of ripening. J Bioenerg Biomembr. 1999, 31 (5): 527-533. 10.1023/A:1005408809619.
58. Navet R, Jarmuszkiewicz W, Almeida AM, Sluse-Goffart C, Sluse FE: Energy conservation and dissipation in mitochondria isolated from developing tomato fruit of ethylene-defective mutants failing normal ripening: The effect of ethephon, a chemical precursor of ethylene. J Bioenerg Biomembr. 2003, 35 (2): 157-168. 10.1023/A:1023750204310.
59. McGuire RG: Reporting of objective color measurements. HortSci. 1992, 27 (12): 1254-1255.

60. Ozogul F, Taylor KDA, Quantick PC, Ozogul Y: A rapid HPLC-determination of ATP-related compounds and its application to herring stored under modified atmosphere. Int J Fod Sci Tech. 2000, 35 (6): 549-554.
61. Liu H, Jiang YM, Luo YB, Jiang WB: A simple and rapid determination of ATP, ADP and AMP concentrations in pericarp tissue of litchi fruit by high performance liquid chromatography. Food Technol Biotech. 2006, 44 (4): 531-534.
62. Bingham IJ, Farrar JF: Activity and capacity of respiratory pathways in barley roots deprived of inorganic nutrients. Plant Physiol Bioch. 1989, 27 (6): 847-854.
63. Wan CY, Wilkins TA: A modified hot borate method significantly enhances the yield of high-quality RNA from cotton (*Gossypium hirsutum* L. Anal Biochem. 1994, 223 (1): 7-12. 10.1006/abio.1994.1538.
64. Kuang JF, Chen JY, Luo M, Wu KQ, Sun W, Jiang YM, Lu WJ: Histone deacetylase HD2 interacts with ERF1 and is involved in longan fruit senescence. J Exp Bot. 2012, 63 (1): 441-454. 10.1093/jxb/err290.
65. Wei YZ, Hu FC, Hu GB, Li XJ, Huang XM, Wang HC: Differential expression of anthocyanin biosynthetic genes in relation to anthocyanin accumulation in the pericarp of *Litchi Chinensis* Sonn. PLoS One. 2011, 6 (4): e19455. 10.1371/journal.pone.0019455.
66. Zhong HY, Chen JW, Li CQ, Chen L, Wu JY, Chen JY, Lu WJ, Li JG: Selection of reliable reference genes for expression studies by reverse transcription quantitative real-time PCR in litchi under different experimental conditions. Plant Cell Rep. 2011, 30 (4): 641-653. 10.1007/s00299-010-0992-8.

CHAPTER 10

Effect of Post-harvest Treatments of Chemical and Plant Growth Regulators on Physical Parameters of Sapota Fruit cv. Kalipatti

Tsomu T* and Patel HC

Department of Horticulture, B.A. College of Agriculture, Anand Agricultural University, Anand 388110, Gujarat, India

ABSTRACT

Investigation was conducted to study the effect of post-harvest treatment of chemical and plant growth regulators on physical characteristics sapota fruit cv. Kalipatti. Fruits were treated with CaCl2 (5000 and 10000 mg/l) followed by plant growth regulators, Gibbrellic acid (GA3) (150 and 300 mg/l), kinetin (100 and 200 mg/l), ethrel (1000 and 2000 mg/l) and control (untreated fruit) and stored at room temperature. $CaCl_2$ (5000 and 10000 mg/l) was proved very effective in reducing loss in weight, spoilage and registered high fruit firmness, increase shelf life and ripening period. There was a significant increase in Physiological loss in weight (PLW), with enhanced storage period irrespective of calcium chloride (5000 and 10000 mg/l) and Gibbrellic acid (GA3) (150 and 300 mg/l) treatments. On the other hand, Physiological loss in weight (PLW), spoilage was increase throughout the storage. However, minimum Physiological loss in weight (PLW) and, total spoilage were noticed under $CaCl_2$ 5000 mg/l and 10000 mg/l treated fruits. The study suggests that calcium chloride (5000 mg/l) as post-harvestdip improves the fruit firmness, shelf life and ripening period of the sapota upto 12 days of storage.

Keywords: Sapota; Post-harvest; Calcium chloride; Gibbrellic acid (GA_3); Kinetin; Ethrel; Storage

INTRODUCTION

Sapota, popularly known in India as chiku, is native to tropical America, belongs to family Sapotaceae. Being climacteric in nature, sapota fruits need ripening treatments after full maturity. During ripening fruit passes through a series of changes in colour, texture and flavor indication that compositional changes are taking place. The ripe fruits have pleasant aroma and are excellent in sweetness without any astringency due to decrease in polyphenols with concurrent increase in sugars, production of ethylene, rate of respiration and catalase activity [1]. The post-harvest losses are very high in tropical countries. Various chemicals have been used to hasten or delay ripening, to reduce losses and to improve and maintain color and quality by slowing down the metabolic activities of fruit. These chemicals arrest the growth and spread of microorganism by reducing the shriveling which ultimately leads to an increased shelf life and maintain the marketability of the fruit for a longer period. Therefore the present investigation is carried out to evaluate the "Effect of post-harvest treatments of plant growth regulators and chemicals on physico-chemical characteristics and shelf life of sapota [*Manilkaraachras*(Mill) Forseberg.] fruit cv. Kalipatti."

MATERIALS AND METHODS

Present investigations were carried out at Anand Agricultural University, Anand. The fruit of sapota cv. Kalipatti were harvested at optimum stage of maturity. The fruits were dipped for 5 min. in solution of calcium chloride (5000 and 10000 mg/l) and ethrel (1000 and 2000 mg/l), GA_3 (150 and 300 mg/l) for 10 minutes, kinetin (100 and 200 mg/l) for 20 minutes and the fruits kept under control were dipped in distilled water for 5 minutes.

The experimental data was analyzed in Completely Randomized Design (CRD) with four repetitions. Fruits were than air dried and packed in Corrugated Fibre Board (CFB) boxes and kept at room storage. The fruits were assessed at 4th, 6th, 8th, 10th and 12th day of storage for physiological loss in weight, spoilage, fruit firmness, shelf life and days taken to ripening.

Physiological loss in weight (PLW%)

The initial weight of fruits was recorded at the time just before giving post-harvest treatments. To assess the loss in weight, observations were recorded at 4th, 6th, 8th, 10th and 12th day and were subtracted from initial weight. The loss in weight was calculated and expressed in percentage. It was calculated by following formula.

$$\text{PLW}\ (\%) = \frac{\text{Initial weight}\ -\ \text{Final weight} \times 100}{\text{Initial weight}\ (g)}$$

Shelf life (days)

The shelf life of fruits was noted by keeping the fruits at room temperature and the days taken from harvesting to optimum eating stage.

Spoilage of fruits (%)

The number of visibly diseased, rotten, over ripe fruits were counted and expressed as percentage over the total number of fruits at 4th, 6th, 8th, 10th and 12th days of storage period.

Number of days taken to ripening (day)

The number of days taken by each treatment to attain eating ripeness by more than 90 percent fruits present in a lot was considered as days taken to ripening.

Fruit firmness (kgcm^{-2})

Sapota fruit firmness (fruit's pressure) was measured by means of a pocket Penetrometer in terms of pressure required by piercing through the fruit. The pressure required was recorded in kilogram per square centimeter.

RESULTS AND DISCUSSION

The physiological loss in weight was significantly diminished by post-harvest $CaCl_2$ treatments (**Table 1** and **Figure 1**). The fruits treated with $CaCl_2$ (5000 and 10000 mg/l) recorded the lowest loss in weight. The reduction in weight loss might be due to the maintenance of firmness of fruits by calcium as it decreased the enzyme activity responsible for disintegration of cellular structure, which decreases the gaseous exchange

[2]. The loss in weight during the storage was significant regardless of $CaCl_2$(5000 and 10000 mg/l) treatment.

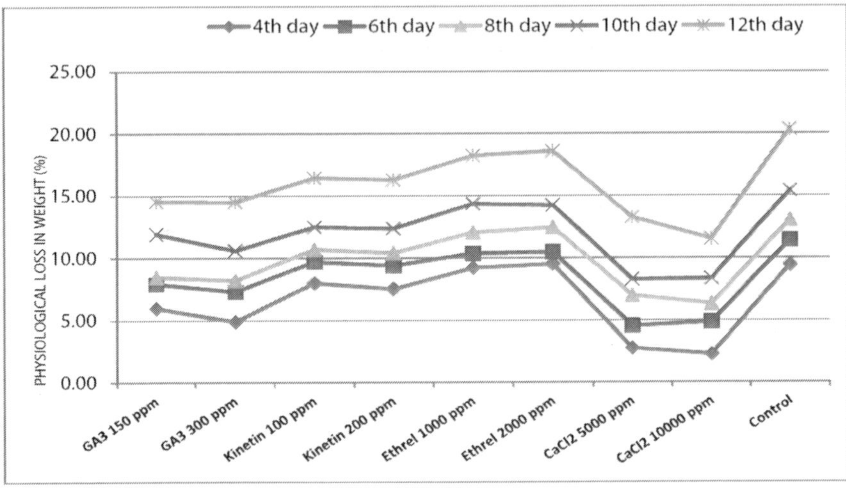

Figure 1: Effect of post-harvest treatments of chemical and growth regulators on physiological loss.

Table 1: Physiological loss in weight due to the effect of different levels of postharvest treatments of chemical and plant growth regulators on sapota fruits cv. Kalipatti.

Sr. No.	Treatments	Physiological loss in weight (%) No. of days (Storage period)				
		4th	6th	8th	10th	12th
1	GA_3 150 mg/l	5.97	7.91	8.48	11.90	14.52
2	GA_3 300 mg/l	4.89	7.30	8.19	10.57	14.50
3	Kinetin 100 mg/l	7.98	9.68	10.67	12.48	16.44
4	Kinetin 200 mg/l	7.50	9.37	10.4	12.35	16.27
5	Ethrel 1000 mg/l	9.20	10.36	12.02	14.34	18.22
6	Ethrel 2000 mg/l	9.50	10.46	12.44	14.22	18.58
7	$CaCl_2$ 5000 mg/l	2.71	4.53	6.96	8.26	13.23
8	$CaCl_2$ 10,000 mg/l	2.23	4.85	6.30	8.32	11.49
9	Control	9.42	11.40	13.00	15.40	20.31
S.Em. ±		0.21	0.33	0.40	0.44	0.49
C.D. (0.05%)		0.62	0.94	1.17	1.28	1.42
C.V. %		6.49	7.71	8.23	7.39	6.14

The shelf life of fruits was significantly prolonged due to calcium chloride and plant growth regulator treatments over control or untreated fruits. Among the different treatments tried, $CaCl_2$ 10000 (11.25 days) and 5000 ppm (10.50 days) followed by GA_3 300 ppm (8.75 days) were found most effective to extend the shelf life of sapota fruits over control (**Table 2** and **Figure 2**).

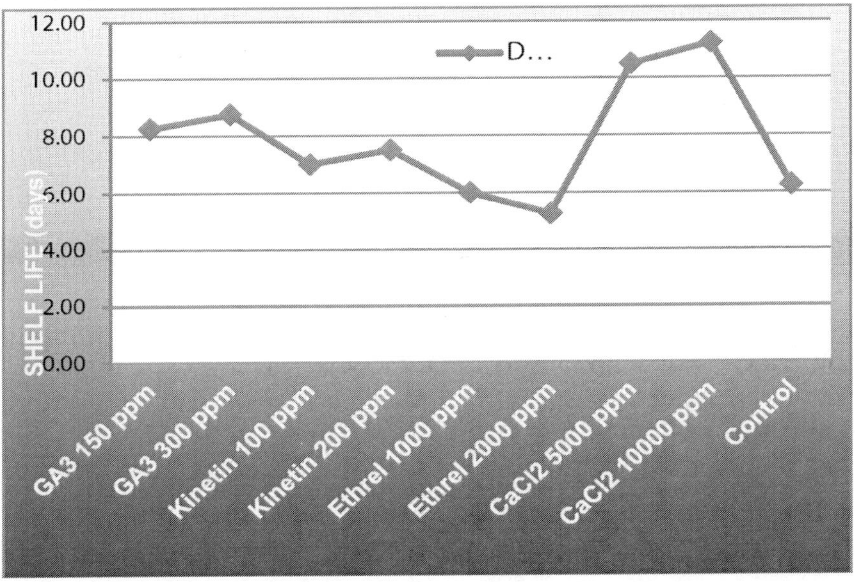

Figure 2: Effect of post-harvest treatments of chemical and growth regulators on Shelf life (days) of sapota fruits cv. Kalipattiin weight (%) of sapota fruits cv. Kalipatti.

The sapota fruits under treatments of calcium chloride were found to have better quality with extended shelf life. Calcium helps in structural integrity of both the cell wall and plasma membrane thus delaying ripening and extending storage life [3].

The fruits treated with $CaCl_2$ 10000 ppm (8.5 days) and 5000 ppm (8.0 days) and GA_3 300 ppm (7.50 days) and 150 ppm (7.25 days) took significantly maximum days for ripening. The untreated fruits ripened in six days and ethrel 2000 ppm treated fruits ripened almost in 4 days earlier than rest of the treatments.

The delay in ripening of the fruits was due to the fact that it slowed down the process of ripening by retarding the pre-climacteric respiration rate and ethylene production and through postponement of their climacteric peak a *vis* a *vis* control (**Table 2** and **Figure 3**). These changes lead to reduced degradative metabolism in terms of catalase and PME activities and thus helpful in extending shelf-life of the fruits.

Table 2: Effect of different levels of post-harvest treatments of chemical and plant growth regulators on shelf life, days taken to ripening and fruit firmness of sapota fruits cv. Kalipatti.

Sr. No.	Treatments	Shelf life (days)	Days taken to ripening (days)	Fruit firmness ($kgcm^{-2}$)
1	GA_3 150 mg/l	8.25	7.25	0.54
2	GA_3 300 mg/l	8.75	7.50	0.55
3	Kinetin 100 mg/l	7.00	6.00	0.41
4	Kinetin 200 mg/l	7.50	6.50	0.44
5	Ethrel 1000 mg/l	6.00	5.00	0.37
6	Ethrel 2000 mg/l	5.25	4.50	0.39
7	$CaCl_2$ 5000 mg/l	10.50	8.00	0.64
8	$CaCl_2$ 10,000 mg/l	11.25	8.50	0.66
9	Control	6.25	6.25	0.30
S.Em. ±		0.36	0.33	0.02
C.D. (0.05%)		1.03	0.95	0.06
C.V. %		9.08	10.30	8.05

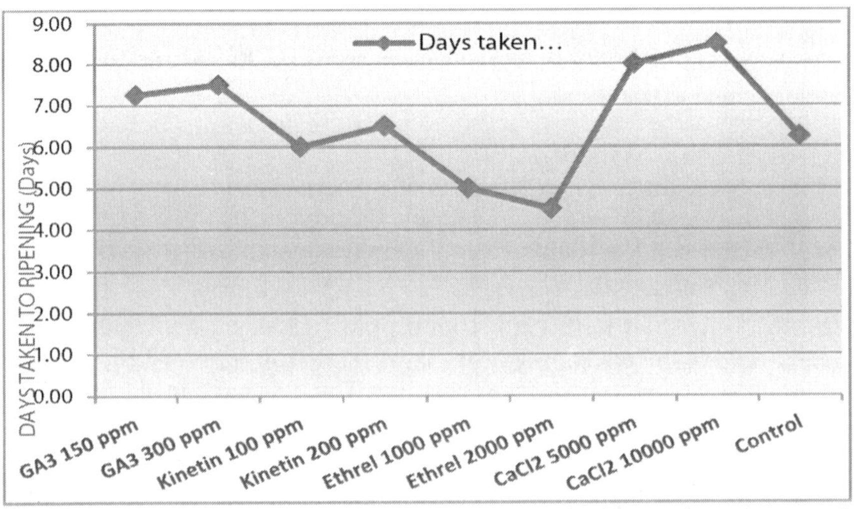

Figure 3: Effect of post-harvest treatments of chemical and growth regulators on days taken to ripeninig of sapota fruits cv. Kalipatti.

The present investigation is in conformity with the results reported by Gautam and Chundawat [4] in sapota cv. 'Kalipatti', Attri and Singh [5] in sapota cv. 'Cricket Ball' Damodaran et al. [6] in sapota cv. 'Cricket Ball' and Sudha et al. [7] in sapota cvs. 'PKM-1' and'CO-2'.

Fruit firmness during the storage is associated with acceleration of hydrolytic enzymes. Post-harvest treatments assisted sapota fruits to maintain firmness up to the end of storage period. Fruits treated with $CaCl_2$ 10000 ppm (0.66 $kgcm^{-2}$) and 5000 ppm (0.64 $kgcm^{-2}$) showed the highest firmness. The fruits treated with $CaCl_2$ retained more firmness in comparison to control (**Table 2** and **Figure 4**). Firmness in many fruits is an important characteristic that is used to determine stability and it is predominantly determined by cell wall composition and structure. Calcium treatments have been known to delay the softening and improve the fruit quality [3].

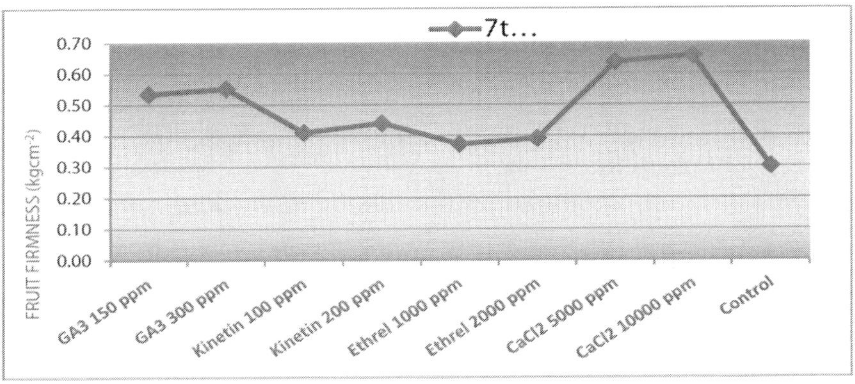

Figure 4: Effect of post-harvest treatments of chemical and growth regulators on fruit firmness (Kgcm-2) of sapota fruits cv. Kalipatti.

The spoilage percent of sapota fruits was increased as the storage period advanced irrespective of any treatment. Calcium treated fruits showed significantly lesser extent of rotting which may be due to the higher fruit flesh and skin calcium content, which resulted in stronger intracellular organization and rigidified cell wall. Similar findings were reported by Singh and Narayana [8] in mango (**Table 3** and **Figure 5**).

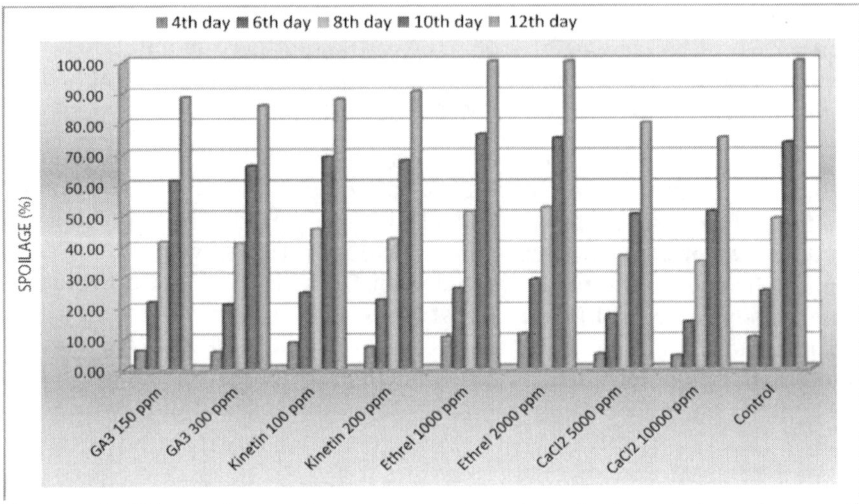

Figure 5: Effect of post-harvest treatments of chemical and growth regulators on spoilage (%) of sapota fruits cv. Kalipatti.

Table 3: Effect of different levels of post-harvest treatments of chemical and growth regulators on spoilage of sapota fruits cv. Kalipatti.

Sr. No.	Treatments	Spoilage (%) No. of days (Storage period)				
		4th	6th	8th	10th	12th
1	**GA_3 150 mg/l**	6.00	21.70	41.33	61.13	88.50
2	**GA_3 300 mg/l**	5.50	20.87	40.83	65.92	85.75
3	**Kinetin 100 mg/l**	8.50	24.78	45.50	68.96	87.75
4	**Kinetin 200 mg/l**	7.00	22.44	42.24	67.67	90.25
5	**Ethrel 1000 mg/l**	10.25	26.14	50.90	76.25	100.00
6	**Ethrel 2000 mg/l**	11.25	28.90	52.32	75.00	100.00
7	**$CaCl_2$ 5000 mg/l**	4.50	17.51	36.57	50.02	80.00
8	**$CaCl_2$ 10,000 mg/l**	4.00	15.02	34.69	50.71	75.00
9	**Control**	10.00	25.03	48.55	73.27	100.00
S.Em. ±		0.37	0.96	1.65	2.94	1.78
C.D. (0.05%)		1.06	2.79	4.80	8.51	5.17
C.V. %		9.84	8.55	7.57	8.97	3.98

CONCLUSION

The present study was conducted to investigate the effect of postharvest treatments of chemical and plant growth regulators on physical parameters of sapota fruit cv. Kalipatti. The various post-harvest treatments included were fruit dip in $CaCl_2$, GA_3, kinetin, ethrel and control fruits were dipped in distilled water. Treated fruits were stored under room temperature. Fruits treated with $CaCl_2$ (5000 mg/l) helps in maintaining the highest fruit firmness, delayed ripening and shelf life. The least spoilage and PLW was again found in fruits treated with $CaCl_2$ (5000 and 10000 mg/l) followed by GA_3 (150 and 300 mg/l) as compared to other post-harvest treatments.

REFERENCES

1. Sankaranarayanan R, Geetharani P, Manivannan MI (2007) Ripening behaviour of sapota. South Indian Hort 55: 276-282.
2. Levy D, Poovaiah BW (1979) Effect of calcium infiltration of senescence of apples. Hort Sci 14: 466.
3. Sharma RM, Yamdagni R, Gaur H, Shukla RK (1996) Role of calcium in horticulture-a review. Haryana J of Hort Sci 25: 205-212.
4. Gautam SK, Chundawat BS (1989) Post-harvest changes in sapota cv. 'Kalipatti': I-Effect of various post-harvest treatments on biochemical changes. Indian J Hort 46: 310-315.
5. Attri BL, Singh DB (1996) Effect of GA3 and Silver nitrate on the physico-chemical characteristics of sapota cv. Cricket ball during storage. Prog Hort 28: 3-4.
6. Damodaran T, Attri BL, Medhi RP, Nair SA, Alex L (2001) Studies on post-harvest of sapota (Achraszapota L.)cv. Cricket Ball during storage. Indian J Hort 58: 342-345.
7. Sudha R, Amutha R, Muthulaksmi S, Rani W, Indira K, et al. (2007) Influence of Pre and Postharvest Chemical Treatments on Physical characteristics of sapota (Achras sapota L.) var. PKM 1. Research J of Agri and Bio Sci 3: 450-452.
8. Singh BP, Narayana CK (1999) An integrated approach for storage of mango. Indian J of Hort 56: 5-9.

CHAPTER 11

Investigating the Effects of Plant Essential Oils on Post-Harvest Fruit Decay

Hazem S. Elshafie[1] and Ippolito Camele[1]

[1] *School of Agricultural, Forestry, Food and Environmental Sciences, University of Basilicata, Potenza, Italy*

ABSTRACT

Essential oils are one of the most important natural products derived from plants, due to their various biological properties and their medicinal and nutritional uses. This chapter provides an overview of several different aspects relating to essential oils including a historical perspective, the uses of essential oils, their main sources and antifungal activity, their bioactive single constituents and their modes of action. The chapter will also give an insight into the chemical measures necessary for controlling plant pathogens and their negative impact on human health and/or the environment. It will also review the different sources of essential oils such as sage, oregano, thyme and marjoram from the Lamiaceae family, vervain from the Verbanacae family, and magnolia from the Magnoliaceae family. The antimicrobial activity of essential oils is reviewed, with particular emphasis on the antifungal properties exhibited against some serious pathogenic fungi and post-harvest disease. Moreover, various antimicrobial tests and techniques, such as various kill-time studies, killing time determination, LD-50 and growth curve recording, poisoned food techniques, spore germination and measurement of metabolic CO_2 are included. Finally, five case studies relating to the antifungal activity of some plant essential oils, either in vitro or in vivo, against post-harvest pathogenic fungi are reviewed at the end of this chapter.

Keywords: Plant essential oils, Antifungal activity, Post-harvest diseases, Chemical measurements, Antimicrobial tests, Fungal pathogenicity

1. INTRODUCTION

An essential oil is a concentrated hydrophobic liquid containing volatile aromatic compounds from plants. Essential oils are also known as volatile oils or ethereal oils. An oil is "essential" in the sense that it carries a distinctive essence of the plant. It does not form a distinctive category for either medical, pharmacological, or culinary purposes.

Essential oils have several applications in the manufacture of perfumes, cosmetics, and soaps. In the food industry, essential oils can be used for flavouring, and for adding scents to incense. The use of some essential oils as alternative antimicrobial agents, replacing chemical treatments, has attracted considerable interest from post-harvest scientists in recent years. Most essential oils and their single constituents have been reported to inhibit post-harvest fungi both *in vitro* and *in vivo*. The use of these volatile compounds has garnered a great deal of interest recently.

There is no doubt that essential oils have important pharmacological properties, which result in their wide use in pharmaceutical practices. The first investigation into the antimicrobial activity of essential oils was a study by Buchholtz [1] who found that thymol has higher growth inhibitory properties on bacteria having been cultivated in a tabac decoction than phenol, which was previously considered to be the best systematic substance for use as surgical antiseptic [2]. Therefore, a great deal of research has been carried out aimed at studying the antimicrobial activities of several essential oils that have already been registered in the pharmacological industry.

The fungitoxic activity of essential oils may be due to synergism among their components, since most of this activity has been reported to be enhanced when combined [3]. Bioactivity in the vapor phase of essential oils was recognized as a characteristic that makes them attractive for use as possible fumigants for the protection of stored product [3–6].

This chapter will address different aspects relating to plant essential oils and fungal pathogenicity ranging from the chemical use in controlling plant pathogenic fungi as well as discussing the main sources of essential oils, and their biological activity. Furthermore, this chapter will provide an insight into the mode of action of single constituents of different essential oils, using different case studies.

2. CHEMICAL CONTROL OF PLANT PATHOGENIC FUNGI

Collectively, fungi and fungal-like organisms cause more plant diseases than any other group of plant pathogens, with over 8,000 species shown to cause disease. The importance of fungi as agents of plant and human diseases, producers of industrial and pharmacological products, and as decomposers, has prompted scientists worldwide to study their biology. Some of the world's great famines and human suffering can be blamed on plant pathogenic fungi [7].

Chemical disease control employs the use of chemicals that are generally toxic to pathogens and characterized by their specific effect, such as many commercial fungicides and antibiotics. Basically, any chemical agent must be effective and act safely, without harming any live organism and must also have minimal or no effect on the environment, microflora and soil.

Chemical fungicides are generally used in the control of fungal diseases. Recently, growing public concern over the health and environmental hazards associated with the increased levels of chemical pesticides and the lack of approval for the renewal of some of the most effective active molecules, has led to the development of safe, alternative, and natural methods of post-harvest disease control [8]. To minimize the development of pest resistance as a result of the overuse of chemical fungicides, it was advisable to use a variety of fungicides belonging to different chemical groups. Throughout the 1980s and 1990s, public concern about the use of agricultural chemical pesticides grew as a result of the high risk of poisoning to all living organisms, beneficial micro-flora and micro-fauna, and the contamination of food products. The risk from chemical pesticides is due to the possibility that they may be absorbed through the skin, inhaled or ingested through consumption of contaminated plants.

To overcome the risks of chemical pesticides, some precautions should be taken with respect to the handling of the chemicals and the management of agriculture practices. Environmental concerns focus mainly on protectant fungicides such as copper and sulfur sprays, which have the potential to affect a broad range of organisms if they are washed off leaves and then accumulate in the soil or are washed into waterways. A lot of countries have already started educational programs for farmers to reduce the overuse of chemical pesticides, and hence increase the productivity of several economic crops.

Recently, there has been a great interest in the use of essential oils and plant extracts as possible natural substitutes for conventional synthetic pesticides. This may be attributed mainly to ecosystem pollution and pesticide resistance in pests, insects and fungal pathogens.

3. PLANT ESSENTIAL OILS

Essential oils are volatile compounds produced in many species of plant. These oils are thought to play a role in plant defense mechanisms acting against phytopathogenic microorganisms [9–13]. Subsequently, plant essential oils were subjected to pharmacological studies and, later, to numerous and frequent tests of their antimicrobial activity.

Several methods have been used to evaluate the *in vitro* antimicrobial activity of different essential oils. Among the most common methods are agar diffusion tests, serial broth or agar dilution tests, and vapor phase tests [14].

Further antimicrobial tests comprise the following:

1. **Various kill-time studies:**

This test deals with determining the activity of a compound relative to the activity of phenol, after 15 min (phenol coefficient).

1. **Killing time determination:**

This test deals with determining the exact time needed for the complete inhibition of the target organism after contact with the test compound, using contaminated silk threads.

1. **LD-50 and recording of growth curves:**

This test records growth curves and determines the amount of a compound effectively inhibiting the growth of 50% of test organisms.

1. **Poisoned food techniques:**

Tests in which the delay of microbial growth is determined in the presence of growth inhibitors.

1. **Spore germination:**

This test is suitable for use with fungi, especially during short contact time studies.

1. **Measuring of metabolic CO_2:**

This test monitors the presence or absence of the growth of yeast, or visualizes growth using indicators such as sulfur salts from sulfur supplemented cow milk as a growth medium, 2,3,5-triphenyltetrazolium chloride, or p-iodo-nitrophenyltetrazolium violet.

3.1. Main Sources of Plant Essential Oils

1. **Family Lamiaceae**

Sage

Sage is considered to be the main genus among the *Lamiaceae* family, which consists of about 900 species widely distributed in the temperate, subtropical and tropical regions all over the world but especially in the Mediterranean region, central Asia, central and South America, and in southern Africa. Globally, the best known species of the family, used in both traditional and modern medicine are *Salvia officinalis*, *S. fruticosa* and *S. divinorum*.

Another important plant is oregano, considered to be the most valued species worldwide. About 60 plant species were listed within this common name. The majority of oregano species belong to the *Lamiaceae* and *Verbenaceae*.

Oregano

Oregano is a perennial herb; its flowers are purple, 3–4 mm long, and produced in erect spikes. It is sometimes called wild marjoram, and its close relative *Origanum majorana* is known as sweet marjoram. Many subspecies and strains of oregano have been developed by humans over the centuries for their unique flavors or other characteristics. It is an important culinary herb, used for the flavor of its leaves, which can actually be more flavorful when dried than fresh. It has an aromatic, warm and slightly bitter taste, which can vary in intensity.

The antifungal and antibacterial activity of oregano essential oil against a number of plant pathogens has been reported by Adebayo et al. [15] These pathogens include *Aspergillus niger* v. Tieghem, *A. flavus* Link, *A.*

ochraceus Wilhelm, *Fusarium oxysporum* Snyder and Hansen, *Penicillium* sp. L., *Pseudomonas aeruginosa* Schroter ATCC 2730, *Staphylococcus aureus* Rosenbach ATCC 6538, *F. solani* var. *coeruleum* (Martius) Saccardo, *Clavibacter michiganensis* S., *Phytophtora infestans* Mont., *Sclerotinia sclerotiorum* Lib. and *Xanthomonas vesicatoria* D.

Thymes

Thyme is an evergreen herb with culinary, medicinal and ornamental uses. The most common variety is *Thymus vulgaris*. Thyme belongs to the genus *Thymus* of the mint family (*Lamiaceae*) and is a relative of the oregano genus *Origanum*.

The essential oil of *T. capitatus* Hoffm. & Link, displayed antifungal activity in stored foods and inhibited the growth of both *B. cinerea* and *Monilinia fructicola* [16]. *T. vulgaris* showed antifungal activity against some post-harvest fungal pathogens such as *B. cinerea*, *P. italicum*, *P. citrophthora* and *Rhizopus stolonifer* [10].

Marjoram

Marjorana hortensis (*Lamiaceae*), commonly known as marjoram, is a perennial herb or undershrub with sweet pine and citrus flavors. It has a long history of medicinal and culinary use.

Camele et al. [10] reported that the last two essential oils of *Lamiaceae* family showed antifungal activity against *P. citrophthora* and *R. stolonifer*. *M. hortensis* showed antifungal activity against *C. acutatum* and *B. cinerea*, and antibacterial activity against two strains of G+ve (*Bacillus megaterium*and *C. michiganensis*) and five strains of G-ve (*Escherichia coli*, *X. campestris*, *B. mojavensis*, *P. savastanoi* and *P. syringae* pv. *phaseolicola*) (Elshafie et al., data not published).

1. Family Verbanacae

Vervain

Verbena officinalis (*Verbenaceae*) commonly known as vervain, is a perennial medicinal plant which grows natively in Europe. It is widely naturalized outside its native range, for example in North America. It has been used traditionally as folk medicine in some countries, and recently was reported to act as an anticancer agent. Despite its widespread uses, the mechanisms of the pharmacological actions of the herb are still unclear.

The antimicrobial activity of vervain essential oil was reported recently by Elshafie et al. [12] who found that this essential oil significantly reduced (*in vivo*) the brown rot lesion diseases of peach caused by *M. laxa*, *M. fructicola* and *M. fructigena*.

1. Magnoliaceae

Magnolia

Magnolia is a large genus of about 210 flowering plant species in the family *Magnoliaceae*. Its trees are very ancient and the flowers and oil have been used in many cultures around the world.

M. liliflora essential oil showed potential *in vitro* and *in vivo* antifungal effects against *B. cinerea*, *Colletotrichum capisci* (Syd.) E.J. Butler & Bisby, *F. oxysporum* Snyder and Hansen, *F. solani*, *P. capisci* Leonian, *Rhizoctonia solani* (Cooke) Wint. and *S. sclerotiorum* [17].

The essential oil of *M. liliflora* also showed a potential *in vivo* antifungal effect against *P. capsici*, and this activity could be attributed to its constituents: α-terpineol, α-bourbonene, β-caryophyllene, 2-β-pinene, α-humulene, farnesene, and caryophyllene oxide components [17].

3.2. Antimicrobial Activity of Plant Essential Oils

Numerous essential oils have shown an antifungal effect against several post-harvest pathogens, e.g. *B. cinerea* Pers. [18–20], *Aspergillus* spp. [21–23], *Fusarium* spp. [24,25], *Penicillium* spp. [26], *R. stolonifer* (Ehrenb.: Fr.) Vuill. [27,28], *C. gloeosporioides* Penz. [29]. The antifungal effect is attributed mainly to the inhibition of both mycelial growth and spore germination. This hypothesis suggests that the expression of disease will be restricted by the impeding of the initial infection and the subsequent mycelial spread beyond the infection site [30].

Clove oil is an essential oil extracted from the clove plant, *Syzygium aromaticum* (L.) Merr. & Perry, and has been reported to act as a bioactive substance — especially its active component monoterpene eugenol [31] — against *B. cinerea*, *M. fructigena* Honey, *P. expansum* Link and *Phlyctema vagabunda*Desm. in apples. Carvacrol (one of the major constituents of oregano essential oil) is a phenol that was reported to show a high inhibition of mycelium growth in *Neofabraea alba* (E.J. Gutrie) Verkley on apples [32].

Essential oils from basil (*Ocimum basilicum* L.), fennel (*Foeniculum sativum* Mill.), lavender (*Lavandula officinalis* Chaix), marjoram (*O.*

majorana L.), oregano (*O. vulgare* L.), peppermint (*Mentha piperita* L.), rosemary (*Rosmarinus officinalis* L.), sage (*Salvia officinalis* L.), savory (*Satureja montana* L.), thyme (*T. vulgaris* L.) and wild mint (*Mentha arvensis* L.) showed a potentially significant antifungal activity higher than that available from chemical treatments in post-harvest treatments against *B. cinerea* and *P. expansum* on apples [3].

The different efficacies of essential oils are due to the differing fungitoxic properties of each single active constituent, as well as the synergic effect [3,32]. This hypothesis suggests that the possible phytotoxic effects of essential oil treatments may be due to the same active components existing in each essential oil. The length of storage may also negatively influence the antifungal activity of the essential oil treatments. Therefore, treatments using essential oils should only be used for short storage times, or they should be repeated after a defined time period has elapsed, depending on the fruit cultivar in question [3]. The selection of an essential oil for a post-harvest treatment must be based mainly on the characteristics of the fruit, desirable storage time and decay.

The antifungal activity of essential oils could be enhanced by the method of application. The potential of using essential oils by dipping or spraying to control post-harvest decay has already been examined in fruit and vegetables [33–35]. The combination of various post-harvest treatments may improve the efficacy of controlling post-harvest pathogens [30,36,37].

Essential oils from thyme (*T. capitatus* L.), spearmint (*M. spicata* L.) and anise (*Pimpinella anisum* L.) exhibited inhibitory effects on the development of *M. fructicola* (G.Winter) Honey [36,37]. The essential oil of Lemon myrtle (*Backhousia citriodora* F.Muell.) has been reported to have antifungal activity against *M. fructicola*, and has been shown to have a strong antimicrobial activity, mainly with respect to its potential application as a topical pharmaceutical product. Its main constituent, citral, had also been reported to exhibit a fungitoxic effect against a range of post-harvest pathogens [38].

Mechanism of antifungal activity of essential oils

Apart from the positive effect of each of the chemical constituents of different essential oils, several different studies have indicated that there seems to be a synergetic effect between the individual chemical constituents. This synergism in the aromatic plants components functions to make them more effective and reduces the developing resistance of any pathogenic fungi.

In particular, some single constituents such as carvacrol, γ-terpinène and p-cymene become more effective when they are combined together and act synergistically [15].

On the other hand, p-cymene is efficient facilitator of the transport of carvacrol across cell wall components and the cytoplasmic membrane of pathogenic fungi. Thymol, eugenol and carvone are widely used in the control of several fungi, particularly those which contaminate various important economic crops [13].

Another hypothesis suggested by Soylu et al. [39,40], is that the observed diameter reduction and lyses of the hypha wall, may be attributed to the enzymatic reactions within the essential oil which act to regulate synthesis of the wall. Furthermore, the lipophilic properties of the above mentioned single components might have the ability to degrade the plasma membrane, and thus to increase the permeability of the cytoplasm.

The following will give an insight into some case studies relating to the antimicrobial activity of some Mediterranean plant essential oils:

1st case study: *Biochemical characterization of oregano essential oil*;

2nd case study: *Antimicrobial activity of oregano, thyme and vervain essential oils*;

3rd case study: *In vivo antifungal activity of thyme and vervain essential oils*;

4th case study: *In vitro antifungal activity of the main components of vervain, thyme and oregano essential oils*;

5th case study: *In vitro and in vivo antifungal activity of single constituents of oregano essential oil.*

3.3. Case Studies

First case study

Biochemical characterization of oregano essential oil

Mancini et al. [11] studied the biochemical characterization of *O. vulgare* ssp. *hirtum* essential oil from the Southern Apennines (Italy). *O. vulgare* is composed mainly of phenolic compounds belonging to the carvacrol/thymol chemotype. The possible inhibitory fungicidal activity of *O. vulgare* was determined against *M. laxa* (Aderh. & Ruhland) Honey, *M. fructigena* and *M. fructicola* as follows: prepare different concentrations of each oil + potato dextrose agar (PDA) + (0.2 %) with 0.2 % Tween 20 and

250, 500, and 1000 ppm of each essential oil under study, then pour 14 ml of PDA+ oil in Petri dish.

Following that, completely dry off the preparation under laminar flow, 0.5 cm disc from the studied fungi 96 h old was inoculated in the centre of each Petri dish. All plates were incubated at 22°C for 96 h in the absence of light, and the diameter of any fungal mycelium growth was measured in mm. PDA plates + Tween 20 without oils were inoculated with the same fungi as a control. Fungitoxicity was expressed as the percentage of growth inhibition (PGI) and calculated according to Zygadlo et al. [41] as follows:

$$PGI\ (\%) = 100 \times (GC\text{-}GT)/GC$$

Where GC represents the average diameter of fungi grown in PDA (control); GT represents the average diameter of any fungi cultivated on the treated PDA containing the essential oil.

The oils tested have shown antifungal activity against *M. laxa*, *M. fructigena* and *M. fructicola* (Table 1) [11] and have neither shown any phytotoxic activity against germination and initial radicle elongation of *Sinapis arvensis* L., *Phalaris canariensis* L., *Lepidium sativum* L. and *Raphanus sativus* L., nor have they shown any haemolysing effect against the cell membrane of bovine erythrocytes [11].

Second case study

Antimicrobial activity of oregano, thyme and vervain essential oils

Adebayo et al. [15] reported that oregano essential oil has antimicrobial activity against a number of plant pathogens such as *A. niger* van Tieghem, *A. flavus* Link, *A. ochraceus* Wilhelm, *F. oxysporum*, *F. solani* var. *coeruleum* (Martius) Saccardo, *Penicillium* spp., *P. aeruginosa* Schroter ATCC 2730, *S. aureus* Rosenbach ATCC 6538, *C. michiganensis* (Smith) Davis, *P. infestans* Mont, *S. sclerotiorum* Lib. and *X. gardneri* Dowson. In addition, different species of oregano such as *O. vulgare* and *O. syriacum*L. showed antifungal activity against *B. cinerea* [33,34].

Thyme and vervain essential oils were obtained from *T. vulgaris*, *V. officinalis* L. have shown *in vitro*potential fungicidal activity against four causal agents of post-harvest orange fruit rot: *B. cinerea*, *P. italicum*, *P. citrophthora* (R.E. Sm. & E.H. Sm.) Leonian and *R. stolonifer* (Ehrenb.: Fr.) Vuill. [10].

TABLE 1. Antifungal activity of *O. vulgare* essential oils from three different localities.

Inhibition Percentage (PGI %)				
Treatments		*M. laxa*	*M. fructigena*	*M. fructicola*
O. vulgare (M)	1000 ppm	100.0 ± 0.0a	100.0 ± 0.0a	100.0 ± 0.0a
	500 ppm	79.6 ± 2.21b	69.2 ± 1.67c	54.3 ± 2.65c
	250 ppm	67.9 ± 1.10c	58.0 ± 0.83d	14.3 ± 6.19e
O. vulgare (MP)	1000 ppm	100.0 ± 0.0a	100.0 ± 0.0a	100.0 ± 0.0a
	500 ppm	74.2 ± 3.31c	62.1 ± 3.35cd	45.6 ± 2.65c
	250 ppm	48.4 ± 2.21d	38.4 ± 3.35e	20.6 ± 2.65de
O. vulgare(SGP)	1000 ppm	100.0 ± 0.0a	100.0 ± 0.0a	100.0 ± 0.0a
	500 ppm	92.9 ± 1.10a	86.9 ± 1.68b	85.00 ± 1.77b
	250 ppm	46.8 ± 4.42d	43.1 ± 3.34e	25.6 ± 0.88d

[i] - Where: M: Mandia sample; MP: Marconia di Pisticci sample; SGP: San Giovanni a Piro sample. Values followed by the same letter in each vertical column are not significantly different according to Tukey test *at* $P < 0.05$, data are expressed as mean (SDS), R=3.

Third case study

In vivo antifungal activity of thyme and vervain essential oils

Elshafie et al. [12] studied the *in vivo* antifungal activity of both thyme oil and vervain oil against post-harvest brown rot disease of peach. They reported that thyme oil was mainly composed of *o*-cymene (56.2%), while the main components of vervain oil were citral (44.5%) and isobornyl formate (45.4%) [10]. In addition, both oils were evaluated for their *in vivo* antifungal activity against the post-harvest pathogen *Monilinia* on peach fruits. All tested peach fruits of cv. "Springcrest", were not treated by either pre- or post-harvest chemical pesticides, and were sterilized superficially with a 2 % sodium hypochlorite solution and were later washed with sterile distilled water. They were finally air dried before being inoculated with the above mentioned three phytopathogenic fungi at room temperature by injuring the fruit surfaces and inoculation with about 10 μL of fungal suspension of 10^6 spore/mL. Spore suspensions were prepared by adding two loopfuls of each fresh fungal mycelium (7–10 days old) to 10 mL of sterile distilled water before each of the suspensions was filtered and the concentration adjusted by serial dilution in sterile distilled water. One day after inoculation, each single fruit group was sprayed with different concentrations of thyme essential oil at 250 and 500 ppm, or at 500 and 1000 ppm of vervain essential oil. Each experiment was repeated twice. Four fruits, after being wounded with a sterile needle, were sprayed only with sterile distilled water and used as negative control, while four fruits for each oil concentration were not inoculated and were used instead as a control to determine the possible phytotoxicity. All the fruit series were kept in a moist chamber at a high relative humidity (about 95%) for 4 days at room temperature before being observed for eventual appearance of symptoms. The fungitoxicity effectiveness was expressed as the diameter of the brown rot lesion in mm on fruit with respect to control. Both oils studied showed a promising fungicidal effect *in vivo* on the post-harvest diseases of peach fruits such as brown rot lesion caused by *M. laxa*, *M. fructigena*, and *M. fructicola* (Figure 1,12).

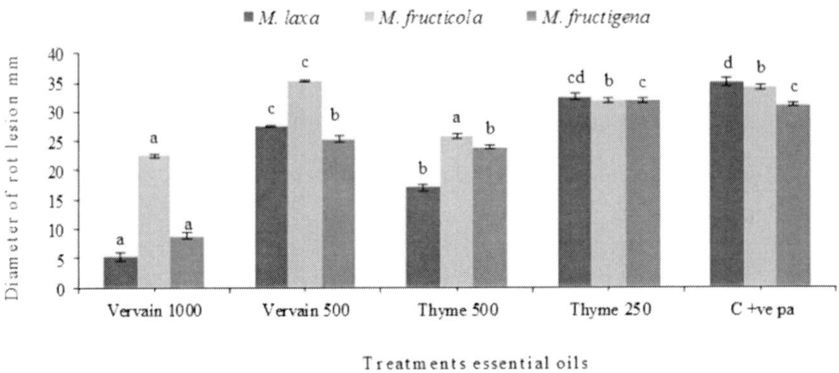

Where: C+ve pa: is positive control where the fruits were challenged only with pathogens.

FIGURE 1. In vivo antifungal activity of thyme and vervain oil against postharvest brown rot disease of peach caused by M. laxa, M. fructicola and M. fructigena. Bars with different letters indicate means values significantly different at $P < 0.05$ according to Tukey test. Data are expressed as mean of three replicates ± SD.

Fourth case study

In vitro antifungal activity of the main components of vervain, thyme and oregano essential oils

Camele et al. [42] investigated the biological activity *in vitro* of the main components "β-fellandrene, β-pinene, camphene, carvacrol, citral, *o*-cymene, γ-terpinene and thymol" extracted from three Mediterranean aromatic plants (*V. officinalis, T. vulgaris* and *O. vulgare*), against five etiological agents of post-harvest fruit decay, *B. cinerea, P. italicum, P. expansum, P. citrophthora* and *R. stolonifer*. The possible fungistatic or fungicidal activity of each studied essential oil components was determined as follows: (A) putting single 3-mm-thick and 0.5-cm-diameter PDA plugs containing fungal mycelium onto the central part of the surface of Petri dishes containing PDA pre-treated with different concentrations (50, 150 or 250 ppm) of each tested single component dissolved in 0.2% Tween-20, (B) dropping, under axenic conditions, 10 μL aliquots of single suspensions containing 1×10^4 conidia/mL of the single target microorganism species onto the central part of surface of Petri dishes containing PDA prepared with the same percentage of Tween-20 and the single three component concentration. Results showed that citral exhibited a fungicidal action against *P. citrophthora*; carvacrol and thymol showed a fungistatic activity against *P. citrophthora* and *R. stolonifer*. Thymol showed fungicidal activity against *P. italicum*. Citral and carvacrol at 250 ppm, and thymol at 150 ppm all stopped the growth of *B. cinerea*.

Fifth case study

In vitro and in vivo antifungal activity of the single constituents of oregano essential oil

Elshafie et al. [13] evaluated the antifungal effect of the single components of *O. vulgare* ssp. *hirtum*essential oil against the post-harvest pathogens *M. laxa, M. fructigena* and *M. fructicola in vitro* and *in vivo*. The chemical characterization of *O. vulgare* reported by Mancini et al. [11] explained that it contains five main single components: carvacrol, thymol, linalool, citral and *trans*-caryophyllene [11].

In vitro antifungal activity

The possible fungicidal activity of the above five standards was determined according to the method of Soylu et al. [40]. Three-mm-thick and 0.5-cm-diameter PDA plugs, axenically taken from the peripheral portion of basic colonies, were inoculated onto the central part of PDA Petri dishes pre-treated with different concentrations of each single component (50, 150 or 250 ppm) dissolved in 0.2% Tween 20. All plates were incubated at 22°C for 96 h in the absence of light. Negative controls comprised either PDA plates without any treatments, or PDA plates treated only with 0.2% Tween 20. The antifungal activity was expressed by measuring the diameter of any mycelium growth in mm [42].

Carvacrol and thymol have exhibited the highest activity during *in vitro* tests against all tested post-harvest *Monilinia* pathogens (Figure 2,13). Citral showed moderate antifungal activity, lower than that of carvacrol and thymol. Linalool and *trans*-caryophyllene showed slight antifungal activity against all studied pathogens. On the other hand, thymol showed fungitoxic inhibition, whereas carvacrol and citral showed fungistatic activity.

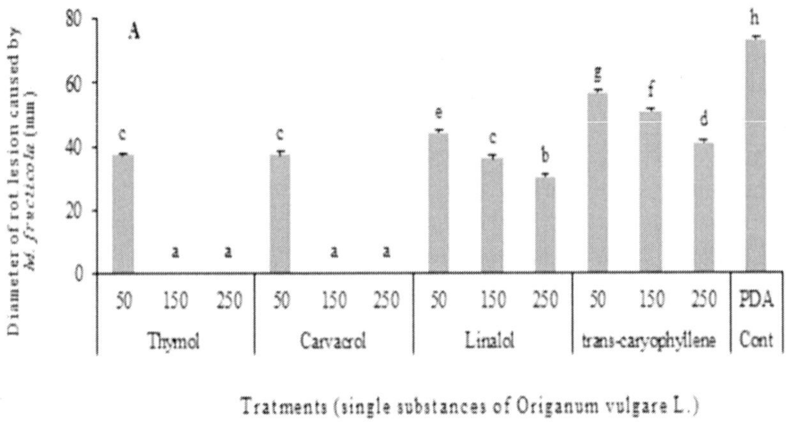

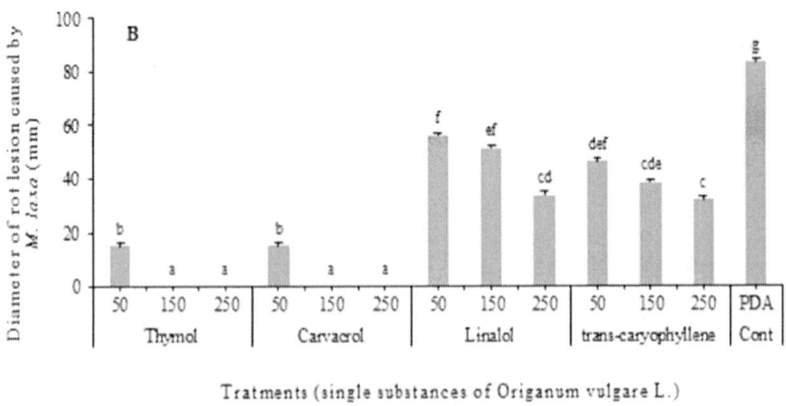

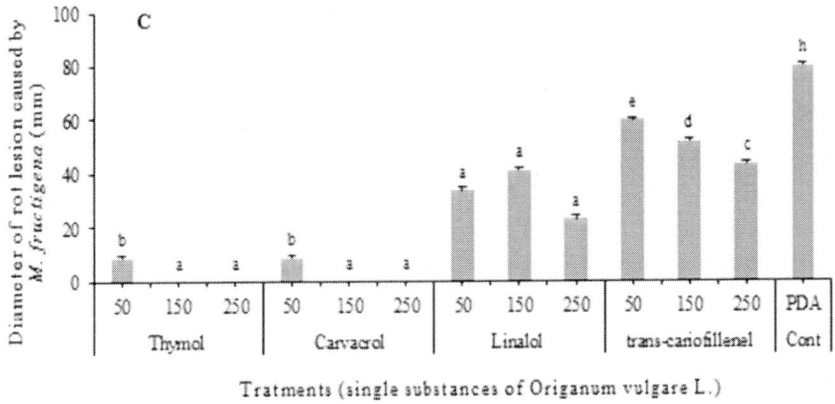

Where: 50, 150 and 250 are the concentrations of each single substance in ppm; PDA is potato dextrose agar.

FIGURE 2. In vitro antifungal activity of the four single substances of O. vulgare oil against M. laxa, M. fructicola and M. fructigena. Bars with different letters indicate means values significantly different at P < 0.05 according to Tukey test. Data are expressed as mean of three replicates ± SE.

In vivo antifungal activity

The bioactive treatments which exhibited *in vitro* activity were selected for evaluation of their i*n vivo*activity against three *Monilinia* species causing brown rot of peach fruits, following the method of Hong et al. [43]. Tested peach fruits cv. "Springcrest", were not treated with either pre- or post-harvest chemical pesticides, and were superficially sterilized with 2 % sodium hypochlorite solution, then later washed with sterile distilled water, before they were finally, air dried and inoculated with the above mentioned three phytopathogenic fungi at room temperature. Each inoculum was performed by injuring the surface of the fruits with a sterile needle and then adding 10 μl of fungal suspension containing 10^6spore /ml. Liquid fungal cultures were prepared by adding two 3-mm-thick and 0.5-cm-diameter (4 days old) fungal discs to 150 ml of sterilized potato dextrose broth (PDB) medium. They were then incubated at 22 °C for 7–9 days. One day after inoculation, each fruit group was sprayed with an emulsion containing different concentrations of each single component at 150 or 500 ppm, dissolved in 0.2% Tween 20. The negative control composed of three groups of fruit sprayed only with sterile distilled water, whereas the positive control composed of three groups of fruits inoculated only with *Monilinia*isolates. The severity of symptoms induced by infection of the single *Monilinia* isolates was determined by measuring the diameter of brown rot lesions in mm after 3–5 days of incubation at room temperature (16–24 °C).

Carvacrol and thymol have shown a promising inhibition of the brown rot of peach fruits caused by *M. laxa*, *M. fructicola* and *M. fructigena in vivo* especially at a dose of 500 ppm (Figure 3,13). In particular, carvacrol showed the highest significant antifungal activity against *M. fructicola*.

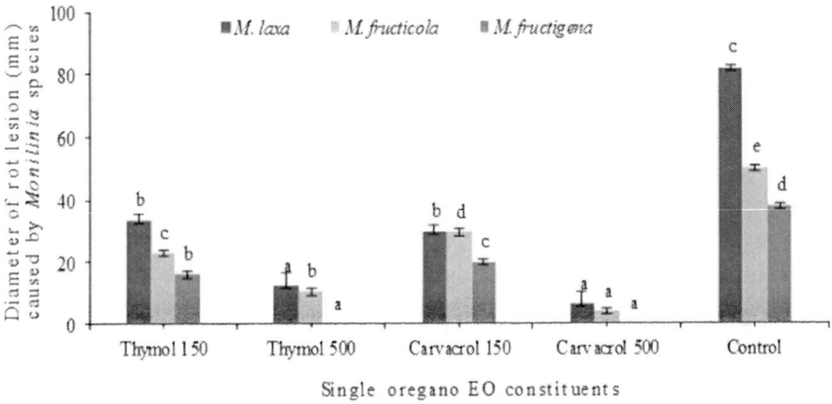

Where: 150 and 500 are the concentrations of each single substance in ppm.

FIGURE 3.In vivo antifungal activity of thymol and carvacrol against brown rot disease of peach fruit caused by M. laxa, M. fructicola and M. fructigena. Bars with different letters indicate means values significantly different at $P < 0.05$ according to Duncan test. Data are expressed as mean of three replicates ± SD.

4. CONCLUSIONS

Plant essential oils are one several promising environmentally friendly alternatives to conventional synthetic pesticides used to control several fungi and fungus-like organisms such as plant pathogens, food contaminants and decays. Most of the essential oils and their formulations mentioned in this review showed a high efficacy — either via direct contact or through incorporation into nutrient media — against different fungal pathogens at very low concentrations when used either *in vitro* or *in vivo*.

Among the essential oils used, oregano, thyme and vervain and their active constituents such as carvacrol, thymol, linalool, citral and their isomers, effectively inhibited mycelial growth and spore germination through fungistatic and/or fungicidal actions. The active ingredients of the essential oils and their isomers can be used effectively as seed and soil treatments — controlling most of the post-harvest decay fungi — and in plant and food protection.

REFERENCES

1. Buchholtz L. Antiseptika und Bakterien. Archivexperim. Pathologie Pharmakologie. 1875;4:1 82.
2. Ashhurst AP. The centenary of Lister (1827–1927). A tale of sepsis and anti-sepsis. Ann. Med. History. 1927;9:205–211.
3. Lopez-Reyes JG, Spadaro D, Gullino ML, Garibaldi A. Efficacy of plant essential oils on post harvest control of rot caused by fungi on four cultivars of apples in vivo. Flavour Fragr. 2010;25:171–177.
4. Hidalgo PJ, Ubera JL, Santos JA, LaFont F, Castellanos C, Palomino A, Roman M. Essential oils in Culamintha sylvatica Bromf. ssp. ascendens(Jordan) P.W. Ball wild and cultivated productions and antifungal activity. J. Essent. Oil Res. 2002;14:68-71.
5. Kordali S, Kotan R, Mavi A, Cakir A, Ala A., Yildirim A. Determination of the chemical composition and antioxidant activity of the essential oil ofArtemisia dracunculus and of the antifungal and antibacterial activities of Turkish Artemisia absinthium, A. dracunculus, Artemisia santonicum, and Artemisia spicigera essential oils. J. Agric. Food Chem. 2005;53:9452-9458.
6. Viuda-Martos M, Ruiz-Navajas Y, Fernandez-Lopez J, Perez-Alvarez J. Antifungal activity of lemon (Citrus lemon L.), mandarin (Citrus reticulataL.), grapefruit (Citrus paradise L.) and orange (Citrus sinensis L.) essential oils. Food Control. 2008;19:1130-1138.
7. Ellis SD, Boehm MJ., Mitchell TK. Fungal and Fungal-like Diseases of Plants. Fact sheet, Agriculture and Natural Resources. 2008; Ohio State University.
8. Lopez-Reyes JG, Spadaro D, Prelle A, Garibaldi A, Gullino ML. Efficacy of plant essential oils on postharvest control of rots caused by fungi on different stone fruits in vivo. J. Food Protect. 2013;76:631–639.
9. Liu WT, Chu CL. Thymol and Acetic acid vapors reduce postharvest brown rot of apricots and plums. HortScience. 2002;37:151–156.
10. Camele I, De Feo V, Altieri L, Mancini E, De Martino L, Rana GL. An attempt of post-harvest orange fruit rot control using essential oils from Mediterranean plants. J. Med. Food. 2010;13:1515–1523.
11. Mancini E, Camele I, Elshafie HS, De Martino L, Pellegrino C, Grulova D, Vincenzo De Feo. Chemical composition and biological activity of the essential oil of Origanum vulgare ssp. hirtum from different areas in the southern Apennines (Italy). Chem. & Biodiver. 2014;11(4):639-651.
12. Elshafie HS, Mancini E, Camele I, Martino LD, De Feo V. In vivo antifungal activity of two essential oils from Mediterranean plants against post-harvest brown rot disease of peach fruit. Indus Crop Prod. 2015a;66:11–15.

13. Elshafie HS, Mancini E, De Martino L, Pellegrino C, De Feo V, Camele I. Antifungal activity of some constituents of Origanum Vulgare L. essential oil against post-harvest disease of peach fruit. J. Med. Food. 2015b;18(8):929-934.
14. Handbook of essential oils, Science, technology, and applications. Edited by: K. Hüsnü Can Baser and Gerhard Buchbauer. CRC Press. Taylor & Francis Group.
15. Adebayo O, Dang T, Bélanger A, Khanizadeh S. Antifungal Studies of Selected Essential Oils and a Commercial Formulation against Botrytis Cinerea. J. Food Res. 2013;2(1):217-226.
16. Tsao R, Zhou T. Antifungal activity of monoterpenoids against post-harvest pathogens Botrytis cinerea and Monilinia fructicola. J. Essent. Oils Res. 2000; 12:113–121.
17. Bajpai VK, Kang SC. In Vitro and In Vivo Inhibition of Plant Pathogenic Fungi by Essential Oil and Extracts of Magnolia liliflora Desr. J. Agr. Sci. Tech. 2012;14:845-856.
18. Wilson CL, Solar JM, Ghaouth EL, Wisniewski ME. Rapid evaluation of plant extracts and essential oils for antifungal activity against Botrytis cinerea. Plant Dis. 1997;81:204–210.
19. Walter M, Jaspers MV, Frampton CM, Eade K, Stewart A. Control of Botrytis cinerea in grape using thyme oil. Australas. Plant Pathol. 2001;30:21–25.
20. Bouchra C, Achouri M, Idrissi LM, Hmamouchi M. Chemical composition and antifungal activity of essential oils of seven Moroccan Labiataeagainst Botrytris cinerea Pers. Fr. J. Ethnopharmacol. 2003;89:165–169.
21. Dubey NK, Kishore N, Shing SN. Antifungal properties of the volatile fraction of Melaleuca leucodendron. Trop. Agric. 1988;60:227–228.
22. Rahman MU, Gul S. Mycotoxic Effects of Thymus Serpyllum Oil on the Asexual Reproduction of Aspergillus Species. J. Essent. Oils Res. 2003;15:166–171.
23. Shashi SK, Patra M, Shukla AC, Dikshit A. Use of essential oil as botanical–pesticides against post-harvest spoilage in Malus pumilo fruits. Biocontrol. 2003;48:223–232.
24. Vaughn FS, Spencer FG. Antifungal activity of natural compounds against thiabendazole-resistant Fusarium sambucinum strains. J. Agric. Food Chem. 1994;42:200–203.
25. Daferera DJ, Ziogas BN, Moschos G. The effectiveness of plant essential oils on the growth of Botrytis cinerea, Fusarium sp. and Clavibacter michiganesis subsp. michiganensis. Crop Prot. 2003;22:39–44.
26. Arras G, Aggabio M, Piga A, D'Hallewin G. Fungicide effect of volatile compounds of Thymus capitatus. Acta Hort. 1995; 379:593–600.
27. Thompson DP. Effect of essential oils on spore germination of Rhizopus stolonifer and Aspergillus species. Mycologia. 1993;78:482–485.
28. Reddy MVB, Angers P, Gosselin A, Arul J. Characterization and use of essential oil from Thymus vulgaris against Botrytis cinerea and Rhizopus stolonifer in strawberry fruits. Phytochemistry. 1998;47:1515–1520.
29. Palhano FL, Vilches TT, Santos RB, Orlando MT, Ventura AJ, Fernandes PM. Inactivation of Colletorichum gloeosporioides spores by high hydrostatic pressure combined with citral or lemongrass essential oils. Int. J. Food Microbiol. 2004;95:61–66.
30. Lazar-Baker EE, Hetherington SD, Ku VV, Newman SM. Evaluation of commercial essential oil samples on the growth of post-harvest pathogen Monilinia fructicola (G. Winter) Honey. Lett. in Appl. Microbiol. 2011;52:227–232.
31. Amiri A, Dugas R, Pichot AL, Bompeix G. In vitro and in vivo activity of eugenol oil (Eugenia caryophylata) against four important post-harvest apple pathogens. Int. J. Food Microbiol. 2008;126(1-2):13-9.

32. Neri F, Mari M, Brigati S, Bertolini P. Control of Neofabrea alba by plant volatile compounds and hot water. Post-harvest Biol. Technol. 2009;51:425-430.
33. Smid EJ, Dewitte Y, Gorris LG. Secondary plant metabolites as control agents of post-harvest Penicillium rot on tulip bulbs. Post-harvest Biol. Technol. 1995;6:303-312.
34. Dixit SN, Chandra H, Tiwari R, Dixit V. Development of a botanical fungicide against blue mould of mandarins. J. Stored Prod. Res. 1995;31:165-172.
35. Arrebola E, Sivakumar D, Bacigalupo R, Korsten L. Combined application of antagonist Bacillus amyloliquefaciens and essential oils for the control of peach post-harvest diseases. Crop Prot. 2009;29(4):369–377
36. Feng W, Zheng XD, Chen JP, Yang Y. Combination of cassia oil with magnesium sulphate for control of post-harvest storage rots of cherry tomatoes. Crop Protect. 2008;27:112-117.
37. Dutta PK, Tripathi GK, Mehrotra J. Perspectives for chitosan based antimicrobial films in food applications. Food Chem. 2009;114:1173-1182.
38. Tsao R, Zhou T. Antifungal activity of monoterpenoids against post-harvest pathogens Botrytis cinerea and Monilinia fructicola. J. Essent. Oils Res. 2000;12:113–121.
39. Soylu S, Yigitbas H, Soylu EM, Kurt S. Antifungal effects of essential oils from oregano and fennel on Sclerotinia sclerotiorum. J. Appl. Microbiol. 2007;103:1021-1030.
40. Soylu EM, Kurt S, Soylu S. In vitro and in vivo antifungal activities of the essential oils of various plants against tomato grey mould disease agent Botrytis cinerea. International J. Food Microbiol. 2010;143(3):183-189.
41. Zygadlo JA, Guzman CA, Grosso NR. Antifungal properties of the leaf oils of Tagetes minuta L. and T. filifolia lag. J. Essent. Oil Res. 1994;6:617–621.
42. Camele I, Altieri L, De Martino L, De Feo V, Mancini E, Rana GL. In vitro Control of Post-Harvest Fruit Rot Fungi by Some Plant Essential Oil Components. Int. J. Mol. Sci. 2012;13:2290-2300.
43. Hong C, Michailides TJ, Holtz BA. Effects of wounding, inoculum density, and biological control agents on post-harvest brown rot of stone fruits. Plant Dis. 1998;82:1210–1216.

CHAPTER 12

Postharvesting Techniques and Maintenance of Seed Quality

Zubaida Yousaf*, Nadia Saleh, Asba Ramazan and Arusa Aftab

[1]*Department of Botany, Lahore College for Women University, Lahore, Pakistan*

ABSTRACT

Healthy seeds and grains are the demanding enterprise of the recent era for the production of high yield in the next season. The seeds must be stored for the maintenance of high-yielding crop. During storage, major losses of seeds are caused by various biological and nonbiological factors. There is a need to examine reasonable factors of these crop losses, which ultimately affect the market value and quality of the seed. The quality of seeds can be maintained by using careful postharvest handling techniques. There is need to establish the well-suited methods to assess the losses during the process and to use the best technique to minimize the loss and to ensure the quality and safety of the crop. The target is to achieve the high-quality seeds of the national and international standards that could meet the demand of the supplier. This chapter emphasizes on the aspects and postharvest techniques that are used to maintain seed quality. A comprehensive review of the better, economical, convenient, and productive methods is provided, focused on the needs of developing countries but also with relevance in more industrialized countries.

Keywords: crop maintenance, handling techniques, storage losses, postharvest

1. INTRODUCTION

Seed industry survives for the sake of better quality of seeds and postharvest storage techniques. It is a seed quality program that seeks the high achievement and gives the quality assurance. Seed quality is one of the most demanding enterprises of present days. The majority of the seeds especially cereals and legumes by small-scale farmers are produced and stored on the farms. The major problem is the damage due to biological factors such as molds, and insects can be alleviated by implicating effective storage techniques [1–4]. It is the foremost priority of the farmers to prevent the crop losses and storage of seeds/grains. Farmers often purchase the new seeds/grains from the market to produce the next crop for higher yield due to the risk of crop loss. There is a need to develop the effective strategies for seed storage that can give the confirmation of better crop yield and reduce the risks of storage losses [5].

1.1. Valuable Seeds: Achieving Seed Quality

The value of the seeds allows the farmers to produce high-yielding crops from good quality healthy seeds. Seed quality is the measure of the potential to produce desired quality, healthy, and high-yielding crops at low planting rates [5]. Seed quality can neither be obtained automatically nor by a permanent process. There is nature's pressure against the quality of the seeds. Efforts are being made to produce best quality seeds for the farmers. Any handling or production stage can reduce the quality of the seed. To alleviate these losses, careful technical management is required for all seed operations [6].

1.2. Principles of Seed Storage

Seeds when stored in natural environment or ambient temperature respond readily to temperature, available oxygen, and relative humidity. Metabolic activities, age, and longevity of seeds can be manipulating by controlling the humidity temperature and oxygen [7]. Reduction of seed moisture content up to an appropriate limit is prerequisite for storage as seed could be damaged because of desiccation. Seeds can be stored for a longer period due to lower level of humidity. According to the thumb rule, the life of the seed doubles by decreasing the moisture content to 1% in case if seed moisture content is between 5 and 14%. Higher moisture content is more affected by higher temperature so seeds need to be stored in cool location. The life of the seed doubles by decreasing the temperature to 5°C and is applicable between 0 and 50°C [8].

Oxygen level can be controlled by hermetic storage in a sealed container which reduces the physiological aging of the grains as well as reduces the physical damage due to insects and fungal growth [8].

2. POSTHARVEST TECHNIQUES OF SEED STORAGE

2.1. Drying

Physiological maturity attained by the cereal and legumes at moisture content between 35 and 45% is crop dependent. Temperature affects the storage of seeds at moisture content between 10 and 14%. For high-quality yield of crops, timely harvesting and drying are necessary. Biologically active seeds deteriorate readily under most of the circumstances due to fungal contamination and attack of insects and other pests. The main purpose of drying is to reduce the respiration in seeds [6, 9]. The process also impedes qualitative damage due to fungi and other insect pests. Drying can itself affect the quality of the seeds. Extensive drying under very high temperature can damage the seed. In summer season, simple drying methods are used through exposure to sun and adequate wind. The alternative drying methods have been devised for high-yielding varieties and improved farming practices and irrigation to deal with increased production or harvesting in wet season in multi-cropping [10, 11].

2.2. Sun Drying

Sun drying of seeds is the drying method in the tropical developing countries. The method is employed when crop is ready for the harvest. Some seeds like maize can be sun dried although during drying, crops get sensitive to insect infestation, rodent or birds' attack, and mold damage. A threshed seed drying by spreading on sheets or tray is a common phenomenon but has a risk of soil or stone contamination. For example, paddy at rice mills is at large quantity [6]. Rice is dried on especially build drying floor that allow easy run off of rain water. The seeds are dispersed in thin layer and turned at regular intervals to facilitate drying and covered at night with the help of sheets. There are some disadvantages of the process that temperature is an uncontrollable factor. In paddy rice, high temperature can cause stress or cracks in the seeds which lead to high level of damage during milling. Yield can be contaminated by dust, atmospheric contamination or insect infestation, or human or animal disturbance [6, 9].

2.3. Solar Drying

It is the modification of sun drying in which sun rays are collected in an especially designed unit for air removal in adequate ventilation system. The unit has 20–30° higher temperature than open drying, and less time is consumed in the process [12]. In solar dryers, solar collector is used to heat air which then allowed to pass to the seed beds. It comprises two basic designs: natural convection dryers use thermal gradients and forced convection dryers force the air through solar collectors and seed layers. These dryers are suitable for farm use. The former design of the Asian Institute of Technology in Bangkok has been used as blueprint for several convection dryers and comprises a drying bin, a solar chimney, and a solar collector [13]. The solar collector consists of black polythene sheet or layer of burnt paddy husk; it is covered with clear polythene sheet. Perforated platform presents in drying bin. The disadvantages of the process are as follows: high structural profile, stability problem in windy condition, and the need of replacement of polythene sheets at regular intervals [6].

2.3.1. Mechanical Dryers

The same principle of drying is used by mechanical dryers as forced convection solar dryer; the dryer forced the air through the seed bed and the air is heated with the help of a flat plate instead of conventional means. In modern automated storage system, drying takes place at one of the two points either in prestorage dryers (prior to loading seeds in freestanding loading) or in store dryer (after loading in final storage compartment) [9]. In prestorage dryers, ambient air is used in continuous flow dryers, and heat is generated by thermostatically controlled furnace which is powered by electricity, diesel, or gas. Heat may be supplied by direct or indirect way. Indirect way is preferred due to the separate outlet for the combustion product not through seeds. In the batch dryers, seeds are fed into properly defined batches, whereas in continuous flow dryers, the grains are flowed into the system and recovered at desired moisture content [6].

2.3.2. Tray Dryers

Tray dryers are batch dryers of flat beds. The seeds are dispersed on the mesh tray at the depth of 600–700 mm, and warm dried air is passed through seeds to sufficiently dry them [6, 14].

2.4. Radial Drying Bin

Radial drying bin comprises two vertical metal mesh cylinders, one inside the other. Seeds are loaded between these two cylinders, and air is blown to the inner cylinder and passed from the inner to the outer mesh cylinder. By reversing air through seeds, air can be sucked from the central cylinder. However, there is the risk of overdrying of seeds in the inner cylinder which are in direct contact with the hot air. Air is wetter and cooler at the leaving side toward outside [6].

2.5. Continuous Flow Dryers

Moisture content of the seeds can be removed by sucking or blowing hot air by top-to-bottom passage through the system. Bin is present at the top of the drying section with cooling system at the base. Seed beds can be horizontal, vertical, or inclined. Seeds are moved by conveyors, scrapers, vibration, or gravity. The degree of drying depends upon the speed, size, and rate of flow of outlet conveyor of the dryer. Continuous flow dryer is varied by relative direction of air stream and seed flow [15]. Several continuous flow dry techniques are described below.

2.5.1. Cross Flow

The seed passes through the two perforated sheets downward to the column by allowing horizontal passage of air through the seeds. The advantage of the dryer is that the moisture gradient can be defined at any stage of drying seeds [6].

2.5.2. Counterflow

A round bin is used to unload seeds with the upward flow of air. Little evaporative cooling takes place when hottest air passes through the driest seeds.

2.5.3. Concurrent Flow

In concurrent flow, wettest seeds are exposed to the hottest air during passage of air through seed bed. High temperature improves the efficiency of the dryer and cools the seeds by moisture evaporation [6, 16–18].

2.5.4. Cross Flow

Cross-flow dryers have been used widely in recent years, but mixed-flow dryers have advantages over cross-flow dryers. Mixed flow dryers, combination of concurrent, counter and cross flow dryers has great advantage of efficient fuel consumption. But reduction in output because of uneven flow of seeds leading to uneven drying is one biggest hindrance in adaption of mixed flow drying [6].

2.5.5. Tower (Mixed Flow)

They consist of tall rectangular bins for storage, and triangular ducts are present along the width of the dryer at horizontal position. Half of the ducts are used for induction of warm air and removal of damped and cooled air is done by remaining ducts. It has multiple directions of seed flow and air flow [16–18].

2.5.6. Louvered Bed Dryer

The seeds are passed through different types of batches; the hot air is blown to the seeds to dry them. The dryers work on the principle of cross-flow dryer. The speech and depth of the drying beds depends upon the degree of drying. The two basic designs of the dryers are conveyor dryer and cascade dryer. The cascade dryers are gravity fed cross-flow dryers. The seed depth is controlled by roller dams and speed is controlled by output elevators. There can be incorporated changes to vary the length of the dryer. In conveyor dryer, air is blown to seeds through inclined louvered bed and the seed flow is controlled by variable speed, roller chain conveyor and heavy duty. These dryer can be one directional, two directional, or multidirectional; the variation in direction assists in removal of waste material and reduction in size of the dryer [6].

3. IN STORE DRYING

This is an alternative method of drying in which seeds are load into bulks floor storage or in bin and then they are dried in stores [19].

3.1. Bulk On Floor Storage

They consist of especially strengthened wall which can bear the weight of seeds. The seeds are loaded in uniform depth. At one side of the building, fan is present for the aeration purpose and the plenum chamber runs along the center of the store or walls with perforated lateral ducts, below or above the floor level under the bulk of seeds [6].

3.2. In Bin Drying

This type of drying comprises one or more bins for drying purpose and other bins for storage. The dryers reduce the chance of physical damage due to the lesser handling. The shallow layer of seed along the bins for drying reduces time consumption and makes the process safer. The semi-dried batches free the space for incoming seeds and consist of lateral ventilated system or ventilated floors about 0.5 m above the base [6, 20].

3.3. Bag Dryers

The drying in bags is difficult because there is not proper insurance of passage of air through the seeds. In sack platform dryers air is blown through the floor of air duct whereas in fan blowers heated air is blown from the floor apertures and sacks are placed on them. Larger bags are stacked in the center of tunnel in moisture extraction unit. Through the air ducts, hot air is blown with the help of a fan. One must be careful of correct dimension to avoid uneven drying. However this system is not appropriate for even drying of seeds because of short circuit over certain areas. [6].

4. STORAGE LOSSES

It is reported that a total of 30% crop losses take place due to harvest. However, this is the "worst case" figure to cite for the crops in priority area of development. Storage losses cannot be predicted before harvest. There can be many biological, climate, handling, harvesting, storage, and distribution and social and cultural factors which can cause crop losses. By careful handling, 50% postharvest and storage losses can be disseminated. There is no such method to predict the exact figures of such losses. Efforts have been made to determine the reliable baseline methods to figure out crop loss activities. A methodology has been made to estimate the standardized postharvest losses by crop activities [21]. Loss assessment and loss reduction programs have been prompted by the Food and Agricultural Organization (FAO) of the United Nations. The focus of these programs was to obviate the reduction of crop losses of staple foods. The methodology for assessment of seed loss during harvest was summarized [22], although there was no universal method to assess storage losses. Sampling procedures, varied handling and storage produce, and irregular movement of batches make the loss assessment phenomena non-generalized especially for perishable commodities. Care should be taken in designing a methodology with positive aims to make it acceptable, economical, and meaningful. The methods are varied for perishable commodities due to their different nature whereas relatively uniform for cereal seeds. To find out standard moisture content and dry matter, the weight loss of undamaged and damaged seeds can be compared. The purpose of storage is to avoid both biological and financial losses of yield. To obviate these losses, we must have known the root causes of these deprivations [9].

4.1. Loss and Damage

The term damage and loss can be confusing sometimes when used synonymously. Loss is defined as the measurable qualitative or quantitative decline in foodstuff. Superficial deprivation of commodities is defined as the damage, in which physical spoilage results in loss of commodity. Damaged commodity can either be used but loss is a permanent decline. [6, 7].

4.2. Categories of Storage Loss

Categories of qualitative or quantitative storage loss have economic impacts. Physical weight or volume loss which is considered as quantitative loss can be figured out readily. Quality losses can be measured by the simple judgment of commodity and comparing with standard quality items. Commodities can be rejected by changes in taste, texture, appearance, loss of nutritional value, and the presence of contaminants. The following categories can be listed to demonstrate the storage losses of crops [23].

The primary cause of losses directly affects the stored crops. They may be biological, chemical, or biochemical in nature.

4.3. Biological Losses

Biological factors responsible for crop deprivation are rodents, insects, birds and microbes (fungi and bacteria). Microbes grow on the crops and cause the weight loss, crop spoilage, and other defects which reduce the market demand of the yield. Development of infestation can make trouble if produce is stored for longer period. The damage or loss caused during storage in the cereal seeds by birds, rodents, and the condition can be harsher by microbial (fungi and bacteria) attack in the field. There will be quality loss, if the disease is superficial; it can be quantitative loss if infections penetrate into deeper layers of seeds. In case of superficial disease, it is possible to remove damaged area and use the rest of the portion [14, 24, 25]. *Chemical losses* involve loss of flavor, color, texture, and nutritional value due to pesticides and chemical reactions [26]. *Biochemical losses* can be softening, discoloration, and bad flavor due to enzyme-activated reactions. *Mechanical losses* are due to bruising, breakage, processing, and injury during handling or harvesting. *Physical losses* are related to climate condition such as low or high temperature, improper storage atmosphere, and high humidity. Physical factors can also mediate chemical and biochemical losses [7]. *Physiological losses* involve weight losses due to respiratory heat loss. Susceptibility of infection and damage by pathogen increases during wilting, senescence, ripening, and wilting.

Microbiological and biological factors are important in seed, whereas mechanical, physiological, and microbial factors cause losses in perishable crops. The factors that encourage primary crop losses are secondary losses by inappropriate handling of equipment, technology, and control. The factors are lack of harvesting equipment, skills, packaging, handling, adequate containers, appropriate transport, drying and storage conditions, proper processing technique, and adequate management [6, 9].

4.4. Weight Loss

Weight loss is not necessarily the sign of loss of crops. Reduction in moisture content can be a reason of weight loss. Shrinkage factor is a tool to recognize for commercial transactions. Moisture loss can lead to economic loss if it is not taken into account by grading for price control. Weight loss can be due to feeding of birds, insects, rodents, and microorganisms. Weight loss can be measured by comparing the weight before and after storage in the sack. There can be also an increase in moisture content due to water production in seed by insect infestation and can lead to an increase in weight. If insect infestation increase the moisture content of seed or insect may consume seed, leaving behind dust, weight loss cannot be detected easily [27]. To detect these losses, a useful mass of infested and noninfested seeds is ground into flour, and their weight is compared. A decrease in flour yield of infested mass than sound mass will be observed. Be aware of malpractices to compensate weight loss by adulteration of stones, earth, or sand and water. So there is a need to assess not only moisture changes but also the quantity of foreign matter present in the yield [23, 28].

4.5. Quality Loss

Quality is an important trait considered by the consumers, and local traders have different criteria for assessment according to circumstances. Biochemical factors (acidity, sugars, flavor, and smell) involve size, shape, and appearance. Quality loss can also be due to contamination and foreign matter (insect fragments, rodent hair, excreta, weed seeds, earth, glass, stones, and parts of plants) content. Contaminants that are difficult to remove comprise soluble excretion of insects, pesticides, oils, toxins produced from fungal infections, and pathogenic organisms spread by rodents. Loss potential will be increased by raising the standard rules by consumers [24, 29].

4.6. Nutritional Loss

The loss depends upon the qualitative and quantitative loss of nutritional value to the human population which affects the nutritional status of that population. This is mainly caused by feeding of pest on selective part of seed. *Plodia* and *Ephestia* feed selectively on the seed embryo and remove the vitamin and protein content. Many pests feed on bran of cereal seeds as a result reduced vitamin content. *Liposcelis* spp. selectively feeds on bran of rice and embryo (Pike, 1994). Weevil is an endosperm feeder and declines the carbohydrate content [6, 30].

4.7. Loss of Seed Viability

The loss of seed viability is linked with. The reasons of the losses may be temperature, excessive respiration, moisture content, infestation, light, and infestation-controlling methods. Insects that are selective to attack the embryo cause great losses in germination as compared to others. Seed loss can be detected by standard germination tests [31].

4.8. Commercial Losses

Commercial losses are either due to direct consequences (foregoing factors) or indirect consequences (cost of preventive action or equipment). There may be a loss of goodwill, monetary loss, and loss caused by legal action. Intercountry trade might be affected by commercial loss. Losses can be rapidly reduced by experience and knowledge. Postharvest losses not necessarily take place due to inappropriate storage. There may be biological, physical, or mechanical factors in deterioration of cereal seeds. There is a need to broaden the intervention techniques to deliver high-quality product from the field to the market. For example, after the Tanzania outbreak of pest in the maize crop, Somalia and Malawi refused to take the maize due to insect spreading [32].

4.9. Temperature-Dependent Injury

High-temperature solar radiation on exposure deteriorates the fresh commodities rapidly. By increase of temperature, respiration also increases; there should be proper ventilation and cooling of the crops to obviate the problem. In similar way, there can be crop injury due to low temperature between 0 and −2°C. However some crops of tropical and subtropical origin showed tolerance for chilling injuries at 12–14°C. Chilling injuries (skin pitting, discoloration, uneven or abnormal ripening, and sensitivity to rapid deterioration) are apparent when commodity is removed from the environment [7, 29].

5. ASSESSMENT OF LOSSES

5.1. Assessment of Seed Losses in Storage

The agents causing seed loss during storage are insects, rodents, and molds. In past this issue gained attention of many scientist but still there is a need to find some more effective technique to avoid seed loss due to insects. Insects can cause both qualitative and quantitative loss by boring or feeding on seeds; weight loss has got more attention [6].

5.2. Weight Losses Caused By Insects

The assessment is made through collection of samples of seed and various intervals after storage and comparing the samples to observe the changes in subsequent samples. Assessing quantity loss with subsequent samples at different intervals will be used to estimate storage losses at different occasions. Sample collection from each batch of seed and their quantity loss are estimated accordingly. Sample should be collected without disturbing the pattern of infestation in the bulk stores. Three samples must be collected when subsequent regular sampling is not possible: first, at time of storage; next, in the midway of storage period; and last, at the final few weeks of seed storage.

The pattern of used seed and quantity loss is noted [22, 27].

5.3. Methods of Estimation of Weight Loss

There are two methods used for estimation of weight loss by the insects, when subsequent sampling is possible: volumetric method and thousand grain mass (TGM) method.

In this case, when subsequent sampling is impossible, count and weight and converted percentage methods are used [6].

5.4. Volumetric Method

Volumetric method is commonly known as the standard volume weight (SVW) or bulk density method. This is used to measure the bulk density of clean sample by the use of equipment. SVW is figured out from the sample of the seeds at the beginning of the storage period, and losses are determined. This method strictly calculates the weight loss by grain boring of insects and moisture difference over a subsequent time period in standard volume container. In stand volumetric method to determine accurate ratio for moisture content and dry weight of seed, moisture can be taken as constant term and the crop as dry matter. However, volume and frictional characters can also be affected by changes in moisture content. There is direct relation between moisture content and volume of the sample therefore seed should be packed loosely. To make the moisture constant, it is necessary to calculate the standard volume of dry matter at different moisture contents. The process is time-consuming and needs care and equipped laboratory [33]. Another factor affect the volume of sample is weight of insecticidal dust, adheres to the seed surface increase the volume of seed and frictional character. Sieving can be a useful phenomenon to remove dust. However, volumetric phenomena are less helpful due to overassessment of losses [6, 34].

5.5. Thousand Grain Mass Method

This method differs from volumetric method in comparison with fixed number of seeds instead of fixed volume. This means that weight of seeds is multiplied by thousand and corrected by a dry matter. It is calculated by weighing and counting the seeds in a given sample. Standard reading is made by taking measurement at the beginning of storage of seeds, and then subsequent readings are compared with baseline reading [35].

5.6. Count and Weight Method

The method, sometimes called gravimetric method, is applied where the baseline readings of seed storage are not obtained in the start of the season. A sample of 1000 seeds and minimal medium are used in this estimation. Weight and number of seeds in each sample fraction is determined after isolation of damaged seeds. The results are then calculated by putting the values in the following equation:

$$\frac{(U \times N_d) - (D \times N_a)}{U\,(N_d + N_a)} \times 100 = \%\ \text{weight loss} \tag{1}$$

where U = weight of undamaged seeds, D = weight of damaged seeds, N_a = number of undamaged seeds, and N_d = number of damaged seeds.

In this method, moisture content of the separate fraction is unnecessary for a single sample, and differences of assumption are likely small. The method does consider hidden infestation in damaged category, and it considers seed infestation that is random by the insect which is not necessarily to be true [36]. The method can cause misleading results for low level of infestation and multiple infestations in large seeds. The method is useful in field level for quick estimation at extremes. To overcome the biased estimation, several refinements have been made. For example, different-sized seeds have different moisture contents and can have hidden infestation. These seeds are graded according to size and categories before counting and weighing [22]; severely attacked grains are separated, and reading of hidden grains is taken after emergence of infestation [37, 38]. There is another way to know the hidden infestation by dissecting seeds, but the method is tedious and has a chance of change in moisture content of seeds, for the sake of calculations that are needed to be made on dry matter.

5.7. A Count and Weight Method Modification

In this modification of count and weight method, highly infected grains have also been taken into account. To avoid the underestimation of losses, the number of infected seeds is counted for sum of the seeds in whole sample [39]. To assess the weight loss, the following equation is used:

$$\frac{\left[\mathrm{TND}(\mathrm{D}+\mathrm{U})\right]+\mathrm{FW}(\mathrm{N_dU}-\mathrm{N_aD})}{\left[\mathrm{TND}(\mathrm{D}+\mathrm{U})\mathrm{U}\right]+\left[\mathrm{FW}(\mathrm{N_d}+\mathrm{N_a})\mathrm{U}\right]}\times 100 \tag{2}$$

where U is the weight of undamaged seeds, D is the weight of damaged seeds, N_a is the number of undamaged seeds, and N_d is the number of damaged seeds.

The weight loss is calculated by averaging weight loss of two subsamples.

The percentage weight loss is calculated by deviation in Eq. (3):

$$\frac{\text{Undamaged weight (UW)} - \text{final weight (FW)}}{\text{Undamaged weight (UW)}}\times 100 \tag{3}$$

The undamaged weight of the sample (in the absence of missing seeds) is figured out by applying assumption in count and weight method. The average unit weight of undamaged grains in the original sample will be equal to the average unit weight in the remaining undamaged grain subsamples. The undamaged weight of the whole original grain sample is figured out as the product of the total number of seed estimated in the original sample and unit weight of undamaged seeds in the subsamples. The total number of seeds is the sum of the damaged and undamaged seeds. The final step will be the ratio of the final sample weight to the average unit weight of the seeds. The method focuses on missing seeds of the sample instead of underestimating lost samples as in conventional count and weight method [39].

5.8. The Converted Percentage Damage Method

The method is suitable for quick estimation of loss by boring insects without using equipment such as during a rapid field appraisal. Weight loss is assessed by taking percentage damage seed as sample. A study has been made to find out the relationship of damage and weight loss [23]. A conversion factor can be used subsequently to estimate the weight loss of same type to seeds in different samples. The conversion factor is calculated by the following formula by using same count and weight method and so faces same type of errors:

$$\frac{\%\text{ damage seeds}}{\%\text{ weight loss}} = \text{Conversion factor} \tag{4}$$

To overcome the error of count and weight method, 10% or more damaged sample seeds are used. The sample size should not be smaller than 500 seeds taken as the percentage of the total seeds present. Rapid loss technique is a rapid weight loss assessment in the field based on the damage and weight loss relationship. Standard graphs are used in this technique to relate percentage damage and weight loss of samples under study [6, 21]. For preparing reference graph, ten samples of 500 seeds are required in the laboratory. Weight loss is estimated by count and weight method, and percentage damage is calculated. The graph is plotted between percentage weight loss and percentage damage. In the field, clean sample of 200 seeds is gathered, percentage damage seed is determined, and loss figure of reference to graph is read off. For a sample of 200 seeds of Bambara or cowpeas, the technique depicts accuracy on ±7%. The method is better than percentage damage method and allows good assessment of weight loss by taking many individual samples in short time period at the field. Visual scales are rapid loss assessment methods. The abovementioned techniques are time-consuming and need appropriate equipment and well-trained technicians. Visual scale has benefit in assessing damage in fields, is ideal for field use and rapidity, and is required only for reference scales and operator unbiased. This technique enables wider coverage in small time and small sample error. On-site loss assessment avoids extra spoilage and anomalous results by double checking on the site before leaving. The visual scale should be calibrated according to the objective of the study, weight loss, and farmer perception of value [6, 21].

5.9. Losses By Vertebra Pests

Vertebra pests such as rodents and birds remove the whole seed from the sample so it is impossible to estimate loss by vertebra pests. The loss can be assessed by comparing the reference percentage of seed loss and average seed weight [40]. Population studies and feeding trials are used to calculate the losses by pests and rodents, but these are often with small accuracy in comparison to efforts expanded [41]. Pests used stored grains only as their part of diet; feeding trial can overestimate the loss of stored seeds. The quantity of seed loss by rodents is questionable. Loss of crops due to rodents goes behind when compared with loss of storage container, personal property, buildings, and potential health risks.

5.10. Weight Loss By Molds

Seeds infected by molds will cause weight loss; the weight loss can be assessed by the same method that is used to calculate weight loss by insect. The moldy seed weight loss increased due to the absorption of moisture so the loss by mold can be compensated. The method is not that good to assess the real loss of seeds, and the seeds may be misled as undamaged seeds due to no apparent indication of infestation on surface. To estimate the weight loss due to mold, damaged seeds are separated from undamaged seeds, and then moldy seeds are discriminated from damaged seeds. The weight loss due to mold will be equal to the weight of mold [21, 27].

5.11. Total Loss In A Season

The above losses are the seed losses at a given point in a time. The picture can be misleading; there must be a relation in the pattern of seed used in a season. In an undisturbed stored crop, if the sample loss is 10% throughout the storage seasons, the total loss will be due to insects. The seeds will loss at different quantities at different intervals of time due to insect exposure at different time lengths throughout the season [27]. The percentage loss of seeds increases gradually with time due to increase in insect infestation. After permitting changes in moisture content, the loss in seeds can be calculated by measuring the weight of seeds at the time when the seeds are still in the store and when taken out from the store. The loss other than insect damage can be obtained by subtracting the loss caused by other insects. The overall losses of the seed after store are considerably lower than the suggested value. Many loss assessment protocols at commercial and farm level have been reported [22]. The key to get the best assessment result is to draw an acceptable methodology for each commodity.

There were small figures for losses in commercial operations and none for cooperative level storage. The condition reflects the buying and selling of seeds in the developing world in short time period. This gives the picture of private sector (market liberation and parastatal marketing) involvement, but there is few information about storage loss. Entrepreneurs could store large quantity of seed for longer time period. However private sector has increased this level for farm storage. Many efforts, time and money was invested to measure the storage losses in farm storage, but the effort was not fruitful like the earlier projects. Moreover, the study should be undertaken with the postharvest sector as a whole, and precise measurements should be avoided. Social survey can be helpful to discover the farmers' problem for loss assessment and appropriate measurement techniques [42].

6. HARVESTING AND MATURITY INDICES

Harvesting and improper handling of commodities can cause bruises and injuries which directly affect their market value and make them unattractive. Injuries give the open space for microbial attack, that cause rotting, respiration increase, and shortage of storage life. Improper harvesting can cause crop loss and severe damage in seeds [9, 43].

6.1. Harvesting and Handling

Harvesting is the first step of postharvest and is the last step of crop production. The method and condition of the harvest affect the further handling, processing, and storage of crops. Premature harvest causes loss of quality of seeds, and due to high water content, they will deteriorate in the store. Overmature crop harvest causes biological and physical losses of crops by consistent wetting and drying of crops [9]. Damp seeds need to be threshed quickly and dried after harvest. Different parts of plant are harvested by different harvesting methods; in case of forage, the whole plant is trimmed off; in case of cereal seeds, partial or part of the plant is threshed and cleaned; and straw or chaff is removed for further processing. Small scale produces performed threshing and harvesting by threshing combines harvesters (equipped through community groups) while in developing countries threshing and harvesting carried out by hand unlikely cause damage or deterioration of crops in store. Large-scale commercial producers use mechanical harvester equipment; their use is limited due to the production of cash crops. Post-harvest risk of crop damage in storage is reduces by manual harvesting Small scale produces performed threshing and harvesting by threshing combines [9, 21]. Conventionally, seeds are beaten with stick or against hard surface (wooden bar, log of wood, stone, and wooden metal or tub) for the sake of threshing. The methods can cause damage or cracks on the seeds, while seeds that are trodden under foot will be a less damaging method for the seeds. Grain heads or ears of sorghum, millet, or wheat are commonly beaten with sticks. However, manual harvesting is laborious, and economical process has the risk of physical damage. Maize cobs are beaten with sticks or shelled by hand, which result in high-level damage. Mechanized threshers are made to reduce the damage of seeds; the models are highly sophisticated.

Seed harvesting is carried out by a combination of various steps of threshing, cleaning, or combine harvester. Large-scale harvesting is carried out by mechanical equipment that is especially designed to harvest cereal seeds [6].

6.2. Storage Structures For Seeds

Seeds/grains are durable crops that usually require simple systems in their storage.

6.3. Farm-Level Storage

For protective outdoor or indoor storage, the seeds must be guarded for physical damage such as high temperature, adverse weather, snow and rain, and biological factors which include microorganisms, birds, rodent, mites, and insects. The process of farm storage is used by various countries to store major part of seeds [44, 45]. The storage structure varies in capacity and has range of 100 kg to few metric tons. Modifications in locally prepared storage structure could be done according to climatic conditions. There are some traditional storage structure. The bins are commonly made of ferro-cement, plywood, metal, and high-density and high-molecular-weight polyethylene. Plywood is the most suitable storage structure and underground structures of different shapes and sizes, which work on the principle of hermetic storage. This builds the higher level of carbon dioxide and low oxygen level which is lethal to pest and microbial attack during seed storage [46]. Traditional methods are cheaper, but they are not effective against pest and microbial attack. Silos or metal bins are also used at farm level to store seeds.

6.4. Bagged Storage

In developing countries, seeds are stored in gunny or woven polypropylene bags in traditional warehouses, whereas silos for bulk storage, seeds elevator, and flat storage structure are used in developed countries [45, 47]. Bag storage is a laborious and costly process and has greater chance of seed spillage and biological losses. There may be water seepage and humidity problem due to inappropriate flooring of warehouses. Bags do not need any fumigation facilities or aeration system. In developing countries, the system will be uneconomical due to small farm size and cheaper manual labor.

6.5. Bulk Storage

There are two types of bulk storage of seeds; it may be vertical (silos or bins) or horizontal (on floor stores). Horizontal stores comprise especially constructed floors of warehouses containing proper ventilation on the floor and walls that are strengthened to withstand the weight of the seeds. Bins and silos are especially designed stores, either round or square, in grouped or freestanding form, and incorporating unloading and loading usually of aeration systems. The alternative forms of bulk storage are belowground or partially belowground storage or enameled, sealed silos, for the storage of high moisture content seeds. The process is appropriate to bulk handling or storage of seeds [48, 49].

6.6. Hermetic Storage

The traditional methods of storage in the natural build of oxygen and lower level of oxygen protect the seeds from biological damage. This traditional storage method is not effective for the seeds that have lesser moisture content, and infestation is less than insects per kilograms of seeds. The controlled atmosphere treatment and fumigation have to be supplemented in hermetic storage [50].

6.7. Outdoor Storage

This is the temporary measure of storage in case of lack of permanent storage. The godowns and silos are built on plinth, and the stacks of seeds are covered with polyethylene covers. The stacks must be properly aerated in a week by raising the cover to the seventh or eighth layer. The cover and plinth (CAP) technique is used commonly for wheat and paddy. However, there is a chance of damage of cover by wind or rain, and effective fumigation cannot be achieved [44].

6.8. Storage Practices For Quality Seeds

Quality of seeds can be maintained by especially designing the small stores to silo or warehouses which play a protective role against adverse temperature conditions, ground water, rain water, pests, and thefts. The structure and contents of the store should be managed [9].

7. MOISTURE CONTROL

A water disposal system and well-designed roof are required (overhung or gutter) to hinder the water running to the store. Drains move the water away from the stores. In large warehouses, side by side joining of stores should be avoided to prevent the water run into the store. Raised floor and proper drainage system protect against the risk of the flood, and water-proof floors and walls guard against ground water. There must be proper ventilation system to control the humidity inside the storage structure [27].

7.1. Protection Against Temperature

It is difficult to control temperature in the storage structures; it needs special design elements to control temperature. Temperature can be ascertained by the use of controlled ventilation. In the cold night, temperature can be manipulated by insulated stores. Stores can be built in east-west direction, and using shiny material outside the stores can be an effective strategy to control heat. Thick wall and wide roofs (for shade) can further reduce the heat of the storage structure. Heater and refrigerators can be installed to modify temperature in stores; the equipment will perform well in case of insulated stores. However in these stores, degree of insulation depends upon atmospheric condition outside of it. [6, 51].

8. MODIFIED ATMOSPHERE STORAGE (MAS) AND CONTROLLED ATMOSPHERE STORAGE (CAS)

In modified atmospheric condition, the gases are added or removed from surrounding to make a controlled atmospheric composition around seeds, which differs from air (78.08% nitrogen, 20.95% oxygen, and 0.03% carbon dioxide). This includes the reduction of oxygen and elevation of carbon dioxide concentration. CAS and MAS are different in level of control although CAS is more exact. The process is used to facilitate whole-store fumigation [52, 53].

8.1. Protection Against Theft

Security guard or fence may be needed to guard the store. Lights are installed to illuminate the stores [6].

8.2. Protection From Birds and Rodents

Bird and rodent contact can be hindered by avoiding the free spaces for the access to the seeds. Rodent guards are commonly tight-fitting shelves, or conventional methods could be tying of thorn bushes to the stores as barrier. Walls of the stores should be smooth, and roof should be rodent proof to avoid their contact inside the storage structure [54].

8.3. Protection Against Insect

The small size insects can easily enters the storage buildings therefore it is difficult to make insect-proof walls and roofs because they can enter from even small cracks. For protection against insects, smooth walls, controlled ventilation, cleaned stores, floor without cracks, and curved floor wall can be effective. For protection against termites, termite-resistant material should be used (concrete, stones, fire bricks, and termite-resistant wood) [55].

9. TRANSPORT

The transport of commodities from field to storage houses can produce some injury, which can later cause deterioration of produce. This main focus is to keep the produce dry and moisture-free. The risk of insect infestation of seeds from the contaminated container has insect residual. Mechanical injury is produced from vibration during transport, poor condition of vehicle and roads, poor driving, insecure container stacking, the use of unsuitable container, and careless handling. Overheating is produced from the sun or vehicle engine that results in moisture loss of produce, and this leads to natural breakdown and decay [6].

10. QUALITY AND SAFETY

The quality of a product defines the class, degree, excellence, or superiority of the crop. The quality of the durable commodity is the combination of characteristics, properties, and attributes that give the commodity value as the food or a source of next crop yield. Foreign material or high moisture content can affect the marketing quality of the crop. High moisture may encourage the shrinkage or biological and biochemical damage in seeds. Low moisture in pulses and paddy rice can break or damage the seeds. The discolored and broken seeds lower the marketing quality and increase the chance to insects and microorganism attack [14, 27]. The main objective of a farmer is to produce apparently good product with few visual defects, and product must score high on yield, disease resistance, ease of harvest, and shipping quality and achieve the national and international quality standards. For the buyer or consumer, the appearance is more important; they are eagerly concerned in healthy seed and lifelong storage. The product should be assured safe for the distribution to suppliers and market [7].

11. POSTHARVEST QUALITY ASSESSMENT TECHNIQUES

There are two methods to assess the quality of crop after harvest such as analytical or objective and subjective or sensory method [56]. There are limitations for subjective method in quality evaluation of hazardous material [57]. Objective method is destructive or nondestructive in nature. Manual sorting of commodities is outdated, costly, unreliable, subjective, laborious, and slow. Food depicts complex and dynamic behavior with respect to varietal and time dependency. Nondestructive methods classified according to physical principles are visual inspection, x-ray imaging and computed tomography, near-infrared spectroscopy, ultrasonic testing, thermographic testing, electromagnetic testing, liquid penetrant testing, magnetic particle testing, acoustic emission testing, infrared and thermal testing, magnetic resonance imaging, electronic nose, etc. [58].

Quality assessment or grading has the following benefits:

i. It helps to facilitate the purchase of an unseen product.

ii. Quality and safety improvement incentives.

iii. Grading makes the market information meaningful.

iv. It helps in facilitating quality and price comparison.

v. It reduces the deception and fraudulent marketing risk.

vi. Diverse marketing mechanism can be enabled through the process such as commodity exchange, future trading, inventory credit, credit letters, and facilitating resolution of disputes regarding composition or quality [6].

12. FUTURE CHALLENGES FOR POSTHARVEST TECHNOLOGY

In increasing urban population and shifting lifestyle in developing countries, demand for food has increased many times. Seed is a unit which not only is used as food but also responsible for production of the next generation. The most attention-demanding perspectives in seed biology are postharvest management and quality maintenance of seed. Although in past years, great success has been achieved in development of novel packaging, storage and transport systems, pests, and seed-borne disease control for market access. However, to achieve future goals, further research and technology development should be dedicated to explore genetic aspects of quality traits like stress resistance, resistance to postharvest diseases, and pest management. Researchers should try to work on integrated approaches for postharvest management of seed. Nanotechnology is an emerging field which is leading to tremendous achievements in crop sciences. Seed biologist should also try to explore this field so that seed could be preserving for longer time without altering the genetics.

13. CONCLUSIONS

Seed quality should be maintained for the better quality crops. Maintenance of the seeds is the major problem in the developing countries these days. There is a need to establish better, economical, convenient, and productive methods for postharvest handling and seed storage. Researchers should focus on translating knowledge into agricultural outputs. The chapter deals with the value of healthy seeds, factors affecting the seed quality, postharvest techniques for the seed storage, and techniques and safety measures for their quality assessment for the maintenance of good quality seeds to meet the international standards emphasizing on the needs of developing countries. This could be achieved by opting more sensitive and advance technologies with special emphasis on seed response to environmental and maternal factors.

REFERENCES

1. Delouche JC, Caldwell WP. Seed vigor and vigor tests. Proceeding of Association of Official Seed Analyst. 1960;50:124–129.
2. Delouche JC. Planting seed quality. In: Proceeding of 1969 Beltwide Cotton Prod.-Mech. Conference. New Orleans. January 1969. p. 16–18.
3. Grabe DF. Agronomic significance of seed deterioration. Agronomy Abstract. 1965;57:40.
4. Woodstock LW. A respiration test for corn seed vigor. Proceeding of Association of Official Seed Analyst. 1966;56:95–98.
5. Gregg BR, Billups GL. Seed Conditioning Technology Part A. Science Publishers. USA. 2010, Vol. 2, p. 1–2.
6. Boxall RA, Brice JR, Taylor SJ, Bancroft RD. Crop post-harvest, science and technology, volume 1, principles and practice. In: Golob P, Farrell G, Orchard E (Eds) Crop Post- Harvest: Science and Technology. Blackwell Science Ltd, Oxford, UK. 2002. p. 141–204. doi: 10.1002/9780470751015.ch5.
7. Mohammed M. Manual on post-harvest management strategies to reduce losses of perishable crops. UWI/CTA/NAMDEVCO Workshop on Strategies to Reduce Post-harvest Losses in Perishable Commodities at NAMDEVCO Conference Facility. Piarco, Trinidad. February 24–25, 2014.
8. Harrington JF. Seed storage and longevity. Cited in: Kozlowski TT, Seed Biology. 1972;3:145–245.
9. Kiaya V. Post-harvest Losses and Strategies to Reduce Them. New York: Action Contre la Faim (ACF International). 2014.
10. López J, Uribe E, Vega-Gálvez A, Miranda M, Vergara J, Gonzalez E, Di Scala K. Effect of air temperature on drying kinetics, vitamin C, antioxidant activity, total phenolic content, non-enzymatic browning and firmness of blueberries variety O Neil. Food and Bioprocess Technology. 2010;3:772–777.
11. Mujumdar AS, Law CL. Drying technology: Trends and applications in post-harvest processing. Food and Bioprocess Technology. 2010;3:843–852.
12. Chen XD, Mujumdaer AS (Eds) Drying Technologies in Food Processing. Oxford: Blackwell. 2008.
13. Boothumjinda S, Exell RHB, Rongtawng S, Kaewnikom W. Field tests of solar rice dryers in Thailand. In: Proc ISES Solar World Forum. Perth, Oxford: Australia (Parkville ISES), Pergamon Press. 1983. p. 1258–1263.
14. Kitinoja L, Gorny JR. Post-harvest Technology for Small-scale Produce Marketers Economic Opportunities, Quality and Food Safety. Post-harvest Horticulture Series. Department of Pomology, University of California. Davis, USA. 1999. p. 21.
15. Radajewski W, Jolly P, Abawi GY. Optimization of solar grain drying in a continuous flow dryer. Journal of Agricultural Engineering Research, 1987;38:127–144.
16. Colliver DG, Brook RC, Peart RM. Optimal management procedure for solar grain drying. American Society of Agricultural Engineers. Chicago. 1978;12:78–35.
17. Heid WG. Solar-assisted Combination Drying: An Economic Evaluation. Agricultural Economic Report, Economics, Statistics and Co-operative Services, US Department of Agricultural, No. AER-453. 1980.
18. Hellickson MA. Cost/performance goals for agricultural solar collectors. American Society of Agricultural Engineers. 1979;54:79–40.
19. Chung DS. Lee C. Physical and thermal properties of grains. In: Champ BR and Highley E (Eds) Preserving Grain Quality by Aeration and In-store Drying.

Proceedings of an International Seminar. Kuala Lumpur, Malaysia. Canberra, Australia: ACIAR. October 9–11, 1986.
20. Bartosik RE. A model-based fan and burner control strategy for the in-bin drying and conditioning of corn. 2005.
21. Harris KL, Lindblad CJ. Post-harvest Grain Loss Assessment Methods. Minnesota: America Association of Cereal Chemist. 1978. p. 193.
22. Boxall RA. A Critical Review of the Methodology for Assessing Farm-level Grain Losses after Harvest. Slough, UK: Tropical Products Institute G191. 1986.
23. Boxall RA. Storage losses. In: Golob P, Farrell G, Orchard JE (Eds) Crop Post-harvest: Science and Technology Volume 1: Principles and Practice. Oxford: Blackwell Sciences Ltd. 2002. p. 143–169.
24. Kader AA. Post-harvest Technology of Horticultural Crops. 3rd ed. Oakland: University California Agriculture and Natural Resources, Publ. 3311. 2002.
25. Gross K, Wang CY, Saltveit ME. The Commercial Storage of Fruit, Vegetables and Florist and Nursery Stocks. USDA Agriculture Handbook 66. 2002. http://www.ba.ars.usda.gov/hb66/index.html.
26. Atanda SA, Pessu PO, Agoda S, Isong IU, Ikotun I. The concepts and problems of post-harvest food losses in perishable crops. African Journal of Food Science. 2011;5:603–6013.
27. Boxall RA. Post-harvest losses to insects – A world overview. International Bioterioration and Biodegradation. 2001,48:137–152. Retrieved from www.scopus.com.
28. Grolleaud M. Post-harvest Losses: Discovering the Full Story. Rome: FAO. 1997. Available from: http://www.fao.org/docrep/004/ac301e/ AC301e04.htm#3. 2.1%20Rice
29. Lipinski B, Hanson C, Lomax J, Kitinoja L, Waite R, Searchinger T. Reducing Food Loss and Waste. World Resources Institute Working Paper. Washington, DC: 2013.
30. Grolleaud M. Post-harvest Losses: Discovering the Full Story. Overview of the Phenomenon of Losses during the Post-harvest System. Rome, Italy: FAO, Agro Industries and Post-Harvest Management Service (AGSI). 2002.
31. ISTA. International rules for seed testing. Proceedings of the International Seed Testing Association. 1966;31:49–91.
32. Tyler PS, Golob P, Compton J, Bickersteth S. Study on Phytosanitary Requirements to Promote Maize Trade in Eastern and Southern African Countries at Risk from the Larger Grain Borer. Report for the European Commission. Chatham, UK: Natural Resources Institute. 1990.
33. Adams JM, Schulten GGM, Harris KL, Lindblad CJ. Post-harvest Grain Loss Assessment Methods. American Association of Cereal Chemists. USA. 1978 p. 83–95.
34. FAO I. WFP. The State of Food Insecurity in the World. Food and agriculture organization of the united nations. Rome. 2013.
35. Reed C. The precision and accuracy of the standard volume weight method of estimating dry weight losses in wheat, grain sorghum and maize, and a comparison with the thousand grain mass method in wheat containing fine material. Journal of Stored Products Research. 1987;23:223–231.
36. Adams JM, Harman GW. The Evaluation of Losses in Maize Stored in a Selection of Small Farms in Zambia with Particular Reference to the Development of Methodology. Slough, UK: Tropical Products Institute G109. 1977.
37. Ratnadass A, Fleurat-Lessard F. Improvement in the practical methods of assessment of losses caused by insects in grain stored at the village level in tropical Africa. In: Fleurat-Lessard F, Ducom P (Eds) Proceedings of the 5th International Working Conference on Stored Product Protection, Bordeaux. Bordeaux, France: INRA. 1991.

38. Ratnadass A, Berté S, Diarra D, Cisé B. Insect losses in sorghum stored in selected Malian villages, with particular emphasis on varietal differences in grain resistance. In: Highley E, Wright EJ, Banks HJ and Champ BR (Eds) Proceedings of the 6th International Working Conference on Stored Product Protection, Canberra. Wallingford, UK: CAB International. 1994.
39. Compton JAF, Floyd S, Ofosu A, Agbo B. The modified count and weigh method: An improved procedure for assessing weight loss in stored maize cobs. Journal of Stored Products Research. 1998;34:277–285.
40. Boxall RA, Gillett R. Farmer Level Storage Losses in Eastern Nepal. Slough, UK: Tropical Products Institute G157. 1982.
41. Hernandez A. Drummond DC. A study of rodent damage to food in some Cuban warehouses and the cost of preventing it. Journal of Stored Products Research. 1984;20:83–86.
42. Goletti F, Wolff C. The Impact of Post-harvest Research. Washington, DC: International Food Policy Research Institute, Markets and Structural Studies Division Discussion Paper, 29. 1999.
43. FAO. Global Food Losses and Waste: Extent, Causes and Prevention. . Food and agriculture organization of the united nations Rome. 2011.
44. Semple RL. Post-harvest technology: Storage development and application in developing Asian countries. In: Semple RL, Hicks PA, Lozare JV, Castermans A (Eds) Grain Storage Systems in Selected Asian Countries. REAPASIA. People Republic of China. 1992. p. 9–59.
45. McFarlane JA. Storage methods in relation to post-harvest losses in cereals. Insect Science Application. 1988;9:747–754.
46. Shejbal J, de Bioslambert JN. Modified atmosphere storage of grains. In: Multon JL (Ed). Preservation and Storage of Grains, Seeds and Their By-Products: Cereals, Oilseeds, Pulses, and Animal Feed. New York: Lavoisier. 1988. p. 749–777.
47. Kennedy L, Devereau AD. Observations of large-scale outdoor maize storage in jute and woven polypropylene sacks in Zimbabwe. In: Highley E, Wright EJ, Banks HJ, Champ BR (Eds) Stored Product Protection: Proceedings of the 6th International Working Conference on Stored-Product Protection. Wallingford: CAB International. 1994. p. 290–295.
48. Multon JL. Spoilage mechanisms of grains and seeds in the post-harvest ecosystem, the resulting losses and strategies for the defence of stocks. In: Multon JL (Ed.) Preservation and Storage of Grains, Seeds and Their By-Products: Cereals, Oilseeds, Pulses and Animal Feed. New York: Lavoisier. 1988. p. 3–63.
49. Bailey JE. Whole grain storage. In: Sauer DB (Ed). Storage of Cereal Grains and Their Products. St. Paul, MN: American Association of Cereal Chemistry. 1992. p. 157–182.
50. Alvindia DG, Caliboso FM, Sabio GC, Regpala AR. Modified atmosphere storage of bagged maize outdoors using flexible liners: A preliminary report. In: Highley E, Wright EJ, Banks HJ, Champ BR (Eds) Stored Product Protection: Proceedings of the 6th International Working Conference on Stored-Product Protection. Wallingford: CAB International. 1994. p. 22–26.
51. Longstaff BC. Temperature manipulation and the management of insecticide resistance in stored grain pests: A simulation study for the rice weevil, Sitophilus oryzae. Ecological Modelling. 1988;42:303–313.
52. Banks HJ, Annis PC. Comparative advantages of high CO2 and low O2 types of controlled atmospheres for grain storage. In: Calderon M, Barkai-Golan R (Ed.) Food Preservation by Modified Atmospheres. Boca Raton: CRC Press. 1990. p. 94–122.
53. Paster N, Calderon M, Menesherov M, Barak V, Mora M. Application of biogenerated modified atmospheres for insect control in small grain bins. Tropical Science. 1991;32:355–358.

54. Smith RH. Rodents and birds as invaders of stored-grain ecosystems. In: Jayas DS, White NDG, Muir WE (Eds) Stored-Grain Ecosystems. New York: Marcel Dekker. 1995. p. 289–323.
55. Hagstrum DW, Flinn PW, Shuman D. Acoustical monitoring of stored-grain insects: An automated system. In: Highley E, Wright EJ, Banks HJ, Champ BR (Ed.) Stored Product Protection: Proceedings of the 6th International Working Conference on Stored-Product Protection. Wallingford: CAB International. 1994. p. 403–405.
56. Ranganna S. Handbook of Canning and Aseptic Packaging. Tata McGraw-Hill Publishing Company. Dehli, India. 2000.
57. Jha SN, Garg R. Non-destructive prediction of quality of intact apple using near infrared spectroscopy. Journal of Food Science and Technology. 2010;47: 207–213.
58. Siddiqui MW. Post-harvest Biology and Technology of Horticultural Crops: Principles and Practices for Quality Maintenance. CRC Press. USA. 2015.

CHAPTER 13

Tissue Biochemical Diversity of 20 Gooseberry Cultivars and The Effect of Ethylene Supplementation on Postharvest Life

Maria Anastasiadi[a], Paul M. Mwangi[a], José J. Ordaz-Ortiz[a,1], Sally P. Redfern[b], Mark Berry[b], Monique S.J. Simmonds[c], Leon A. Terry[a,*]

[a] *Plant Science Laboratory, Cranfield University, Bedfordshire, MK43 0AL, UK*

[b] *Unilever R&D Colworth, Sharnbrook, Bedfordshire, MK44 1LQ, UK*

[c] *Royal Botanic Gardens, Kew, Surrey, TW9 3AB, UK*

ABSTRACT

The European gooseberry (Ribes uva-crispa) is still an understudied crop with limited data available on its biochemical profile and postharvest life. A variety of polyphenols were detected in the skin and flesh of 20 gooseberry cvs, representing mainly flavonol glycosides, anthocyanins and flavan-3-ols. In contrast, gooseberry seeds were for the first time characterised by the presence of considerable amounts of hydroxycinnamic acid glycosides tentatively identified by UPLC-QToF/MS. All cvs examined represented a good source of vitamin C while being low in sugar. Furthermore, the postharvest stability of bioactives was explored by supplementation of exogenous ethylene in air at 5 °C. Results suggest a low sensitivity of gooseberries to ethylene. The overall quality of gooseberries remained stable over two weeks, showing potential for extended bioactive life.

Keywords: Ribes uva-crispa; Ascorbic acid; QToF/MS; Ethylene; Storage; Bioactive life

1. INTRODUCTION

Gooseberries are deciduous shrubs being members of the Grossulariaceae family and genus Ribes like blackcurrants, redcurrants, whitecurrants and jostaberries (Bordonaba and Terry, 2011). The genus is comprised of more than 150 diverse species with currants and gooseberries being the most popular (Barney and Hummer, 2005). Ribes uva-crispa L. (synonym Ribes grossularia L.) is a European species and the most prevalent species among the gooseberries found across the world. It is native to United Kingdom, Caucasus Mountains and North West Africa (Barney and Hummer, 2005). The size of gooseberries varies as does their skin colour ranging from green to pink, red, purple, white, and yellow (Hummer and Dale, 2010). The commercial value of gooseberries is limited at present, mainly due to low demand and high cost of production especially during harvesting (Barney and Hummer, 2005; Dale, 2000). Other influencing factors include prevalence of crop diseases such as powdery mildew (Barney and Hummer, 2005) and lack of high yielding cvs (Pluta et al., 2010).

In recent years there is a rising trend in domestic cultivation of Ribes berries both in Europe and other regions (Barney and Hummer, 2005; Mitchell et al., 2011). Reasons for the increasing interest include small agricultural requirements, resistance to cold winters and the development of improved cvs with better disease resistance, colour, flavour and yield. Quality of gooseberries is primarily based on its visual, textural, organoleptic and nutritional characteristics (Terry et al., 2009). The later attribute has attracted considerable interest over the years, as some bioactive components in berries have been associated with potential health-promoting properties (Bordonaba and Terry, 2011; Folmer et al., 2014; Wang and Stoner, 2008).

Despite the amount of information on the qualitative and quantitative content of bioactives in berries, the nutritional quality of gooseberries has not yet been sufficiently explored due to limited commercial interest. Only a few gooseberry cvs have been studied thus far in which, polyphenols have been extracted from the fruit as a whole (Jordheim et al., 2007; Määttä-Riihinen et al., 2004; Pantelidis et al., 2007; Wu et al., 2004). There is limited information about the spatial contribution across tissues in the final polyphenolic content of gooseberries. The main phenolic compounds reported in gooseberries include anthocyanins (Jordheim et al., 2007; Määttä-Riihinen et al., 2004; Pantelidis et al., 2007; Wu et al., 2004), flavonol glycosides and proanthocyanins (Chiang et al., 2013; Häkkinen et al., 1999a; Mikulic-Petkovsek et al., 2012a; Russell et al., 2009).

The limited commercial value of gooseberries is also depicted in the scarce data available on the postharvest life of gooseberries and the stability of their presumed bioactive components during storage. At present, only a few reports exist on the physical, physiological and biochemical changes occurring in gooseberries during different storage conditions (Harb and Streif, 2004; Kampuse et al., 2015; Muizniece-Brasava et al., 2015). In addition, a better understanding is needed on the role of ethylene in the postharvest life of gooseberries. Gooseberries are classified as non-climacteric fruits, although they are able to produce ethylene in low amounts (0.035–0.35 ng kg^{-1} s^{-1} at 20 °C) (Cantwell, 2002; Thompson, 2002). The role of ethylene on Ribes berries, however, has not been thoroughly investigated with reports often contradictory regarding their sensitivity (Cantwell, 2002; McKay and Van Eck, 2006).

The objective of this study was thus two-fold: to explore the biochemical profile of different tissues of a wide selection of gooseberry varieties grown in the UK, and elucidate postharvest changes in biochemistry and quality characteristics for two gooseberry varieties held for 15 days at low temperature with or without application of exogenous ethylene.

2. MATERIAL AND METHODS

2.1. Chemicals

All HPLC and LC–MS grade solvents were obtained from Fisher Scientific (Loughborough UK). (+)-catechin, (−)-epicatechin, procyanidin B1, procyanidin B2, neochlorogenic acid, caffeic acid, sinapic acid, p-coumaric acid, quercetin-rutinoside, quercetin-glucoside, isorhamnetin-glucoside, isorhamnetin-rutinoside were purchased from Sigma–Aldrich (Dorset, UK). Cyanidin-3-glucoside and cyanidin-3-rutinoside, were purchased from Extrasynthese (Genay Cedex, France). Metaphosphoric acid (Bioxtra ≥ 33.5%), L-ascorbic acid and D-fructose were obtained from Sigma–Aldrich (Dorset, UK). D-glucose and sucrose, were purchased from Fisher Scientific (Loughborough UK).

2.2. Plant material and sample preparation

Gooseberry fruits from 20 cvs of R. uva-crispa were obtained at optimum maturity, from The National Fruit Collection (Brogdale, Kent, UK) on the 13th of July 2012 for biochemical analysis (Fig. 1). Based on the biochemical results obtained, two cvs 'Careless (Kent)' and 'Scotch Red Rough' were selected and harvested again the following year (12th of July 2013) for the purposes of the postharvest trial. Approximately 100–200 berries were harvested from two plants per cv at optimum maturity stage and transported to Cranfield University in cool boxes within 2 h from collection and immediately snap-frozen in liquid nitrogen. The samples were divided into triplicates and stored at −80 °C before analysis. Each sample was further divided into two subsamples. Half of the material (approximately 100 g) was freeze-dried, the seeds were manually removed and the berries were ground into a fine powder for the extraction of phenolics and sugars. The second subsample was kept fresh frozen and powdered in a mortar grinder (RM 200, Retsch Ltd., Derbyshire, UK) with liquid $N_2$and used for the extraction of ascorbic acid to avoid potential degradation of ascorbic acid during the freeze-drying process.

2.3. Postharvest trial

Samples from the two selected gooseberry cvs ('Scotch Red Rough' and 'Careless (Kent)') were transferred inside polystyrene boxes with ice-packs, from The National Fruit Collection to Cranfield within 3 h from harvest. Upon arrival at Cranfield, gooseberries from each cv were split into two batches. The four batches of gooseberries were placed in plastic stackable crates inside water sealed, air-tight polypropylene chambers (88 cm _ 59 cm _ 59 cm) fitted with two 8 _ 8 cm electric fans (Nidec beta SL, RS Components Ltd., Northants, UK) to circulate the ethylene gas during treatment. Two boxes per cv were injected with 11.69 mg L^{-1} of standard ethylene gas (100% ethylene; SIP analytical) and the other two boxes were used as control samples (untreated). Temperature of the storage room was set at 5 _C and the treatment time was 24 h. The concentration of the ethylene gas in the chambers was confirmed after 30 min of injection and after 24 h of storage by withdrawing air from the chambers (including controls), using a tapped 20 mL plastic syringe. The headspace of the sampled air was injected into a gas chromatograph (GC—Model 8340, DP800 integrator, Carlos Erba Instruments, Herts, UK, analytical column, Porapak) fitted with flame ionisation detector (FID) and ethylene gas present quantified.

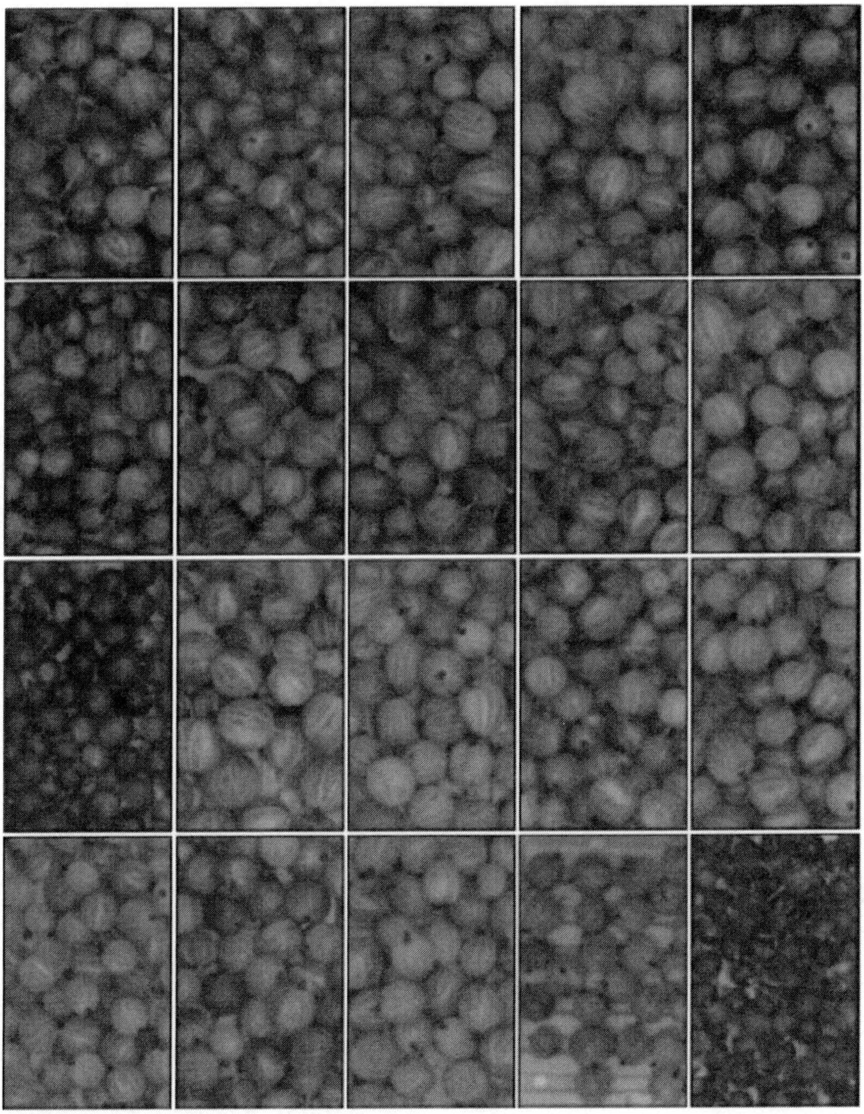

Figure 1. Phenotypic differences of 20 gooseberry cvs, harvested from the National Fruit Collection in 2012. From left to right: 'Careless 3024', 'Guy seedling', 'Jolly Amylers', 'Jubilee Careless', 'Heart of Oak', 'Victoria', 'Ajax', 'Goutray's Earliest', 'Rubuste Nool', 'Lord Elco (Scotland)', 'May Dulle', 'White Eagle', 'Careless (Kent)', 'Nailer', 'Mitre', 'Cousen's Seedling', 'Lord Audley', 'Careless VT 512', 'Bedford Yellow', 'Scotch Red Rough'.

After the treatment was completed (24 h), every batch was split into three replicates and stored in ventilated propylene (15 cm × 22 cm × 8 cm) containers. The containers had an inlet and outlet that provided ventilation to the fruits by way of pumping air into the boxes at a flow rate of about 3.33 mL s^{-1} from a flowmeter controlled unit attached to a ICA 6000 (International Controlled Atmosphere Ltd., Kent, UK). The RH inside the boxes was maintained at 80 ± 10% by placing a beaker of water inside the containers. The RH and temperature in the boxes was continuously monitored using Tinytag Ultra 2 TGU-4500 data loggers (−95% RH, −25 to 85 °C, Gemini Data Loggers Ltd., West Sussex, UK).

2.3.1. Sampling during storage period

Upon arrival (Day 0), five gooseberries were randomly selected from each cv in triplicate. After the treatment (Day 1), five gooseberries were sampled randomly from each box (3 boxes per treatment). Sampling was thereafter repeated at regular intervals (Day 4, Day 7, Day 11, Day 13, Day 15). All samples were subjected to colour and ethylene production measurements before being snap-frozen in liquid nitrogen. The snap-frozen berries were stored at −80 °C until analysis. Prior to extraction the plant material was freeze-dried and powdered as before and analysed for individual soluble sugars and phenolics. Water content of gooseberries was calculated by recording the weight of all samples before and after freeze-drying.

2.4. HPLC-ELSD analysis of non-structural carbohydrates

Extraction of non-structural carbohydrates was performed according to a previous method with slight modifications (Terry et al., 2007). Prior to analysis, the sugars extracts were diluted (1:9, v/v) with HPLC grade water and injected into an Agilent 1200 HPLC fitted with a prevail carbohydrate ES 5 µm size of 250 nm × 4.6 mm diameter and a guard column of the same type. The mobile phase consisted of solvent A (water) and solvent B (acetonitrile) and the elution gradient was as follows: 0–15 min, 80–50% B, 15–20 min, 50–20% B, 20–25 min, 80% B. The eluted compounds were detected by evaporative light scattering detector (ELSD) and quantification was based on external calibration curves of commercial standards.

2.5. Phenolic compounds

2.5.1. Extraction of phenolic compounds

Extraction of phenolic compounds from berries and seeds was performed according to a previous method with slight modifications (Giné Bordonaba and Terry, 2008). Freeze–dried berry powder (150 mg) and freeze–dried seed powder (50 mg) were extracted with 3 mL and 1 mL, respectively, of methanol/water/HCl (70:29.5:0.5 v/v) in a shaking water bath at 20 °C for 1.5 h. The samples were vortexed every 10 min to re-suspend the solids. The extracts were subsequently filtered with 0.20 μm nylon filters and stored at −40 °C before analysis.

2.5.2. HPLC-DAD analysis of phenolic compounds and fraction collection

Analysis of flavan-3-ols, dimeric procyanidins, flavonol glycosides, anthocyanins and neochlorogenic acid in gooseberry extracts (flesh/skin) was performed on a Agilent 1200 HPLC system with a diode array detector (DAD) detector, equipped with a quaternary pump, a fraction collector, a thermostated column compartment operating at 30 °C and a cooled autosampler set at 6 °C. Separation was achieved with an Eclispse XDB-C18 column (150 mm × 4.6 mm; 5 μm, Agilent Technologies) and a 1 mm OPTI-Guard column (mobile phases consisted of 3% formic acid in water) (solvent A) and acetonitrile (solvent B) and the flow rate was 17 $\mu L\ s^{-1}$. The elution gradient was as follows: 0–5 min, 3–5% B, 5–10 min; 5–10% B, 10–20 min, 10–30% B; 20–25 min, 30–60% B; 25–26 min, 60–100% B, 26–30 min, 100% B, followed by 5 min re-equilibration time. The DAD detector was operating at 4 different wavelengths, 280 nm, 320 nm, 360 nm and 520 nm. Quantification was based on external calibration curves of commercial standards. Six anthocyanins detected at 520 nm for which commercial standards were not available, were quantified based on the calibration curve obtained for cyanidin-3-glucoside. Furthermore the fraction collector was used to isolate those peaks and subject them to LC/MS analysis for further elucidation.

2.5.3. UPLC/QToF/MS analysis of unknown anthocyanins

The six fractions isolated at 520 nm with UV spectra similar to the anthocyanin standards, were further analysed with an Agilent 1290 infinity UPLC system, comprised of a binary pump with a jet weaver V35 mixer, a thermostated column, set at 30 °C, and a cooled autosampler set at 6 °C. The

UPLC system was coupled with an Agilent 6540 Ultra High Definition (UHD) Accurate Mass Q-TOF–MS system (Agilent Technologies) equipped with an electrospray ionization source (Agilent Dual Jet Stream). The fractions were evaporated using a miVac centrifugal vacuum concentrator (Genevac) and reconstituted with 100 µL 0.1% formic acid in water and 5 µL were injected into a ZORBAX SB-C18 Rapid resolution, HD 2.1 × 150 mm, 1.8 µm column (Agilent Technologies). Mobile phases consisted of 0.1% formic acid in water (solvent A) and 0.05% formic acid in acetonitrite (solvent B) and the flow was 6.67 µL s^{-1}. The elution gradient was as follows: 0.5–6.5 min, 5% B, 6.5–8 min; 5–15% B, 8–15.5 min, 15–20% B; 15.5–15.55 min, 20–45% B; 15.55–17.55 min, 45–100% B, 17.55–17.60 min, 100–5% B; 17.60–20.0 min, 5% B. Source conditions for the MS were set as follows: nebulizer gas (N_2) temperature 150 °C, at a flow rate of 0.13 L s^{-1} sheath gas (N_2) temperature 400 °C, at a of flow rate of 0.18 L s^{-1} and fragmentor voltage 165 V. MS spectra acquired in full scan, positive ionisation mode in the m/z 50–1500 range.

MS/MS experiments were subsequently carried in positive mode; to obtain the daughter ion spectrum of the selected precursor ions. Product ions were produced by collision-induced dissociation using targeted MS/MS experiments, at three different collision energies (10, 20, 40 eV) with an isolation window of 4 m/z (medium) for all compounds. The LC-MS system and data acquisition were controlled by an Agilent MassHunter Data Acquisition software B.04.00.

2.5.4. UPLC/QToF/MS analysis of hydroxycinnamic acid derivatives

The gooseberry extracts obtained from different tissues (skin/fresh and seeds) were further analysed with the Agilent 1290 infinity UPLC system and the Agilent 6540 UHD Accurate Mass Q-TOF-MS system described above. All extracts were further diluted (1:4) with 0.1% formic acid in water and 5 µL were injected into a ZORBAX SB-C18 column (Rapid resolution, HD 2.1 × 150 mm, 1.8 µm; Agilent Technologies) with the chromatographic method described in Section 2.5.3. Commercial standards of different phenolic compound groups were used as reference compounds (hydroxycinnamic acids, flavonols, flavan-3-ols, dimeric procyanidins).

MS/MS experiments in positive mode were subsequently carried on selected precursor ions to assist with compound identification. Product ions were produced by collision-induced dissociation using targeted MS/MS experiments, at three different collision energies (10, 20, 40 eV) with an isolation window of 4 m/z (medium) for all compounds.

Quantification of the hydroxycinnamic acid derivatives identified with the above method was based on external calibration curves of the corresponding aglycons (p-coumaric acid, caffeic acid, ferulic acid, sinapic acid).

2.6. Extraction and quantification of total ascorbic acid

The total ascorbic acid of the 20 gooseberry cvs harvested in 2012, was determined as total vitamin C using a previously described method with slight modifications (Planchon et al., 2004). The method determined the total vitamin C in the gooseberries as the sum of ascorbic acid and its oxidised form dehydroascorbic acid. Briefly, 500 mg of fresh frozen gooseberries were mixed with 2.5 mL of degassed metaphosphoric acid solution (0.01 M) and the mixture was extracted in a shaking water bath at 25 °C for 10 min. The extract was filtered through 0.20 μm nylon filters and immediately injected in the HPLC. Ascorbic acid was monitored at 248 nm and the total vitamin C was quantified using an external calibration curve of L-ascorbic acid. The assay was performed in triplicate.

2.7. Colour measurement

The colour of gooseberries was assessed using an 8 mm aperture handheld tristimulus colourimeter (Minolta CR- 400, Osaka, Japan). The mean colour space coordinates of Lightness (L*), Chroma (C*) and Hue angle (H°) were taken from three positions on each gooseberry for every sampling point.

2.8. Ethylene production measurement

The ability of the gooseberries to produce ethylene gas was measured with an ETD-300 Ethylene Detector system (Sensor Sense BV, Nijmegen, The Netherlands). In order to determine the amount of ethylene in the cuvettes after 1 h incubation, air was pumped at a continuous flow rate of 1.1 $\mu L\ s^{-1}$. The air from the cuvettes was passed through CO_2 and water scrubbers. Measurements from every cuvette were made for a period of 12 min and expressed in $ng\ kg^{-1}\ s^{-1}$. The equipment was equilibrated before measurement by measuring an empty cuvette for 1 h. A baseline measurement of 3 min was maintained between the samples to eliminate cross contamination.

2.9. Statistical analysis

Analysis of variance (one-way and two-way ANOVA) was performed using Genstat for Windows, Version 14 (VSN International Ltd., Herts., UK). The differences between means of data were compared through Least Significance Difference (LSD) and they were considered to be statistically significant at the 95% confidence level ($p \leq 0.05$). Logarithmic transformations were employed where needed in order to ensure the assumption of equal variability.

3. RESULTS AND DISCUSSION

3.1. Phenolic composition of gooseberries

3.1.1. Characterization and quantification of non-coloured compounds in skin/flesh

Gooseberries (flesh/skin) contained a variety of phenolic compounds including flavonol glycosides, flavan-3-ols, dimeric procyanidins, and hydroxycinnamic acid derivatives (Tables 1 and 2). The major class of polyphenols present in all cvs was flavonol glycosides with the rutinosides of quercetin and isorhamnetin being the predominant glycosides. Quercetin and isorhamnetin glucoside on the other hand were found in much lower amounts. Mikulic-Petkovsek et al. (2012a) also observed that quercetins in gooseberries existed as glucosides and rutinosides and that these are the main form in which they exist in berries from Glossulariaceae family. The same authors have also reported high concentration of isorhamnetin-3-rutinoside in gooseberries (73.9 mg kg^{-1} in red and 48.3 mg kg^{-1} in white gooseberries for FW) when compared to other common berries such as bilberry, blackcurrant and strawberry. Similarly, Määttä-Riihinen et al. (2004) have also reported high concentration of isorhamnetin glycosides in gooseberries. Our results are in good agreement with these previous reports, although the concentration range of quercetin glycosides among the 20 gooseberry cvs analysed within this study were higher than those previously reported (Häkkinen et al., 1999a,b; Määttä-Riihinen et al., 2004; Mikulic-Petkovsek et al., 2012a). These differences could reflect that, to date, only a limited number of gooseberry cvs have been studied. However, the overall trend is in line with Määttä-Riihinen et al. (2004) observation that quercetins are high in berries from Glossulariaceae family. Other flavonol glycosides reported in gooseberries, include glycosides of myricetin and kaempferol, typically present in the form of rutinoside representing minor constituents of the flavonol content of gooseberries (Määttä-Riihinen et al., 2004; Mikulic-Petkovsek et al., 2012a). Furthermore Mikulic-Petkovsek et al. (2012a), identified the presence of small amounts of lacitrin and syringetin rutinoside in a white gooseberry cv and other berry species.

Table 1. Flavonoid composition (mg kg^{-1}) of the flesh/skin of 20 gooseberry cvs on a fresh weight (FW) and dry weight (DW) basis (values inside brackets).[a]

Cultivar	(+)-Catechin		(−)-Epicatechin		ProcB1[b]		ProcB[c]		Q-3-rut[d]		Q-3-gluc[e]		Isorh-3-rut[f]		Isorh-3-gluc[g]	
Scotch Red Rough	20.1	(112.8)	4.2	(23.7)	8.8	(49.5)	2.5	(14.3)	102.5	(767.9)	13.7	(102.7)	81.0	(606.4)	2.7	(20.4)
Lord Audley	26.5	(224.8)	2.2	(18.7)	9.3	(79.2)	2.4	(20.2)	78.2	(585.6)	7.0	(52.8)	38.3	(287.2)	1.6	(12.3)
Heart of Oak	44.3	(264.5)	2.4	(13.9)	12.9	(75.4)	2.3	(13.6)	55.3	(414.1)	7.1	(52.8)	26.8	(200.4)	1.2	(8.7)
Nailer	37.0	(219.3)	2.2	(13.0)	13.1	(70.3)	2.4	(14.4)	49.9	(373.8)	4.8	(35.9)	29.3	(219.1)	1.4	(10.1)
Jolly Amylers	56.5	(364.0)	2.4	(15.6)	16.4	(105.4)	2.3	(15.0)	25.8	(192.9)	3.4	(25.3)	24.0	(179.6)	1.7	(12.7)
Bedford Yellow	46.6	(326.2)	2.7	(18.5)	18.4	(128.5)	4.5	(31.2)	37.5	(280.6)	2.9	(21.9)	13.9	(104.0)	0.9	(6.6)
Cousen's Seedling	14.0	(112.6)	2.0	(16.1)	10.3	(82.5)	3.9	(31.4)	44.7	(334.5)	5.0	(37.3)	40.6	(304.2)	2.5	(18.6)
Careless (Kent)	10.9	(83.2)	1.0	(7.4)	7.0	(53.7)	1.0	(7.8)	43.1	(322.8)	5.3	(39.9)	28.7	(215.0)	2.3	(17.0)
Victoria	34.6	(285.2)	2.3	(18.8)	12.5	(102.2)	2.6	(21.0)	25.7	(192.4)	3.0	(22.5)	16.4	(123.2)	0.5	(3.9)
White Eagle	32.6	(209.6)	2.1	(13.4)	12.4	(79.3)	2.6	(16.8)	16.2	(121.4)	3.3	(25.1)	24.5	(183.2)	2.2	(16.6)
Jubilee Careless	28.5	(192.9)	1.7	(11.6)	10.5	(71.2)	1.6	(10.9)	24.6	(184.1)	2.5	(18.6)	19.5	(145.9)	1.2	(9.3)
Goutray's Earliest	29.6	(258.0)	3.0	(25.8)	10.4	(90.9)	3.7	(32.1)	20.6	(154.4)	3.0	(22.4)	18.1	(135.2)	1.4	(10.6)
May Dulle	15.1	(106.9)	2.6	(18.5)	9.7	(68.6)	1.9	(13.7)	30.6	(229.5)	5.8	(43.5)	21.4	(160.0)	1.5	(11.0)
Mitre	17.8	(126.5)	1.3	(09.5)	8.2	(58.2)	1.2	(8.9)	33.9	(254.0)	3.7	(27.6)	18.8	(141.1)	1.2	(8.9)
Lord Elco	40.1	(320.3)	1.4	(10.9)	13.0	(103.3)	1.7	(13.6)	9.6	(71.6)	0.6	(4.1)	18.3	(137.3)	1.4	(10.9)
Rubuste Nool	30.1	(206.1)	1.6	(10.9)	11.1	(76.0)	1.5	(10.0)	18.1	(135.9)	1.9	(14.5)	18.3	(137.3)	1.4	(10.3)
Careless 3024	26.9	(201.5)	2.3	(17.0)	10.7	(80.4)	2.4	(17.9)	20.2	(150.9)	1.4	(10.6)	15.0	(112.1)	0.6	(4.4)
Careless (VT)	19.9	(130.7)	1.5	(9.7)	8.3	(55.2)	1.5	(10.1)	25.4	(190.3)	2.6	(19.4)	16.8	(125.8)	1.2	(8.8)
Ajax	19.4	(132.1)	1.7	(11.5)	10.3	(69.8)	1.9	(13.2)	19.4	(145.6)	2.3	(16.9)	16.1	(120.8)	1.3	(9.6)
Guy seedling	8.8	(44.7)	1.8	(9.1)	8.7	(43.8)	3.4	(17.2)	15.4	(115.1)	2.6	(19.2)	29.0	(217.4)	2.0	(15.1)
Mean	28.0	(196.1)	2.1	(14.7)	11.1	(77.2)	2.4	(16.7)	34.8	(260.9)	4.1	(31.2)	26.0	(192.8)	1.5	(11.3)
LSD[h]	8.25	(53.48)	0.49	(2.99)	2.27	(12.33)	0.48	(2.77)	16.34	(122.38)	2.10	(15.9)	8.40	(68.4)	0.50	(3.2)

[a] The results are presented in descending order according to the sum of flavonoids for FW.
[b] ProcB1 = procyanidin B1.
[c] ProcB2 = procyanidin B2.
[d] Q-3-rut = quercetin-3-rutinoside.
[e] Q-3-gluc = quercetin-3-glucoside.
[f] Isorh-3-rut = isorhamnetin-3-rutinoside.
[g] Isorh-3-gluc = isorhamnetin-3-glucoside.
[h] LSD = least significant difference at the 95% confidence level.

Table 2. Hydroxycinnamic acid derivative composition (mg kg^{-1}) of the flesh/skin of 20 gooseberry cvs on a FW and DW basis (values inside brackets).[a]

Cultivar	Coumaroyl-hexose1		Coumaroyl-hexose2		Caffeoyl-hexose		Feruloyl-hexose1		Feruloyl-hexose2		Neochlorogenic acid	
Guy Seedling	45.5	(229.8)	15.3	(77.3)	39.9	(201.6)	2.3	(11.9)	11.8	(5.97)	10.4	(52.6)
Heart of Oak	41.5	(241.5)	18.2	(105.8)	33.4	(194.2)	1.5	(8.8)	8.2	(4.93)	7.7	(45.4)
Nailer	33.3	(198.8)	12.8	(76.5)	33.5	(199.8)	1.6	(9.9)	4.7	(2.88)	9.1	(54.5)
Jolly Amylers	38.2	(245.4)	16.7	(107.2)	24.0	(159.7)	1.7	(10.9)	7.0	(4.53)	3.8	(24.4)
Bedford Yellow	35.1	(244.5)	18.3	(127.9)	16.5	(114.7)	1.3	(8.9)	3.4	(2.44)	12.4	(87.0)
Rubuste Nool	32.3	(221.1)	13.3	(91.0)	24.9	(169.9)	2.8	(18.8)	6.9	(4.74)	4.2	(28.7)
Lord Elco	35.9	(286.0)	15.5	(123.7)	20.6	(164.6)	1.9	(15.2)	6.2	(4.99)	3.8	(30.3)
White Eagle	26.4	(168.8)	7.5	(48.0)	29.2	(186.4)	2.1	(13.7)	6.2	(3.97)	3.2	(20.8)
Careless VT	28.5	(218.9)	11.6	(88.9)	21.6	(166.0)	1.4	(11.0)	3.6	(2.76)	0.7	(5.4)
Scotch Red Rough	8.6	(48.5)	4.9	(27.3)	26.1	(145.6)	4.5	(25.0)	12.0	(6.68)	11.7	(66.8)
Mitre	25.6	(182.0)	9.3	(66.1)	20.8	(147.2)	1.4	(10.2)	3.4	(2.43)	0.8	(5.6)
Cousen's Seedling	26.5	(212.1)	9.6	(76.1)	11.2	(89.4)	2.0	(16.0)	5.5	(4.46)	3.4	(27.9)
Careless (Kent)	27.1	(179.7)	10.3	(68.1)	14.3	(94.8)	1.5	(10.3)	3.9	(2.58)	0.7	(4.8)
Careless 3024	17.2	(129.0)	7.2	(54.2)	20.0	(150.1)	1.5	(11.2)	4.5	(3.38)	5.6	(41.8)
Lord Audley	19.3	(163.7)	6.1	(51.2)	15.0	(126.6)	1.3	(11.2)	3.4	(2.84)	8.8	(73.9)
Jubilee Careless	27.6	(186.2)	10.1	(68.0)	11.4	(76.3)	1.1	(7.5)	2.6	(1.78)	0.7	(4.8)
May Dulle	6.4	(45.7)	5.5	(39.5)	22.3	(158.8)	2.3	(16.1)	5.1	(3.64)	8.4	(69.1)
Victoria	14.5	(118.6)	6.9	(56.0)	17.0	(138.4)	1.0	(8.2)	3.8	(3.04)	5.1	(42.2)
Goutray's Earliest	12.3	(107.2)	5.3	(46.7)	17.8	(155.5)	1.1	(9.6)	3.4	(2.95)	6.7	(58.1)
Ajax	17.2	(116.5)	5.8	(39.4)	13.5	(91.7)	1.4	(9.6)	4.0	(2.70)	3.9	(26.5)
Mean	26.0	(177.2)	10.5	(71.9)	21.7	(146.6)	1.8	(12.2)	5.5	(3.68)	5.6	(38.5)
LSD	5.31	(24.87)	2.59	(12.36)	5.75	(29.94)	0.51	(3.63)	0.98	(0.532)	1.61	(11.50)

[a] The results are presented in descending order according to the sum of hydroxycinnamic acid derivatives for FW.

Flavan-3-ols and dimeric procyanidins were present in gooseberries in small to moderate amounts with (+)-catechin being the predominant with an average concentration of 28.0 mg kg^{-1} (FW). These results are comparable with previous reports, with the exception of (+)-catechin which was in concentrations higher than that previously reported (Häkkinen et al., 1999b). Very few reports are available describing the levels of procyanidins in gooseberries, however, according to Wu et al. (2004), procyanidins and prodelphinidins are present in gooseberries mainly in polymeric form.

Of the phenolic acids, only hydroxycinnamic acid derivatives were detected in low to moderate amounts in the skin and flesh of gooseberries. The presence of neochlorogenic acid (trans-5-O-caffeoylquinic acid) was verified through the use of a commercial standard and accurate mass measurements. Several hydroxycinnamic acid glycosides were also tentatively identified with QToF/MS/MS, based on accurate mass measurements and the fragmentation pattern acquired with MS/MS in three different voltages (10, 20, 40 eV) which matched the fragmentation pattern obtained for the corresponding aglycons. The most abundant fragments for each compound are presented in Table 3. The sugar moiety was a hexose and the most abundant pseudo-molecular ion in positive mode was in the form of a [M – Na]$^+$ adduct. The following glycosides were identified in gooseberry skin/flesh extracts: p-coumaric, caffeic, ferulic and sinapic acid hexosides (Table 3).

Table 3. Tentative identification of anthocyanins and hydroxycinnamic acid glycosides in gooseberry cvs with QToF/MS/MS.

Peak	Ret. time (min)		*m/z*	Fragments MS/MS	Tentative identification	Molecular formula
Anthocyanins						
3	7.13	[M+]	463.1303	301.0995	Peonidin-3-glucoside	$C_{22}H_{23}O_{11}$
3	7.27	[M+]	419.1190	287.055	Cyanidin-3-xyloside	$C_{20}H_{19}O_{10}$
4	7.42	[M+]	609.2129	301.0994	Peonidin-3-rutinoside	$C_{28}H_{33}O_{15}$
6	10.11	[M+]	595.1447	287.055	Cyanidin-glycoside	
7	10.36	[M+]	595.1480	(287.0814, 147.0570, 163.1326)	Cyanidin-coumaroyl-glucopyranoside	$C_{30}H_{27}O_{11}$
8	10.96	[M+]	609.1594	301.705	Peonidin-glycoside	$C_{28}H_{33}O_{15}$

Ret. time (min)		*m/z*	Fragments MS/MS	Tentative identification	Molecular formula
Hydroxycinnamic acid derivatives					
5.15, 6.10	[M – Na]+	349.0893	(165.0545, 147.0570)	Coumaric acid hexoses	$C_{15}H_{18}O_8$
4.20	[M – Na]+	365.0862	(181.0508, 163.0402, 145.0283, 135.0437, 117.0333, 89.0383)	Caffeic acid hexose	$C_{15}H_{18}O_9$
4.86, 6.42	[M – Na]+	379.0999	(195.0651, 177.0546)	Ferulic acid hexoses	$C_{16}H_{20}O_9$
5.62	[M – Na]+	409.1109	(225.076, 207.0652, 175.0391)	Sinapic acid hexose	$C_{17}H_{22}O_{10}$

Among the hydroxycinnamic acid glycosides detected in gooseberry flesh/skin extracts, coumaric and caffeic acid glycosides were the most abundant, followed by ferulic acid glycosides (Table 2). Sinapic acid glycosides were also detected in trace amounts (Table 3). 'Guy Seedling' was the cv with the highest content in hydroxycinnamic acid glycosides with a sum of 114.8 mg kg^{-1} (FW), while 'Scotch Red Rough' cv which exhibited the highest flavonoid content among all cvs, had low levels of hydroxycinnamic acid glycosides (56.1 mg kg^{-1}). The levels of hydroxycinnamic acid glycosides reported here are in good agreement with those reported in berries from other species in the Grossulariaceae family (Määttä-Riihinen et al., 2004). Neochlorogenic acid was present in small amounts in gooseberry flesh/skin with a mean concentration of 5.6 mg kg^{-1} (FW) in 'Scotch Red Rough'. Its concentration was almost 2-fold higher in coloured varieties when compared to green varieties. This is the first report of neochlorogenic acid in gooseberries. The presence of several phenolic acids has been previously reported in the literature including protocatechuic acid, p-hydroxybenzoic acid, p-coumaric acid, ellagic acid, ferulic acid, chlorogenic acid, sinapic acid, caffeic acid, syringic acid, vanillic acid and gallic acid (Filipiak-Szok et al., 2012; Mattila et al., 2006; Russell et al., 2009). It is notable that none of the above phenolic acids were detected in free form in gooseberries in our study and no hydroxybenzoic acids were detected in the berries of the gooseberry cvs analysed. Similarly Määttä-Riihinen et al. (2004) reported the presence of hydroxycinnamic acids in gooseberries and other berry species in conjugated form, mainly as caffeoyl- and p-coumaroylglucosides.

3.1.2. Phenolic profile of gooseberry seeds

Unlike the flesh/skin extracts which contained a variety of different phenolic compounds, the only type of phenolic compounds detected in gooseberry extracts of seeds were hydroxycinnamic acid derivatives, which were tentatively identified as glycosides of caffeic, coumaric, ferulic and sinapic acids as described in Section 3.1.1. Two distinct peaks with the same spectral characteristic attributed to coumaric acid glycosides and ferulic acid glycosides were present in all seed extracts, indicating the presence of two isomers of coumaric acid and ferulic acid glycosides (Table 3). Their concentrations are presented in Table 4. Gooseberry seeds were observed as being particularly high in phenolic acid glycosides, with concentrations being 10 times higher compared to the flesh/skin extracts. There was some variability among different varieties with the highest concentrations observed for 'Heart of Oak'. The qualitative profile was similar to the flesh/skin extracts, with coumaric and caffeic acid glycosides being the most abundant, followed by ferulic and sinapic acid glycosides. No phenolic acids were present in free form in the seed extracts, as it was confirmed by comparison with the corresponding aglycon standards. It is also notable that neochlorogenic acid, which was detected in the flesh/skin extracts, was absent in gooseberry seeds. Other compounds detected in the gooseberry seed extracts included trace amounts of (+)-catechin and (−)-epicatechin, quercetin-3-rutinoside, isorhamnetin-3-rutinoside and cyanidin-3-rutinoside and cyaniding-3-glucoside in some red cvs.

Table 4. Concentration of hydroxycinnamic acid glycosides (g kg^{-1}) in the seeds of 20 gooseberry cvs on a DW basis.[a]

Cultivar	Coumaroyl-hexose1	Coumaroyl-hexose2	Caffeoyl-hexose	Feruloyl-hexose1	Feruloyl-hexose2	Sinapoyl-hexose	ΣHCAD[b]
Heart of Oak	4.55	1.09	6.65	0.15	1.10	0.20	13.74
Guy seedling	2.26	0.64	4.19	0.12	0.64	0.19	8.04
Victoria	1.85	0.67	3.99	0.19	0.68	0.33	7.71
Careless VT	2.21	0.75	3.03	0.15	0.76	0.31	7.21
Nailer	1.77	0.64	3.33	0.12	0.65	0.33	6.84
Mitre	1.81	0.65	3.26	0.15	0.66	0.28	6.81
Jolly Amylers	1.70	0.57	3.23	0.12	0.58	0.37	6.57
Bedford Yellow	1.97	0.76	1.74	0.17	0.76	0.24	5.64
May Dulle	1.35	0.46	2.47	0.15	0.47	0.18	5.08
Ajax	0.91	0.35	2.89	0.12	0.36	0.30	4.93
Jubilee Careless	1.52	0.51	1.99	0.13	0.51	0.25	4.91
Careless 3024	1.29	0.50	1.95	0.14	0.50	0.31	4.69
Careless (Kent)	1.55	0.49	1.82	0.11	0.49	0.17	4.63
Cousen's Seedling	1.53	0.45	1.75	0.16	0.45	0.29	4.63
White Eagle	1.14	0.31	2.30	0.10	0.32	0.32	4.49
Goutray's Earliest	0.82	0.33	2.48	0.13	0.33	0.20	4.29
Rubuste Nool	0.98	0.28	1.81	0.08	0.28	0.25	3.68
Lord Elco	0.99	0.30	1.60	0.10	0.30	0.31	3.60
Lord Audley	0.71	0.24	1.75	0.11	0.24	0.15	3.20
Scotch Red Rough	0.61	0.21	1.72	0.16	0.21	0.05	2.96
Mean	1.58	0.51	2.70	0.13	0.51	0.25	5.68
LSD	0.356	0.15	0.917	0.036	0.151	0.077	1.393

[a] The results are presented in descending order according to the sum of hydroxycinnamic acid derivatives for DW.
[b] ΣHCAD = Sum of hydroxycinnamic acid derivatives.

To our knowledge no other reports exist on the phenolic composition of gooseberry seeds, but some comparisons can be made with other Ribes species. Građevac et al. (2012) have tentatively identified 35 compounds in black, red and white currant seed extracts, including phenolic acids derivatives such as sinapoyl glucoside and chlorogenic acid and several phenolic acids in free form, small amounts of flavonols and flavonol glycosides as well as two nitrile containing derivatives of p-coumaric and ferulic acid. In contrast Bakowska-Barczak et al. (2009), have reported flavonols as the main phenolic group present in blackcurrant seed residue in the form of kaempferol, myricetin and quercetin glycosides. The same authors also reported the presence of amounts of p-coumaric acid along with small amounts of p-coumaroyl-glucoside in the seeds (0.487 g kg^{-1} and 0.179 g kg^{-1}, respectively, for DW), although the presence of the aglycon was possibly a degradation product (Bakowska-Barczak et al., 2009). Lu et al. (2002) isolated six glucosides of quercetin, kaempferol and myricetin from blackcurrant seeds along with two nitrile containing derivatives of p-coumaric acid. Furthermore, Lu and Foo (2003) isolated and characterised by ^{13}C NMR two phenolic acid conjugates namely 1-p-coumaroyl-β-D-glucopyranoside and 1-cinnamoyl-β-D-glucopyranoside from blackcurrant seeds along with several anthocyanins and flavonoids.

The presence of phenolic acid glycosides has also been reported in the genus Fragaria, with several hexoses of coumaric, ferulic, caffeic and sinapic acids reported primarily in the strawberry receptacle and in small amounts in the achenes (Fait et al., 2008). These differences in the diversity of compounds and their concentrations could be attributed to differences among species, cvs, extraction and detection methods.

3.1.3. Characterization and quantification of anthocyanins

A total of eight anthocyanins were detected in the pink and red gooseberry varieties. Cyanidin-3-rutinoside and cyanidin-3-glucoside were identified by comparison with authentic standards. The other anthocyanins were tentatively identified based on their accurate mass and the fragmentation pattern acquired with MS/MS in different voltages. (Table 3). The major daughter ion produced was the aglycon after the loss of the sugar moiety. Peak 3 corresponded to peonidin-3-glucoside, with small amounts of cyanidin-3-xyloside, also co-eluting under the same peak as revealed by the mass profile obtained with Q-ToF/MS/MS. Peak 4 was tentatively identified as peonidin-3-rutinoside. Peaks 6 and 7 had accurate mass at m/z 595.1447, and 595.1480, respectively, and after fragmentation they both yielded a major daughter ion with m/z287.055 which was attributed to cyanidin aglycon. Peak 7 additionally had a daughter ion with m/z 147.0570 which is a characteristic fragment of p-coumaric acid. Jordheim et al. (2007), have characterised two isomers of cyanidin-3-O-p-coumaroyl-glucopyranoside in gooseberries by HPLC-DAD and NMR spectroscopy. Peak 7, furthermore had the same UV/Vis spectra (Fig. S1) as cyanidin-3-O-β-(6″-E-p-coumaroylglucopyranoside) as reported by Jordheim et al. (2007). Therefore it was tentatively identified as cyanidin-p-coumaroyl-glucopyranoside. Peak 8 exhibited a similar mass profile as peak 4 and it was tentatively identified as peonidin-3-glycoside. Peak 5 was present in very low amounts and it was not possible to acquire a clear LC/MS profile.

The anthocyanin concentrations in pigmented gooseberry cvs are presented in Table 5. Cyanidin-3-rutinoside (pigment 2) was the predominant anthocyanin identified in all the cvs followed by cyanidin-3-glucoside (pigment 1) and pigment 7. 'Scotch Red Rough' exhibited the highest total anthocyanin content with a sum of 265.8 mg kg^{-1}(FW), which is comparable with the levels of anthocyanins reported for other red gooseberry cvs and similar to the levels found in red currants (Djordjevic et al., 2010; Jordheim et al., 2007; Määttä-Riihinen et al., 2004).

Table 5. Concentration of anthocyanins (mg kg^{-1}) in 20 gooseberry cvs on a FW and DW basis (values inside brackets). The results are presented in descending order according to the sum of anthocyanins per FW.[a]

Cultivar	1		2		3		4		5		6		7		8	
Scotch Red Rough	48.3^{d}	(267.5)	128.8^{h}	(714.6)	3.3^{c}	(18.5)	14.5^{c}	(82.9)	$6.0b^{a}$	(31.9)	5.7^{c}	(31.7)	57.3^{e}	(319.8)	1.9^{a}	(10.9)
May Dulle	33.6^{d}	(239.0)	36.3^{g}	(256.6)	2.0^{b}	(14.0)	2.9^{b}	(20.6)	1.5^{b}	(10.7)	2.6^{b}	(18.2)	21.1^{d}	(149.9)	0.4^{b}	(3.1)
Careless 3024	8.2^{ab}	(61.6)	19.5^{f}	(146.4)	t^{b}		0.8^{a}	(6.2)	t		0.8^{a}	(5.7)	4.6^{c}	(34.4)	t	
Victoria	9.4^{b}	(83.9)	13.5^{c}	(109.7)	nd^{c}		t		t		0.5^{a}	(4.4)	2.8^{a}	(23.1)	nd	
Goutray's Earliest	6.1^{ab}	(53.6)	8.3^{abc}	(72.0)	0.8^{a}	(6.9)	0.9a	(7.7)	0.6c	(5.0)	t		3.0^{a}	(26.1)	t	
Ajax	5.3^{ac}	(36.1)	9.3^{bc}	(63.1)	t		1.1a	(7.7)	t		t		2.1^{a}	(14.2)	nd	
Lord Audley	3.6^{c}	(30.5)	4.8^{a}	(40.9)	t		t		t		nd		1.4^{b}	(11.8)	nd	
Heart of Oak	1.0^{e}	(6.2)	5.4^{ab}	(31.7)	nd		t		nd		t		t		nd	
Bedford Yellow	t		1.8^{e}	(12.2)	t		t		t		t		t		t	
Nailer	t		1.2^{de}	(7.3)	nd		t		nd		nd		t		t	
Jolly Amylers	t		0.6^{d}	(4.3)	nd		t		nd		t		t		nd	
Mean	14.6	(9.73)	20.9	(132.6)	2.0	(13.2)	4.0	(25.0)	2.7	(15.9)	2.4	(15.0)	13.2	(82.8)	1.2	(7.0)
LSD					0.58	(4.05)	1.17	(2.75)	0.38	(2.76)	0.68	(4.22)			0.69	(4.59)

[a] The cultivars for which the levels of anthocyanins were below the detection limit (DL) or quantification limit (QL) have been omitted from the table.
[b] t = traces.
[c] nd = not detected. Means within the same column with no letters in common are significantly different. Data for which LSD is not provided have been log transformed and the significant differences are based on the analysis of variance (ANOVA) for the logarithmic data.

3.2. Presence of non-structural carbohydrates

The concentration of total sugars on a FW basis, ranged between 31.4 g kg^{-1} to 80.2 g kg^{-1} (Table 6), which is consistent with previous reports (Mikulic-Petkovsek et al., 2012b; Viljakainen et al., 2002). Fructose and glucose were the predominant sugars in all cvs accounting for ~49% and ~40.5%, respectively, of total sugars. The mean ratio of sucrose was ~10% of the total sugars in gooseberry cvs. These results are in agreement with Giné Bordonaba and Terry (2008) who reported the same ratios of sugars in 17 UK-grown blackcurrant cvs, indicating a similar sugar profile of different species belonging to the Grossulariaceae. The ratio of fructose to glucose reported here varied from 0.9 to 1.3 and is within the range reported in red and blackcurrants (Milivojevic et al., 2012). The relative proportion of individual sugars in fruits can influence the taste properties such as sweetness (Milivojevic et al., 2012).

3.3. Total ascorbic acid

Ascorbic acid was determined as total ascorbic acid which comprised ascorbic acid and its oxidised form (dehydroascorbic acid). It was detected as a single peak at 3.1 min. The vitamin C content varied from 0.38 g kg^{-1} in 'White Eagle' to 0.85 g kg^{-1}(FW) in 'Ajax' (Table 7). These concentrations are up to 3-fold higher than previous reported for gooseberries (Häkkinen et al., 1999b; Pantelidis et al., 2007; USDA, 2014), although they are similar to the concentrations reported in blackcurrants (Djordjevic et al., 2010), red currant, strawberry and lingonberry (Skrede et al., 2012) and within the range found in common berries (<1 g kg^{-1} FW) (Giné Bordonaba and Terry, 2011). The differences between these and earlier studies could reflect differences in sample preparation and analytical methods or more importantly differences among varieties.

Table 6. Concentration of non-structural carbohydrates (g kg^{-1}) in 20 gooseberry cultivars on a FW and DW basis (values inside brackets).[a]

Cultivar	Fructose		Glucose		Sucrose		Total	
Guy seedling	37.5^{h}	(189.8)	31.7^{f}	(160.6)	10.9^{e}	(55.0)	80.2^{g}	(405.4)
Jubilee Careless	34.1^{gh}	(228.4)	29.0^{ef}	(193.5)	9.4^{de}	(61.4)	72.5^{fg}	(483.3)
Ajax	33.2^{fgh}	(225.2)	28.2^{def}	(191.7)	6.4^{cd}	(43.5)	67.8^{fg}	(460.3)
Scotch Red Rough	30.9^{fgh}	(173.9)	24.5^{cdef}	(138.2)	11.5^{e}	(63.4)	66.9^{fg}	(375.5)
Careless (Kent)	32.1^{fgh}	(212.9)	28.0^{def}	(185.9)	6.3^{cd}	(41.8)	66.4^{fg}	(440.6)
Rubuste Nool	32.5^{fgh}	(218.2)	27.4^{cdef}	(183.8)	6.0^{cd}	(44.0)	65.9^{fg}	(442.4)
White Eagle	30.8^{fgh}	(196.8)	27.7^{cdef}	(176.3)	6.6^{cd}	(41.9)	65.1^{fg}	(415.0)
Careless VT	31.6^{fgh}	(242.4)	27.3^{cdef}	(209.3)	4.5^{bc}	(34.8)	63.5^{fg}	(486.4)
Bedford Yellow	31.0^{fgh}	(218.2)	25.7^{cdef}	(180.7)	5.7^{c}	(40.0)	62.4^{fg}	(439.0)
Nailer	27.8^{efgh}	(166.1)	22.1^{bcde}	(132.1)	10.8^{e}	(64.0)	60.7^{f}	(362.2)
Heart of Oak	24.7^{cdef}	(143.5)	20.3^{bcd}	(118.1)	14.3^{e}	(82.5)	59.3^{f}	(344.2)
Cousen's Seedling	29.3^{efgh}	(235.4)	24.9^{cdef}	(200.2)	3.1^{ab}	(25.0)	57.4^{ef}	(460.7)
Mitre	27.3^{efg}	(195.0)	23.3^{cdef}	(166.4)	5.3^{c}	(37.7)	55.9^{def}	(399.1)
May Dulle	26.1^{defg}	(185.7)	22.0^{bcde}	(156.3)	4.6^{c}	(32.8)	52.7^{cdef}	(374.8)
Jolly Amylers	25.0^{bcdef}	(159.4)	20.7^{bc}	(131.9)	5.5^{c}	(35.2)	51.3^{bcdef}	(326.6)
Victoria	21.9^{abcde}	(178.8)	16.4^{ab}	(133.9)	2.9^{a}	(23.5)	41.2^{abcde}	(336.2)
Lord Elco	19.5^{abcd}	(155.7)	16.3^{ab}	(129.0)	2.9^{a}	(22.8)	38.6^{abcd}	(308.5)
Goutray's Earliest	19.3^{abc}	(168.5)	14.2^{a}	(123.4)	3.1^{a}	(27.2)	36.6^{abc}	(319.0)
Lord Audley	17.8^{ab}	(153.0)	14.0^{a}	(118.4)	2.8^{a}	(23.9)	34.6^{ab}	(292.6)
Careless 3024	16.6^{a}	(124.1)	12.0^{a}	(90.0)	2.8^{a}	(21.2)	31.4^{a}	(235.3)
Mean	27.5	(188.4)	22.8	(156.0)	6.3	(40.9)	56.5	(385.4)

[a] The results are presented in descending order according to total sugar content for FW. Means within the same column with no letters in common are significantly different. Results are based on the analysis of variance (ANOVA) for the log transformed data for all the sugars.

3.4. Postharvest treatment of gooseberries

The two cvs selected for the storage trial in the following year, included the red cv 'Scotch Red Rough' which exhibited the most diverse profile of phenolics and had a high concentration of anthocyanins as well as the green cv 'Careless (Kent)' which is a popular commercial variety of gooseberry. For both cvs (per DW), the phenolic and sugar content at harvest point was compared between the two successive years. The composition of non-structural sugars between the two seasons was similar, with only sucrose in 'Careless Kent' dropping from 41.8 g kg^{-1} in the first year to 26.5 g kg^{-1} in the second ($p < 0.05$) (Table S1). Overall, the profile of phenolics was similar for both years; the most notable difference was the 2-fold drop in anthocyanin content in 'Scotch Red Rough' during the second year from 1477.8 mg kg^{-1} to 749.5 mg kg^{-1}(Table S1).

Table 7. Concentration of total ascorbic acid (g kg^{-1}) in 20 gooseberry cvs on a FW basis.[a]

Cultivar	Total ascorbic acid
Ajax	0.85
Careless VT	0.83
Jubilee Careless	0.81
Lord Elco	0.74
Careless (Kent)	0.74
Nailer	0.68
Scotch red rough	0.64
Heart of Oak	0.60
Victoria	0.56
Bedford yellow	0.55
Guy Seedling	0.54
Mitre	0.53
Cousen's Seedling	0.53
Careless 3024	0.52
May Dulle	0.51
Lord Audley	0.48
Robuste Nool	0.45
Goutrey Earliest	0.44
Jolly Amylers	0.43
White Eagle	0.38
Mean	0.59
LSD	0.049

[a] The results are presented in descending order according to total ascorbic acid content.

3.4.1. Biochemical and physiology changes during storage

In the present study the changes in major phenolic compounds, non-structural sugars, colour and ethylene emission were recorded for both cvs, in order to determine the biochemical and physiology changes in gooseberry fruit during air storage and the effect of external ethylene exposure. The water content was also monitored for both cvs throughout the storage period. 'Careless Kent' had a consistently higher water content compared to 'Scotch Red Rough' with average values at baseline of 87.7% and 81.8%, respectively. The results indicated a marginal water loss of ≤2% for both cvs, irrespective of treatment, which occurred after 7 days of storage (Table S2). Also there was no leakage recorded in any of the cvs.

3.4.2. Ethylene

The change in ethylene production rates for treated and control samples is presented in Fig. 2. Ethylene emission remained at low levels throughout the storage period with a total mean of 5.8×10^{-3} ng kg^{-1} s^{-1}, which is significantly lower compared to other soft fruits such as blackberries and raspberries (0.035–0.35 ng kg^{-1} s^{-1} at 5 °C) (Mitcham et al., http://postharvest.ucdavis.edu/PFfruits/Bushberries/). An increasing trend was observed for the red cv with the control samples showing a sharp increase towards the middle of the storage period, but subsequently fell in the same levels as the treated samples. Ethylene emission levels remained relatively stable for the green variety throughout the storage period.

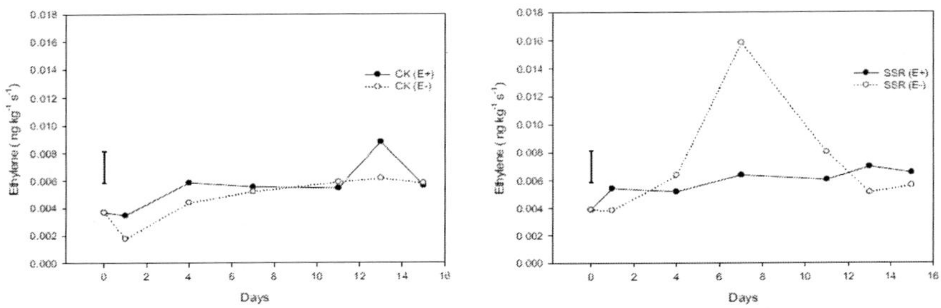

Figure 2. Ethylene emission recorded during the storage period for the two gooseberry cvs, (E +) = ethylene treated samples, E(−) = control samples, CK = 'Careless (Kent)', SSR = 'Scotch Red Rough'. The error bars represent the least significant difference (LSD) at 95% confidence level.

3.4.3. Phenolic compounds

The biochemical results, on a dry weight basis, showed that flavonol glycosides, which represent the major group of phenolic compounds in gooseberries, remained relatively stable for both treated and control samples throughout the storage period. 'Scotch Red Rough' had consistently higher concentrations of flavonol glycosides during the trial, as it is characteristic of the cv (Table S3).

Anthocyanins on the other hand exhibited an increasing trend in 'Scotch Red Rough' over time irrespective of treatment (Fig. 3). A marginal increase observed for the ethylene treated samples was not recorded towards the end of the storage period with both treated and control samples, reaching the same anthocyanin concentrations. This could indicate an initial positive effect of exogenous ethylene on the accumulation of anthocyanins which was counteracted by the sudden increase in ethylene production observed in the control samples towards the middle of the storage period. Other non-climacteric fruits, such as grapes, have exhibited a clear response to postharvest ethylene treatment with anthocyanins positively affected at the end of the treatment period (Becatti et al., 2010, 2014). These differences could be due to species differences and also differences in the treatment protocols applied. Grape berries were exposed to very high ethylene concentrations (1000 mg L^{-1}) and for a longer period of time (36 h), which could explain the acute metabolic responses to ethylene.

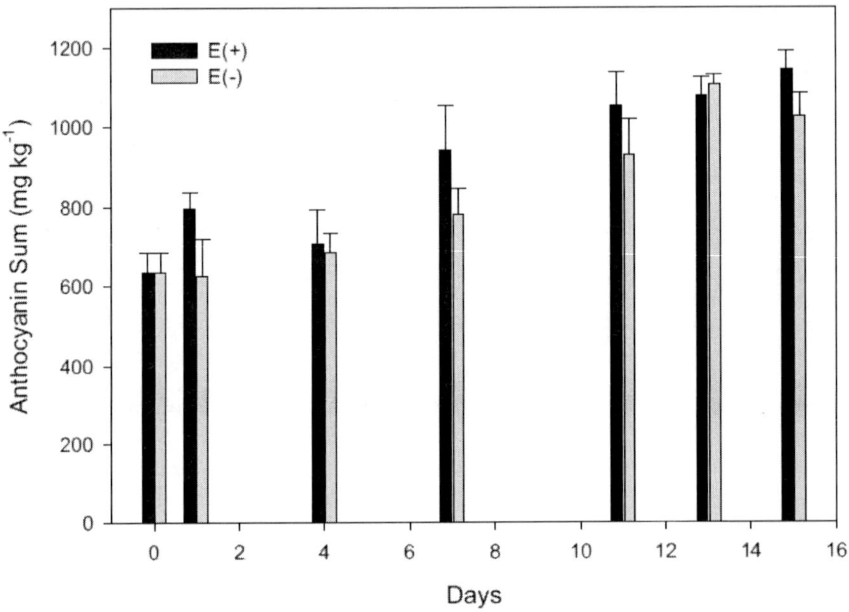

Figure 3. Concentration of total anthocyanins on a DW basis, in 'Scotch Red Rough' during the storage period.E(+) = Ethylene treated samples, E(–) = control. The error bars represent the standard error (SE) of the mean.

On the other hand, the accumulation of anthocyanins under air-storage, observed for the red gooseberry cv, seems to be a common trend among berry species, Kim et al. (2015), recently have demonstrated an increase in anthocyanin content in blackberries during storage at 1 °C, which was more pronounced if the berries were transferred to 20 °C towards the end of the storage period.

Earlier reports, suggest that gooseberry exposure to ethylene, can cause some yellowing/browning of the skin (Cantwell, 2002). In the current study, colour measurements revealed a decrease in L* and C* values over time, for both treated and control gooseberries (Table S2), which was not associated with any perceived colour deterioration. The decrease in L* values could be attributed to several reasons including the observed accumulation of anthocyanin and perhaps gradual water loss. The change in C* values, early in the storage period, may be related to an initial response to exogenous ethylene for the treated samples and the increase in ethylene emission for the control samples, which was further enhanced by water loss later in the storage period.

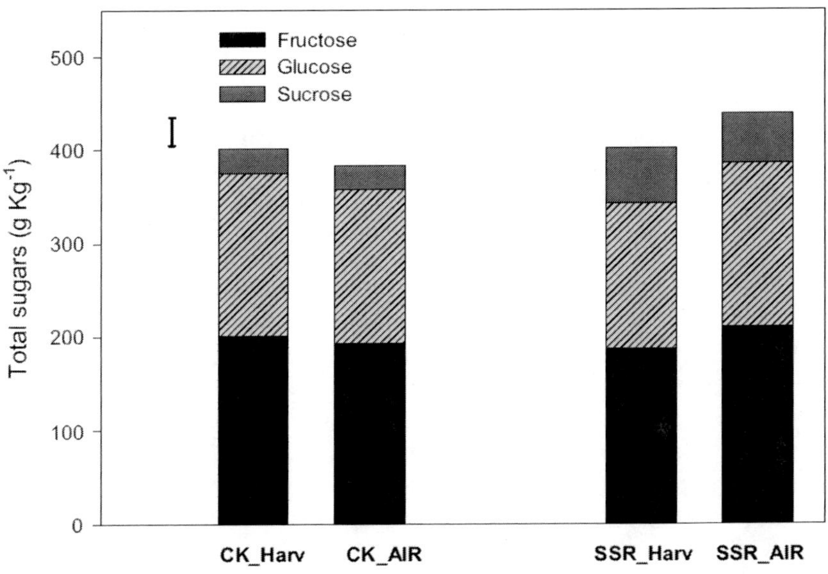

Figure 4. Average concentrations of non-structural carbohydrates on a DW basis, before and during storage for the two gooseberry cvs. CK_Harv = 'Careless (Kent)' at harvest, CK_AIR = 'Careless (Kent)' under air storage, SSR_Harv = 'Scotch Red Rough' at harvest, SSR_AIR = 'Scotch Red Rough' under air storage.

3.4.4. Non-structural carbohydrates

The concentrations of non-structural carbohydrates, for both cvs examined, remained relatively stable for both treated and control samples. The main difference in sugar composition between the two cvs was in the levels of sucrose, which were particularly elevated in 'Scotch Red Rough' throughout the storage period. The average concentration of total sugars for the two cvs, at baseline and during the storage period, is shown in Fig. 4. A previous study on 'Achilles' gooseberries stored under air and different CA conditions, showed that total soluble solids, remained stable for 7 weeks (Harb and Streif, 2004).

The biochemistry and physiology results overall, reveal that gooseberries did not show any clear response to the application of exogenous ethylene and are in broad agreement with previous reports classifying gooseberries as fruits with low sensitivity to ethylene (Cantwell, 2002).

4. CONCLUSIONS

The present study reveals the potential for the use of gooseberries as a rich source of bioactive compounds and demonstrates the distinguishing variability across cultivars and tissue types. Furthermore, it enhances our understanding of their postharvest behaviour especially in relation to their response to exogenous ethylene and the preservation of their nutritional quality under cold storage.

ACKNOWLEDGEMENTS

This study was supported by Unilever U.K. Central Resources Ltd, TSB 101125 (Innovate UK), and the Biotechnology and Biological Sciences Research Council (BBSRC). The authors are grateful to Dr M. Ordidge from the University of Reading for providing access to the National Fruit Collection and Farm Advisory Services Team (FAST) for helping with the sample collection.

REFERENCES

1. Bakowska-Barczak, A.M., Schieber, A., Kolodziejczyk, P., 2009.Characterization of Canadian Black Currant (Ribes nigrum L.)Seed Oils and Residues. J. Agric. Food Chem. 57, 11528–11536.
2. Barney, D.L., Hummer, K.E., 2005. Currants, Gooseberries, and Jostaberries.A Guide for Growers, Marketers, and Researchers in North America. The Haworth Press, Binghamton, NY, USA, pp. 266.

3. Becatti, E., Ranieri, A., Chkaiban, L., Tonutti, P., 2010. Ethylene and wine grape berries: metabolic responses following a short-term postharvest treatment. Acta Hortic.223–228.
4. Becatti, E., Genova, G., Ranieri, A., Tonutti, P., 2014. Postharvest treatments with ethylene on Vitis vinifera (cv Sangiovese) grapes affect berry metabolism and wine composition. Food Chem. 159, 257–266.
5. Cantwell, M., 2002.Summary Table of Optimal Handling Conditions for Fresh Produce. In: Kader, A. (Ed.), Postharvest Technology of Horticultural Crops, 3. UCANR Publications, Oakland, California, USA, pp. 511–518.
6. Chiang, C., Kadouh, H., Zhou, K., 2013. Phenolic compounds and antioxidant properties of gooseberry as affected by in vitro digestion. LWT – Food Sci. Technol. 51, 417–422.
7. Dale, A., 2000. Potential for Ribes cultivation in North America. Hort. Technol. 10, 548–554.
8. Djordjevic, B., Šavikin, K., Zdunic, G., Jankovic, T., Vulic, T., Oparnica, C., Radivojevic, D., 2010.Biochemical properties of red currant varieties in relation to storage.Plant Foods Hum.Nutr. 65, 326–332.
9. Fait, A., Hanhineva, K., Beleggia, R., Dai, N., Rogachev, I., Nikiforova, V.J., Fernie Aharoni, A.R.A., 2008. Reconfiguration of the achene and receptacle metabolic networks during strawberry fruit development. Plant Phys. 148, 730–750.
10. Filipiak-Szok, A., Kurzawa, M., Szlyk, E., 2012.Determination of anti-oxidant capacity and content of phenols, phenolic acids, and flavonols in Indian and European gooseberry. Chem. Pap. 66, 259–268.
11. Folmer, F., Basavaraju, U., Jaspars, M., Hold, G., El-Omar, E., Dicato, M., Diederich, M., 2014. Anticancer effects of bioactive berry compounds.Phytochem.Rev. 295– 322.
12. Giné Bordonaba, J., Terry, L.A., 2008. Biochemical profiling and chemometric analysis of seventeen UK-grown black currant cultivars.J. Agric. Food Chem. 56, 7422–7430.
13. Giné Bordonaba, J., Terry, L.A., 2011. Ribes and Rubus [blackberry, currants and raspberry, etc.]. In: Terry, L.A. (Ed.), Health-Promoting Properties of Fruits and Vegetables, 1. Cabi, Oxford, UK, pp. 260–290.
14. Gopevac, D., Teševi _c, V., Vajs, V., Milosavljevi _c, S., Zduni _c, G., Orpevi _c, B., Stankovi _c, M., 2012.Chemical composition of currant seed extracts and their protective effect on human lymphocytes DNA. J. Food Sci. 77, C779–C783.
15. Häkkinen, S., Heinonen, M., Kärenlampi, S., Mykkänen, H., Ruuskanen, J., Törrönen, R., 1999a.Screening of selected flavonoids and phenolic acids in 19 berries. Food Res. Int. 32, 345–353.
16. Häkkinen, S.H., Kärenlampi, S.O., Heinonen, I.M., Mykkänen, H.M., Törronen, A.R., 1999b. Content of the flavonols quercetin, myricetin, and kaempferol in 25 edible berries.J. Agric. Food Chem. 47, 2274–2279.
17. Harb, J., Streif, J., 2004. Quality and consumer acceptability of gooseberry fruits (Ribes uva crispa) following CA and air storage. J. Hortic. Sci. Biotech. 79, 329– 334.
18. Hummer, K.E., Dale, A., 2010.Horticulture of Ribes.For.Path. 40, 251–263.
19. Jordheim, M., Måge, F., Andersen, Ø.M., 2007.Anthocyanins in berries of Ribes including gooseberry cultivars with a high content of acylated pigments.J. Agric. Food Chem. 55, 5529–5535.
20. Kampuse, S., Muizniece-Brasava, S., Kruma, Z., Sabovics, M., Savri, S., Ruse, K., 2015.The influence of pretreatment technology and packaging material on the quality of fresh gooseberries during storage.Acta Hortic. 1071, 405–412.
21. Kim, M.J., Perkins-Veazie, P., Ma, G., Gina Fernandez, G., 2015.Shelf life and changes in phenolic compounds of organically grown blackberries during refrigerated storage. Postharvest Biol. Technol. 110, 257–263.

22. Lu, Y., Foo, L.Y., 2003. Polyphenolic constituents of blackcurrant seed residue. Food Chem. 80, 71–76.
23. Lu, Y., Foo, L.Y., Herbert Wong, H., 2002. Nigrumin-5-p-coumarate and nigrumin-5-ferulate, two unusual nitrile-containing metabolites from black currant (Ribes nigrum) seed. Phytochemistry 59, 465–468.
24. Määttä-Riihinen, K.R., Kamal-Eldin, A., Mattila, P.H., González-Paramás, A.M., Törrönen, R., 2004. Distribution and contents of phenolic compounds in eighteen scandinavian berry species. J. Agric. Food Chem. 52, 4477–4486.
25. Mattila, P., Hellström, J., Törrönen, R., 2006. Phenolic acids in berries, fruits, and beverages.J. Agric. Food Chem. 54, 7193–7199.
26. McKay, S., Van Eck, A., 2006. Red currants and gooseberries: extended season and marketing flexibility with controlled atmosphere storage. N. Y. Fruit Q. 14, 43– 45.
27. Mikulic-Petkovsek, M., Slatnar, A., Stampar, F., Veberic, R., 2012a.HPLC-MSn identification and quantification of flavonol glycosides in 28 wild and cultivated berry species.Food Chem. 135, 2138–2146.
28. Mikulic-Petkovsek, M., Schmitzer, V., Slatnar, A., Stampar, F., Veberic, R., 2012b.Composition of sugars, organic acids, and total phenolics in 25 wild or cultivated berry species. J. Food Sci. 77, C1064–C1070.
29. Milivojevic, J., Slatnar, A., Mikulic-Petkovsek, M., Stampar, F., Nikolic, M., Veberic, R., 2012.The influence of early yield on the accumulation of major taste and health-related compounds in black and red currant cultivars (Ribes spp.).J. Agric. Food Chem. 60, 2682–2691.
30. Mitcham, E., Crisosto, C.H., Kader, A.A., 2015. Bushberries: Blackberry, Blueberry, Cranberry, Raspberry: Recommendations for Maintaining Postharvest Quality. . (accessed 20-10-2015) http://postharvest.ucdavis.edu/PFfruits/Bushberries/.
31. Mitchell, C., Brennan, R.M., Cross, J.V., Johnson, S.N., 2011. Arthropod pests of currant and gooseberry crops in the U.K.: their biology, management and future prospects. Agric. For. Entomol. 13, 221–237.
32. Muizniece-Brasava, S., Dukalska, L., Kampuse, S., Murniece, I., Savri, S., Dabina- Bicka, I., 2015. The evaluation of physical parameters of gooseberries using different pretreatment technologies and packaging materials.Acta Hortic. 1071, 381–388.
33. Pantelidis, G.E., Vasilakakis, M., Manganaris, G.A., Diamantidis, G., 2007.Antioxidant capacity, phenol, anthocyanin and ascorbic acid contents in raspberries, blackberries, red currants, gooseberries and cornelian cherries. Food Chem. 102, 777–783.
34. Planchon, V., Lateur, M., Dupont, P., Lognay, G., 2004.Ascorbic acid level of Belgian apple genetic resources.Sci. Hortic. 100, 51–61.
35. Pluta, S., Broniarek-Niemiec, A., Zurawicz, E., 2010.Productive value of eighteen gooseberry (Ribes grossularia L.) cultivars of different origin evaluated in central Poland. J. Fruit Ornam. Plant Res. 18, 197–209.
36. Russell, W.R., Labat, A., Scobbie, L., Duncan, G.J., Duthie, G.G., 2009. Phenolic acid content of fruits commonly consumed and locally produced in Scotland. Food Chem. 115, 100–104.
37. Skrede, G., Martinsen, B.K., Wold, A., Birkeland, S., Aaby, K., 2012.Variation in quality parameters between and within 14 Nordic tree fruit and berry species. Acta Agric. Scand. Sec. B Soil Plant Sci. 62, 193–208.
38. Terry, L.A., Chope, G.A., Giné Bordonaba, J., 2007. Effect of water deficit irrigation and inoculation with Botrytis cinerea on strawberry (Fragaria x ananassa) fruit quality. J. Agric. Food Chem. 55, 10812–10819.
39. Terry, L.A., Crisosto, C.H., Forney, C.F., 2009. Small fruit and berries. In: Yahia, E.M. (Ed.), Modified and Controlled Atmospheres for the Storage, Transportation, and Packaging of Horticultural Commodities. CRC Press, Boca Raton, USA, pp. 363– 395.

40. USDA, 2014.National Nutrient Database for Standard Reference Release 27. . (accessed 05-02-2015) http://ndb.nal.usda.gov/ndb/foods/show/2268? fg=Fruits+ and+Fruit+Juices&man=&lfacet=&format= Abridged&count=&max=25&offset=75&sort=&qlookup=.
41. Thompson, J.F., 2002. Transportation. In: Kader, A. (Ed.), Postharvest Technology of Horticultural Crops, 3. UCANR Publications, California, USA, pp. 259–269.
42. Viljakainen, S., Visti, A., Laakso, S., 2002. Concentrations of organic acids and soluble sugars in juices from nordic berries. Acta Agric. 52, 101–109.
43. Wang, L.-S., Stoner, G.D., 2008.Anthocyanins and their role in cancer prevention.Cancer Lett. 269, 281–290.
44. Wu, X., Gu, L., Prior, R.L., McKay, S., 2004. Characterization of anthocyanins and proanthocyanidins in some cultivars of Ribes,Aronia, and Sambucus and their antioxidant capacity. J. Agric. Food Chem. 52, 7846–7856.

CHAPTER 14

Postharvest Practices and Farmers' Perception of Cocoa Bean Quality in Cameroon

Lewis Dopgima Levai[1,2*], Henry Dilonga Meriki[1], Amayana Adiobo[2], Sama Awa-Mengi[3], Jane-Francis Tatah Kihla Akoachere[1] and Vincent P. K. Titanji[4]

[1]*Department of Microbiology and Parasitology, Faculty of Science, University of Buea, PO Box 63, Buea, Cameroon.*

[2]*JP Johnson Biotechnology Laboratory, Institute of Agricultural Research for Development (IRAD) Ekona, PMB 25, Buea, Cameroon.*

[3]*Department of Sociology and Anthropology, Faculty of Social and Management Sciences, University of Buea, PO Box 63, Buea, Cameroon.*

[4]*Cameroon Christian University Bali, PO Box 05, Bali, Cameroon.*

ABSTRACT

Background

After the closure of Cameroon's National Produce Marketing Board in 1990, rigorous measures no longer existed to control production of quality cocoa. This study assesses farmers' knowledge of postharvest practices that could influence cocoa bean quality (CBQ), their perception of quality cocoa and the factors that hinder good farming and postharvest practices (GFPHP). We also make suggestions on how quality control can be incorporated in the production chain to improve on CBQ produced.

Methods

In a cross-sectional, purposive study in Fako Division, South West Region of Cameroon, 100 farmers were enrolled from 10 villages (10 from each village) which make up the major production area in Agro-ecological Zone IV of Cameroon. Farmers were administered a semi-structured questionnaire in the language they best understood.

Results

Cocoa farming in this area is a tradition and our results show that 97 % of farmers are males and acquired farming practices mainly through their family member (43 %) and/or friends (34 %). Farms are self-owned, 76 and 85 %, respectively, are less than 5Ha and have been cultivated for more than 20 years. Only 10 % of farmers could perceive quality cocoa using physical characterization. Although more than 95 % of farmers knew GFPHP, only 48 % undertake these practices. Farmers' focus was on bean weight and not on CBQ and 64 % of farmers think quality has little impact on the price. Most farmers (71 %) sell to "buyam-sellams" who care little about quality. The proportion of respondents who acquired knowledge through formal learning was significantly higher than those who did so from informal learning with regards to both good cocoa production practice ($p = 0.036$) and appreciation of quality attributes ($p = 0.043$). Subsidizing inputs, access to credits, creating cooperatives, regulating price of cocoa and training on postharvest management in decreasing order could enhance practices that will ensure quality cocoa production.

Conclusion

GFPHP is important in the production of quality cocoa. Left alone, farmers will not pay attention to these quality attributes especially as premium quality is not rewarded. Organizing farmers into cooperatives will reduce the link between farmers and companies exporting cocoa and alleviate most bottlenecks in the sector.

Keywords: Cocoa bean, Perception, Postharvest, Quality

BACKGROUND

Cocoa is one of the main cash crops in Cameroon. Cameroon produced 264,077 million tons of cocoa in 2010 [1]. This production dropped to 209,905 million tons in 2013 and rose a little to 232,530 million tons in

2014. This makes Cameroon the fifth world cocoa producing nation after Ivory Coast, Ghana, Indonesia and Nigeria [2]. Presently, the Government of Cameroon projects that with an increase in availability of planting material and the opening of new farms with the help of National Cocoa Development Board that will assist farmers with acquisition of planting materials and accompany them in planting this amount will swell to 600,000 million tons by 2020. This will increase the overall quantity produced but the amount per hectare still remains low. Despite all these, the cocoa sector in Cameroon is still not fully controlled. In 1990, the National Produce Marketing Board (NPMB) was closed down. This board was a government structure that acted in the control of quality and marketing of cocoa beans. After this closure, government involvement was greatly reduced in this sector. Farmers were abandoned to themselves and since they could now sell to anyone ready to buy, stringent measures are no longer taken to guarantee the production of quality cocoa. Licensed Buying Companies (LBC) no longer respected bean quality and buying operations have been left to middlemen ("buyam-sellam") who most of the time are not trained and always in a haste to meet their required monthly tonnage. These forces have greatly compromised the quality of cocoa leaving Cameroon for foreign markets.

The cocoa tree is cultivated principally for cocoa beans. These beans are the principal raw material for the manufacturing of chocolate and other cocoa-based beverages. Inside the cocoa pod (the fruit) is a mucilaginous pulp which represents approximately 40 % of the fruit fresh weight [3]. It is composed of a spongy mucilaginous tissue containing cell sap.

The pulp is the substance that is seen immediately after the pod is opened. It is white, juicy and rich in fermentable sugars such as glucose, fructose and sucrose. It also contains salts (8–10 %), pentosans (2–3 % pectin), organic acids (1–2 %) and 0.6 % proteins [3, 4, 5]. The pulp has a high concentration of acids, especially citric acid, 2.1–2.4 % (w/w), and also has smaller amounts of other organic acids present such as lactic acid, 0.03 % (w/w) and acetic acid, 0.04 % (w/w) [6, 7]. During on-farm processing, this mucilaginous pulp is degraded by microbial fermentation, producing a running liquid known in the industry as 'sweating' [5]. Fermentation, was considered as simply an easy way to remove the pulp to facilitate drying, but its importance to cocoa quality has been well established [4].

The ethanol and acetic acid that diffuse through the seed coat, combined with the heat produced by the exothermic bioconversion of ethanol, causes the death of the seed embryo and marks the end of the fermentation [8, 9]. These changes induce an array of biochemical reactions within the beans and generate the chemical precursors of chocolate flavour, aroma and colour [10, 11, 12, 13]. Other factors such as agronomic management, soil and weather characteristics, as well as postharvest technology, also contribute to chocolate flavour [14].

Cocoa bean is the seed found within the fruit pod embedded in the mucilaginous pulp. The bean is whitish when young and turns brown as it gets matured. Raw cocoa beans have an astringent, unpleasant taste and flavour. The final chocolate flavour is influenced by the origin and cultivar of the cocoa beans, on-the-farm fermentation, drying, roasting and further processing performed by the cocoa and chocolate manufacturer [10, 15, 16, 17, 18].

Seating in its sixteenth meeting in Berlin in 2008, Consultative Board on the World Cocoa Economy agreed that cocoa of merchant quality must be fermented, thoroughly dry, and free from smoky or broken beans, abnormal or foreign odours and any evidence of adulteration, reasonably uniform in size, reasonably free from broken beans, fragments and pieces of shell, and be virtually free from foreign matter. In Cameroon, because of pressure by "buyam-sellams" on the farmers to sell their produce and the fact that quality is not usually considered when buying and selling is done, farmers hardly respect these norms.

According to previous reports [19] market-determined premiums in commodity markets are based among other factors on quality perceptions. Folayan [20] explained that the root cause of quality anomalies in cocoa could be traced to poor farm management, infestation and other diseases, poor handling, bad fermentation, inadequate drying leading to high moisture content which makes the produce vulnerable to mold and bacterial growth. Other causes of low quality cocoa are poor and long storage time which result in fat degradation and pest infestation in the absence of fumigation and other forms of quality maintaining measures.

In this study, we assess farmers' knowledge of postharvest practices that contribute to quality cocoa bean. Their perception of good quality cocoa and the factors that make it difficult for them to practice these technologies have also been examined. Suggestions have also been made on how quality control can be included in the production chain to improve on the quality of cocoa bean produced in the South West Region of Cameroon.

METHODS

This study was conducted in Fako Division of the South West Region, Cameroon. The South West Region is the major production area in Agro-ecological Zone IV characterized by a forest and a mono-modal rainfall pattern. Ten villages (Bafia, Ikata, Likoko Mile 15, Lilale, Malende, Maumu, Muea, Munyenge, Owe and Yoke) were purposively selected for this study.

These villages are amongst those with high cocoa production in this production basin and are closer to the Institute of Agricultural Research for Development (IRAD). It is believed that farmers would be more accessible to research and other technologies that have been developed by IRAD.

The questionnaire was designed to collect information on demography, experience and how knowledge of the cocoa production process was acquired, what kind of postharvest practice they carry out, what limitations they have and suggestions to overcome these limitations. One hundred farmers (10 per village) were randomly enrolled and each of them was administered a semi-structured questionnaire in the language they best understood after an informed consent. Each participant signed a consent form. Authorisation (Ref: 0012/MINRESI/RCSR/01/2014) for this study was obtained from Southwest Regional Centre of the Ministry of Scientific Research and Innovation.

The responses were keyed in an MS excel spread sheet and analysed using the Statistical Package for Social Sciences, SPSS software. Results have been presented as frequency and percentages in tables and figures. Categorical comparisons were performed with Chi-square test. Odds ratios (ORs) and nominal 95 % confidence intervals (CIs) were presented. A two-sided p value < 0.05 was considered significant for all analyses.

RESULTS

Demographic and socioeconomic characteristics of respondents

Cocoa farming in this area is an activity of the middle aged with 70 % of farmers in the age group 31–50. More than 80 % of the farmers are married and have attended school to at least junior secondary (Table 1). The farms are mostly self-owned with, 85 % less than 5Ha. The majority (62 %) have been under cultivation for more than 10 years. Cocoa farming in study area is a tradition and results show that they learned farming practices mainly from their family (43 %) or friends (34 %).

Table 1 Demographic and socioeconomic characteristic of respondents

Characteristics	Category	Frequency	Valid percent
Marital status	Married	84	86.6
	Single	13	13.4

Age group (years)	21–30	7	7.1
	31–40	35	35.7
	41–50	35	35.7
	50–60	12	12.2
	>60	9	9.2
Level of education (years of school)	<7	6	6.0
	7–11	64	64.0
	12–16	24	24.0
	>16	6	6.0
Size of farm (hectares, Ha)	<5 Ha	85	85.0
	>5 Ha	15	15.0
Farm owner	Family	14	14.0
	Landlord	10	10.0
	Self	76	76.0
Occupation	Applicant	1	1.0
	Business	8	8.0
	Civil servant	2	2.0
	Farming	89	89.0
Age of farm (years)	≤10	22	22.4
	11–20	39	39.8
	21–30	28	28.6
	>30	9	9.2
Duration in farming (years)	<5	8	8.0
	5–10	30	30.0
	>10	62	62.0

Acquisition of farming knowhow	Cooperative	17	17.0
	Family	43	43.0
	Farmers field	5	5.0
	School	34	34.0
	Friend	1	1.0
Formal training on postharvest technology	No	49	49.0
	Yes	51	51.0

Perception of quality cocoa based on physical characteristics

Farmers were asked to describe the characteristics based on what they know to be quality cocoa. For each quality attribute (Colour, shell, bean and odour) a table with correct and incorrect responses was provided for them to choose. Based on the present understanding of what is required for the physical characterization of CBQ, most farmers concentrate on the colour, the bean and odour (whether it is smoky). Farmers easily could tell the odour (94 %), external colour (70 %) and internal colour (64 %) of good quality bean but only 20 % could appreciate the quality of the shell as shown in Fig. 1.

To perceive quality cocoa bean, all these attributes are supposed to be correct. When they were put together, only 10 % (Fig. 2) would come out with premium quality beans. Fifty-seven (57) percent had less than three attributes correct (poor bean).

Good farming practices to ensure quality cocoa

Some of the cultural practices were listed and farmers asked to comment on their level of importance to obtaining quality cocoa. Based on their responses as shown in Table 2, apart from use of fertilizer, all of the practices were very important to ensure the production of quality cocoa bean.

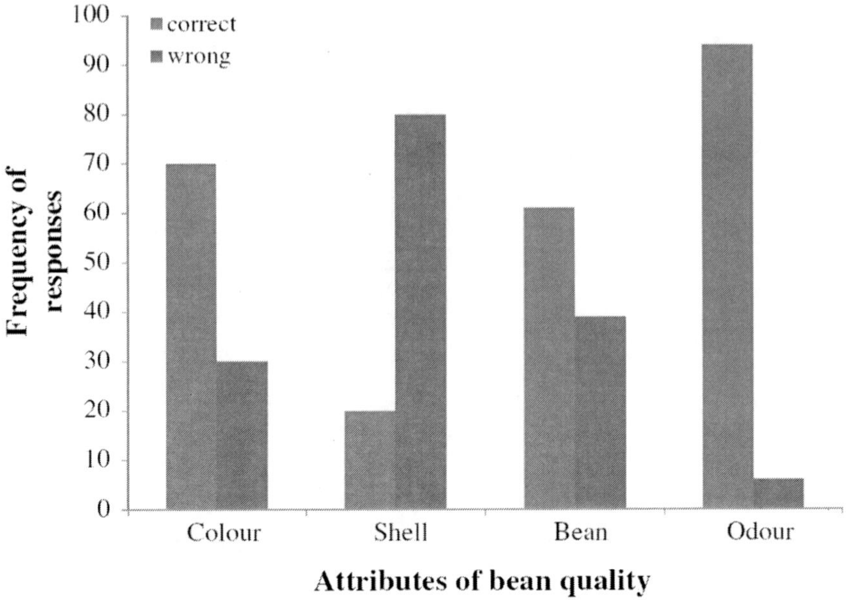

Figure. 1 Frequency of responses to questions on attributes of bean quality

Table 2 Appreciation of agricultural practice to ensure production of quality cocoa

Practice	**Frequency of category**		
	Not important	**Not so important**	**Very important**
Plant good variety	2	8	90
Use of fertilizer	9	20	71
Spray farm	0	1	99
Prune farm	0	0	100
Remove diseased pods	1	4	95
Ferment bean	0	0	100
Dry bean	0	0	100

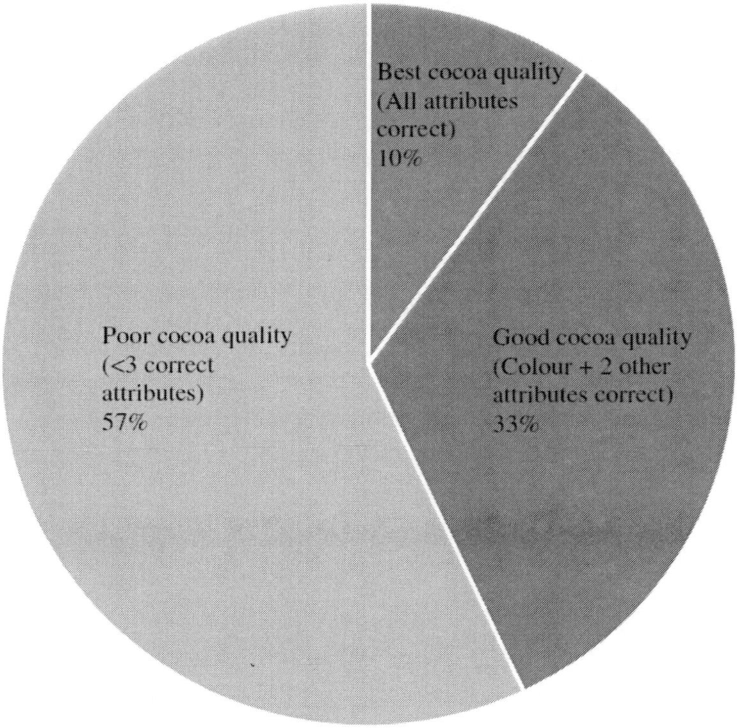

Figure. 2 Appreciation of cocoa quality based on physical attributes (colour, shell, bean and odour) by respondents

We went ahead and asked farmers whether they carry out these practices in their farms to ensure that they always have quality cocoa at the end of each cycle. From the responses we realized that more than 95 % of farmers always spray the pods, prune the branches, ferment and dry their beans. However, 31 % of farmers will not plant improved varieties either because they do not yet have access to these varieties, do not have new land, no knowledge, or the fact that these improved varieties fall after 3 years due to a production that cannot be supported by the root system. Of the 74 % of farmers who do not use fertilizers in their farms more than half (62 %) attribute it to the high cost and 18 % think that it is of no use since their soil is fertile.

Due to the fact that quality is not checked, some farmers (10 %) will not remove diseased pods from the pile. They explained that diseased beans contribute to weight during sales and they do not sort them out as this process is laborious. Whether sorting is done or not does not affect the price, they explained.

Survey data indicate that more than half (52 %) of farmers do not ferment their bean. Fermentation, when done, is mostly carried out for 4–6 days in bags. To achieve this, the pods are opened and pod content filled in 50 kg bags. This permits farmers to easily transfer them from the field on motorbikes to either their homes or to ovens for drying. Some farmers (7 %) carry out fermentation for less than 4 days either because of their turn to use the oven or to get the beans out quickly to a waiting buyer.

Farmers were further asked to list in order of importance three things that would allow them obtain good quality cocoa. According to the results (Fig. 3), farmers would prefer to spray, prune trees and clear their farms in that order. Fermentation and drying only comes afterwards. Upon further questioning (not included in the questionnaire), they indicated that their intention is to have a good harvest and high weighted beans and not how good looking these beans are. Of all the respondents, 64 % said quality will not affect price while only 11 % considered quality cocoa to attract a higher price.

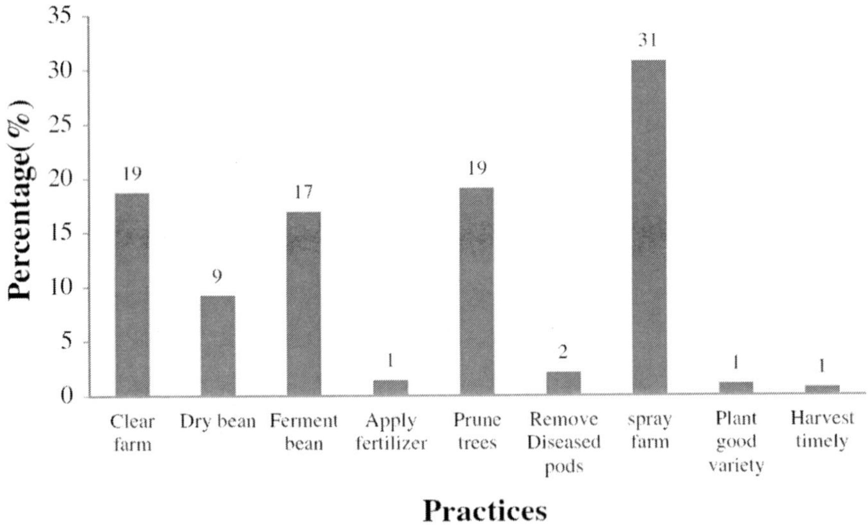

Figure. 3 Frequency of responses on agricultural practices carried out by farmers to ensure production of quality cocoa

Most farmers (71 %) market their beans to middlemen while very few (5 %) sell directly to exporters. In some of the areas, cooperatives have been developed and 24 % of the respondents now sell directly to their cooperatives. This contributed to the fact that 52 % of farmers would not grade their cocoa due to the fact that buyam–sellams do not have a grading scheme and they propose the same price irrespective of. However, 35 % would always grade to check humidity and 13 % will do so only sometimes.

Based on farmers' responses, we could observe that only 48 % undertake practices that could ensure quality cocoa bean as shown in Fig. 4.

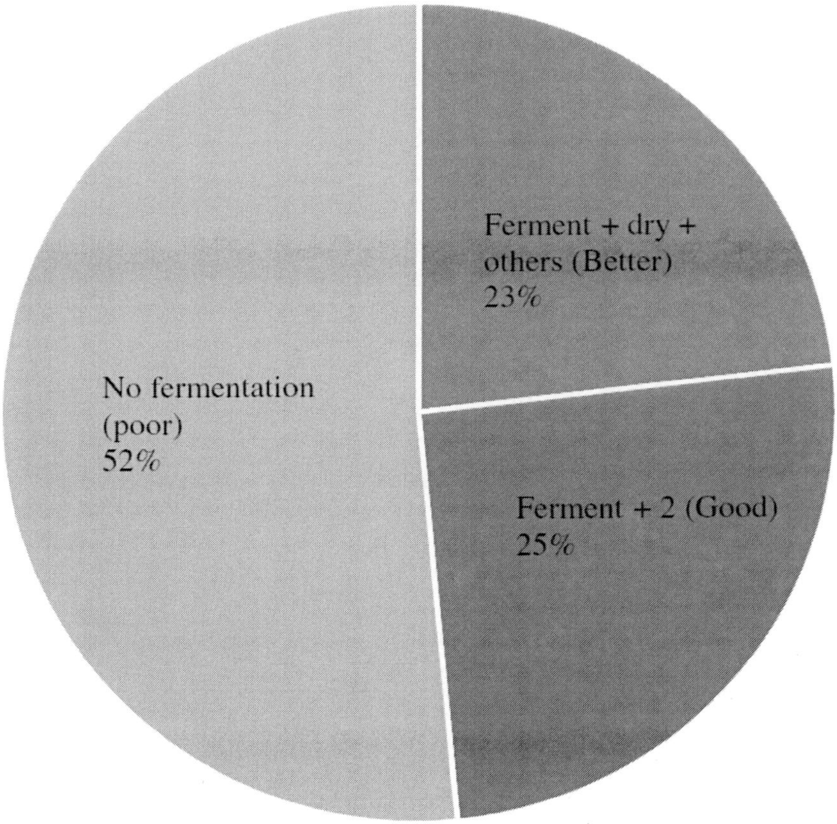

Figure. 4 Use of fermentation and other postharvest practices by farmers to ensure production of quality cocoa

Factors that influence good cocoa practices and appreciation of quality attributes

The proportion of respondents who acquired knowledge through formal [cooperatives and Farmer Field School (FFS)] learning was significantly higher than those who did so from informal (family, friends and self) learning with regards to both good cocoa practice (OR 2.86, 95 % CI 1.05–7.80, $p = 0.036$) and quality attributes (OR 2.68, 95 % CI 1.10–7.06, $p = 0.043$) as shown in Table 3. This was observed when a farmer is under a landlord/family land as opposed to when farm was self-owned (OR 2.70, 95 % CI 1.02–7.14, $p = 0.041$) and when training on postharvest techniques was formal as to informal (OR 2.86, 95 % CI 1.05–7.80, $p = 0.036$). While the other factors did not show a significant difference, only farm ownership significantly influenced good cocoa practice.

Table 3 Factors that influence good cocoa practices and appreciation of quality attributes among farmers

Characteristics	**Categories**	**Good cocoa practice (%)**		**Odds ratios (CI, p value)**	**Quality attributes (%)**		**Odds ratios (CI, p value)**
		Good	**Poor**		**All 4 correct**	**Less than 4 correct**	
Age group (years)	<40	16 (39.0)	25 (61.0)	0.56 (0.25–1.26, 0.156)	18 (45.0)	22 (55.0)	1.30 (0.57–2.96, 0.528)
	≥40	30 (53.6)	26 (46.4)	1	22 (38.6)	35 (61.4)	1
Level of education (years of formal education)	≤12	31 (44.9)	38 (55.1)	0.62 (0.263–1.48, 0.283	31 (44.9)	38 (55.1)	1.90 (0.76–4,75, 0.164)
	>12	17 (56.7)	13 (43.3)	1	9 (30.0)	21 (70.0)	1
Farm ownership	Others (Landlord and family)	16 (66.7)	8 (33.3)	2.70 (1.02–7.14, 0.041)	12 (52.2)	11 (47.8)	0.54 (0.21–1.37, 0.189)
	Self	32 (42.7)	43 (57.3)	1	28 (36.8)	48 (63.2)	1
Duration in farming (years)	≤10	17 (44.7)	21 (55.3)	0.78 (0.35–1.77, 0.556)	16 (43.2)	21 (56.8)	1.21 (0.53–2.76, 0.657)
	>10	31 (50.8)	30 (49.2)	1	24 (38.7)	38 (61.3)	1
Acquisition of farming knowledge	Formal	15 (68.2)	7 (31.8)	2.86 (1.05–7.80, 0.036)	13 (59.1)	9 (40.9)	2.68 (1.01–7.06, 0.043)
	Informal	33	44	1	27	50	1

		(42.9)	(57.1)		(35.1)	(64.9)	
Formal training on postharvest	Yes	25 (48.1)	27 (51.9)	1.04 (0.47–2.28, 0.932)	19 (36.5)	33 (63.5)	1.40 (0.63–3.14, 0.410)
	No	23 (48.9)	24 (51.1)	1	21 (44.7)	26 (55.3)	1

Farmers' recommendations to ameliorate cocoa production

Farmers were finally asked to make recommendations as to what they think could be done to permit them produce quality cocoa all the time. The most prominent suggestions were that there be subsidies for inputs (33.85 %), access to credit (18.08 %) and the formation of cooperatives (10.77 %) (Fig. 5).

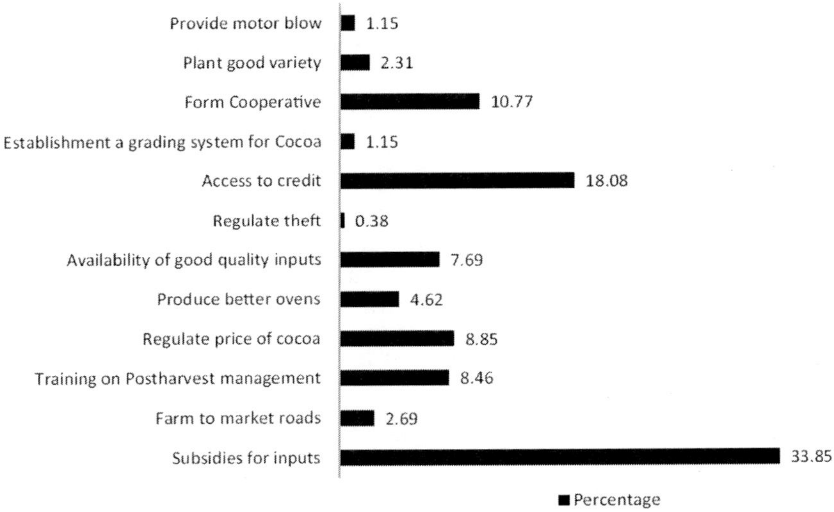

Figure. 5 Recommended factors that could ameliorate cocoa production

DISCUSSION

The dream of every cocoa farmer is to sell their bean at a price that enables them to make maximum gain. This is only possible in a system that is well organized and where premium quality is rewarded. This is presently not the situation that prevails in the cocoa sector in Cameroon. There is available information at the level of IRAD that permits farmers to produce quality cocoa. The process continues to be conducted in a traditional manner as has been noticed in earlier studies [7]. Some of the reasons that hinder them to exploit these technologies have been highlighted in this study.

Fermentation is best conducted in trays, cages or on leaves for 6–7 days [15]. This allows for a single turning of the beans during the second or third day to allow for a proper diffusion of heat and acids. This guarantees a more uniform fermentation and quality bean. In study area, fermentation is carried out in bags and the beans are not turned. Beans from such fermentations are usually not homogenous. Those from the middle of the bag are usually over-fermented while those at the surface are less fermented. This is attributed to the fact that the heat generated during fermentation is less at the surface of the bag and the beans at that location are exposed to acids produced during fermentation only for a short while.

From the responses, we observed that most of the farmers lacked a financial base because production demands a lot in terms of availability and timely application of inputs. With the increase in subsidies and/or availability of credit facilities, more than 50 % of the bottlenecks involved in production can be overcome. These subsidies can be provided to farmers as either production inputs that ranked first (33.85 %) and/or access to credits that ranked second (18.08 %).

To achieve this, farmers themselves have realized that they need to create cooperatives. Cooperatives on the one hand will reduce the long chain that exist between producers and exporters and on the other hand put in place a valid control system for monitoring production of cocoa bean. If this is done, grading of bean can be instituted such that premium quality bean can thus fetch higher prices.

In the days of the NPMB, production capacity of farms of affiliated members could be evaluated and farmers given loans in terms of inputs and/or credit (mostly as inputs) and this amount deducted from their sales. The board could equally assist the farmers when they needed money at very low or no interest and keep their produce to sell when market prices are good because they had good storage facilities. When the NPMB closed down, most farmers fell their trees as it was no longer profitable to farm cocoa.

When middlemen are involved, the marketing chain is long. This places the farmer far from the LBCs and reduces their bargaining power. These middlemen are interested in getting their tonnage for the required LBC as fast as possible and they would hurry the farmers not to even ferment or properly dry beans. Poor resource farmers equally want to make fast money and solve problems or repay an outstanding loan. Faced with this pressure from the middlemen, they heed and this affects the quality that gets to the market.

Lack of training on postharvest technologies was equally raised as a factor that affects the quality of cocoa produced. We noticed that 52 % (Fig. 4) of the farmers we interviewed do not ferment their beans. When fermentation is carried out, it is done in bags and this does not allow for turning of the beans thus some end up over-fermented (in the middle of the bag) and some poorly fermented (close to the surface of the bag). The recommended process is to carry out the fermentation either on the ground or in cages or trays covered with banana leaves for 6–7 days. Farmers use 3–5 days when they ferment as equally noticed by Takrama [21] and Bariah [22]. Duration of fermentation is also influenced by the turn of the farmer to use the oven for drying or the pressure to sell to a waiting buyam–sellam.

Farmers have learned over the past few years to carry out fermentation in bags from their families (43 %) and friends (34 %) Table 3. Most of those who are close to cooperatives and FFS of the Ministry of Agriculture and Rural Development (MINADER) have started using these structures. There was a significant difference in the way the farmers who used the service of formal structures would appreciate and undertake GFPHP when compared to others. It is therefore important that the government through MINADER multiplies these FFS to be close to or present in production basins. Most of the farmers have farming as their principal activity (89 %) and with more than 90 % being able to read and write, the transfer of technologies becomes easier with the use of technical bulletins and practical knowledge in FFS. When farmers were recruited in the Cocoa High Technology Programme in Ghana, 88 % applied the good cocoa practice in their farms to produce quality beans [23].

The fact that farmers always need some form of control was confirmed by the fact that those who worked in farms that they did not own, significantly ($p = 0.041$) undertake good production practices as opposed to those working on their farms (Table 3). This can be explained by the fact that farm owners controlled the way the cocoa production was carried out.

CONCLUSION

Good farming and postharvest practice is important in the production of quality cocoa. Left on their own, farmers will not pay attention to these quality attributes especially as premium quality is not rewarded. Organizing farmers into cooperatives is important in reducing the chain that exists between farmers and LBCs that export cocoa. These cooperatives when established will help alleviate the majority of bottlenecks that exist in the sector and restore quality cocoa production in the cocoa farming sector.

ABBREVIATIONS

CBQ: cocoa bean quality

FFS: farmer field school

GFPHP: good farming and postharvest practices

IRAD: Institute of Agricultural Research for Development

LBC: Licensed Buying Companies

MINADER: Ministry of Agriculture and Rural Development

NPMB: National Produce Marketing Board

ACKNOWLEDGEMENTS

We wish to thank the farmers immensely for their participation and Marcus AJECHA and Rauwitta OMABIT who helped in the survey. There was no funding for this study.

REFERENCES

1. Food and Agriculture Organization of the United Nations. http://faostat3.fao.org/download/Q/QC/E. Accessed 18 July 2015.
2. Top 10 Cocoa Producing Countries—WorldAtlas.com. http://www.worldatlas.com/articles/top-10-cocoa-producing-countries.html.
3. Schwan RF, Wheals AE. The microbiology of cocoa fermentation and its role in chocolate quality. Crit Rev Food Sci Nutr. 2004;44:205–21.
4. Lopez AS, Dimick PS. Biotechnology. In: Reed G, Nagodawithana TW, Weinheim E editors. Biotechnology. 2nd ed. 1995. p. 561–77.

5. Leal GA, Gomes LH, Efraim P, De Almeida Tavares FC, Figueira A. Fermentation of cacao (*Theobroma cacao* L.) seeds with a hybrid *Kluyveromyces marxianus* strain improved product quality attributes. FEMS Yeast Res. 2008;8:788–98.
6. Schwan RF. Cocoa fermentations conducted with a defined microbial cocktail inoculum. Appl Environ Microbiol. 1998;64:1477–83.
7. Ardhana MM, Fleet GH. The microbial ecology of cocoa bean fermentations in Indonesia. Int J Food Microbiol. 2003;86:87–99.
8. Camu N, De Winter T, Verbrugghe K, Cleenwerck I, Vandamme P, Takrama JS, Vancanneyt M, De Vuyst L. Dynamics and biodiversity of populations of lactic acid bacteria and acetic acid bacteria involved in spontaneous heap fermentation of cocoa beans in Ghana. Appl Environ Microbiol. 2007;73:1809–24.
9. Nielsen DS, Teniola OD, Ban-Koffi L, Owusu M, Andersson TS, Holzapfel WH. The microbiology of Ghanaian cocoa fermentations analysed using culture-dependent and culture-independent methods. Int J Food Microbiol. 2007;114:168–86.
10. Hansen CE, Del Olmo M, Burri C. Enzyme activities in cocoa beans during fermentation. J Sci Food Agric. 1998;77:273–81.
11. Hashim P, Selamat J, Muhammad SKS, Ali A. Changes in free amino acid, peptide-N, sugar and pyrazine concentration during cocoa fermentation. J Sci Food Agric. 1998;78:535–42.
12. Cleenwerck I, Gonzalez Á, Camu N, Engelbeen K, De Vos P, De Vuyst L. Acetobacter fabarum sp. nov., an acetic acid bacterium from a Ghanaian cocoa bean heap fermentation. Int J Syst Evol Microbiol. 2008;58:2180–5.
13. Thompson SS, Miller KB, Lopez AS. Cocoa and coffee. In: Doyle MP, Beuchat LR, Montville TJ, editors. Food microbiology: fundamentals and frontiers. 2nd ed. Washington: ASM Press; 2013. p. 881–99.
14. Moreau Cruz JF, Bacelar Leite P, Eduardo Soares S, da Silva Bispo E. Assessment of the fermentative process from different cocoa cultivars produced in Southern Bahia, Brazil. African J Biotechnol. 2013;12:5218–25.
15. Baker DM, Tomlins KI, Gay C. Survey of Ghanaian cocoa farmer fermentation practices and their influence on cocoa flavour. Food Chem. 1994;51:425–31.
16. De Brito ES, García NHP, Gallão MI, Cortelazzo AL, Fevereiro PS, Braga MR. Structural and chemical changes in cocoa (*Theobroma cacao* L) during fermentation, drying and roasting. J Sci Food Agric. 2001;81:281–8.
17. Jinap S, Wan Rosli WI, Russly AR, Nordin LM. Effect of roasting time and temperature on volatile component profiles during nib roasting of cocoa beans (*Theobroma cacao*). J Sci Food Agric. 1998;77:441–8.
18. Kresnowati MTAP, Suryani L, Affifah M. Improvement of cocoa beans fermentation by LAB starter addition. J Med Bioeng. 2013;2:274–8.
19. Anang BT, Fordjour E, Boateng VF. Farmers' management practices and the quality of cocoa beans in upper Denkyira district of Ghana. Asian J Agric Sci. 2011;3:487–91.
20. Folayan JA. Nigerian cocoa and cocoa by-products: quality parameters, specification and the roles of stakeholders in quality maintenance. Pakistan J Nutr. 2010;9:915–9.
21. Takrama JF. Production of Ghana premium cocoa: primary processing operations. Cocoa Res Inst Ghana. 2006;2:1–6.
22. Bariah K. Impact of fermentation duration on the quality of Malaysian cocoa beans using shallow box. KKU Res J. 2014;19(supplement):74–80.
23. Bosompem Martin, Kwarteng JA, Ntifo-Siaw Edward. Perceived impact of cocoa innovations on the livelihoods of cocoa farmers in Ghana: the sustainable livelihood framework (Sl) approach. J Sustain Dev Africa. 2011;13:225–33.

CHAPTER 15

Factors Determining The Adoption and Impact of a Postharvest Storage Technology

Raushan Bokusheva & Robert Finger & Martin Fischler & Robert Berlin & Yuri Marín & Francisco Pérez & Francisco Paiz

Swiss Federal Institute of Technology ETH Zurich, Zurich, Switzerland

HELVETAS Swiss Intercooperation, Bern, Switzerland

Syngenta Foundation for Sustainable Agriculture, Basel, Switzerland

Institute of Applied Research and Local Development (Nitlapan), Central American University (UCA), Managua, Nicaragua

ABSTRACT

This paper evaluates the determinants and impact of adopting the metal silo—a postharvest storage technology for staple grains—which was disseminated by the Swiss Agency for Development and Cooperation (SDC) from 1983 to 2003 in four Central American countries. The aim of the SDC program was to diminish smallholder farmers' postharvest losses by facilitating the manufacture and dissemination of metal silos and thereby to improve regional food security. Our empirical analysis is based on a unique data set obtained from a survey of 1,600 households from El Salvador, Guatemala, Honduras and Nicaragua. We employed a double-hurdle model to identify factors that contributed to the adoption of metal silos and used Tobit and standard regression models to assess the impact of adopting the silos on food security and well-being of households. Our results show that both the household demand for metal silos and the impact of their adoption varied across the four countries, demonstrating the relevance of regional policies for their adoption, as well as their impact. Furthermore, our results indicate that, in addition to achieving household self-sufficiency in maize, the main determinants of adoption were household socio-economic characteristics such as age, land ownership, completion of a training course and quality of basic infrastructure. Finally, when considering a group of economic and social indicators of household well-being, we found that, compared to the silo non-adopters, the adopter households experienced a significant improvement in their food security and well-being between 2005 and 2009.

Keywords: Food security Impact assessment Investment decision Postharvest grain losses Central America

INTRODUCTION

Approximately 16 million people, or 47% of the total population of Guatemala, El Salvador, Honduras and Nicaragua, still live in rural areas (World Bank 2008). Of this population, 62%, i.e. ca. 10 million people, are producers of staple grains (i.e. maize, beans, rice and sorghum) (Baumeister 2010). The majority of staple grain producers are small to medium-sized family farms. Between 39% (Nicaragua) and 92% (Guatemala) of staple grain producers in the above-mentioned countries possess less than 2.1 ha (which corresponds to 3 "manzanas"[1]) of land (Baumeister 2010). Poverty (defined as not having sufficient means to cover basic needs) is widespread for rural staple grain producer families, ranging from 56% in El Salvador to 91% in Honduras (Baumeister 2010).

Staple grains, primarily maize and beans, play crucial roles in food security and income generation, as well as the livelihoods of rural inhabitants of Central America. Maize is the main staple food and beans are an important additional source of protein that complements the maize-based diet. While white maize is mainly used for human consumption, yellow maize is primarily used as fodder. Annual food requirements of a household with an average size of 5.4 persons amounts to about 810 kg of maize and 240 kg of beans (Baumeister 2010).

Postharvest damage (i.e. physical alteration caused by biotic or abiotic agents) and loss (i.e. the difference between totally damaged and recoverably damaged grain still fit for human consumption) of staple grains due to insect pests, rodents and birds are common problems in developing countries. However, precise information on postharvest losses of maize and beans in Central America is scarce. A two-year study (production cycles 1980–81 and 1981–1982) conducted in Honduras by

Raboud et al. (1984) found that postharvest damage and losses of stored maize amounted to 12.5% and 8.1%, respectively (averaged for the two study years). Similarly, Abeleira et al. (2008) found postharvest bean losses in Mexico were 10%.

Given the importance of postharvest management of staple grains in Central America, the program "Postcosecha" ("postharvest" in Spanish) was launched in 1983 by the Swiss Agency for Development and Cooperation (SDC) in Honduras. Later, the program was extended to Guatemala (1990), Nicaragua (1992) and El Salvador (1994). The program consisted of the production and dissemination of metal bins (silos) for the postharvest storage of staple grains. Between 1983 and 2009, almost 670,000 metal silos were produced and disseminated in Central America (SDC 2011). This postharvest technology was developed for the storage of maize and, to a more limited extent, beans. Currently, more than 400,000 mostly rural households, i.e. approximately 2.4 million rural people in Honduras, Guatemala, Nicaragua and El Salvador, use the metal silo for grain storage (Table 1); this number represents 24% of the rural households producing staple grains. Table 1 shows that 46% of the metal silos were disseminated after the direct support by SDC ended in 2003 (SDC 2011), which confirms a successful continuation of the program, particularly in the case of Guatemala.

The means by which metal silos were disseminated after 2003 have evolved differently in the four countries and need to be considered in order to explain the pattern of silo adoption in each country. Most farmers purchase their silos directly from tinsmiths or through a governmental program, or through non-governmental organizations (NGOs). In Honduras, the main means of acquisition was direct purchase from tinsmiths (86%). In Nicaragua, besides direct purchase from tinsmiths (57%), acquisition was through NGOs (23%). To increase food security in rural areas, the Guatemalan government introduced a large subsidy program in 2000, targeting poor rural families; the program subsidizes 62% of the fixed selling price of 58 USD (as of 2009) for a 12 quintal (545 kg) silo by providing the galvanized iron sheets to the contracted tinsmiths. Approximately 75% of the disseminated metal silos fall under the subsidy program, which explains the large increase of disseminated metal silos in recent years. In El Salvador, about 54% of the metal silos were handed over to farmers either as donations (mostly through NGOs), or "in concession", which implies that the farmer obtains the silo free as part of an agricultural subsidy package that included seeds, fertilizer and other relevant materials. However, the silo remained government property and the farmer was not allowed to sell the silo.

Table 1 Adoption of plain metal silo for grain storage according to countries and periods

Country	Number of metal silos disseminated			Number of households using silos in 2009[b]
	1983–2003	2004–2009	Total 1983–2009	
Honduras (1983)[a]	147,427	81,381	228,808	133,850
Guatemala (1990)[a]	103,374	137,994	241,368	158,430
Nicaragua (1992)[a]	59,618	60,785	120,403	68,710
El Salvador (1994)[a]	46,190	30,188	76,378	52,880
Total	357,339	310,348	667,687	413,870

[a]Refers to the official start of the Postcosecha Program by SDC. Some pilot activities producing a small number of metal silos occurred beforehand. Official support by SDC ended in December 2003. [b]assessed considering: a) the number of the metal silos used per household; and b) a lifespan of 15 years for the metal silo.

Source: SDC 2011

A metal silo is a cylindrical structure (one standard design), constructed from high quality galvanized iron sheet (gauge No. 26 or 24) with a top inlet and a smaller lateral outlet at the bottom. The silos are locally constructed by trained artisanal tinsmiths with simple tools (for detailed descriptions and procedures how to fabricate the metal silo see SDC 2008a and Bravo 2009. Metal silos generally hold between 100 and 3000 kg of grain. Rural families in Central America most commonly use either the 12 quintal[2] (545 kg) or 18 quintal (820 kg) size, corresponding to the annual grain consumption of an average family of 5–6 persons. The metal silo can be hermetically sealed, allowing farmers to fumigate the stored grain by using pellets containing phosphine compounds (e.g. aluminum phosphide, "phostoxin" (Bravo 2009)). An important aspect is that the grain must be properly dried (maximum of 13% moisture content) before filling the silo in order to avoid moulds. The metal silo technology has proved to be effective in protecting harvested grains from attack, not only from storage insects but also from rodent pests, birds, insects, and fungal invasion (moulds) (SDC 2008b; Tefera et al. 2010). Users of the metal silos indicated that they were more effective in the control of postharvest losses than traditional storage methods such as granaries, barns, sacks, and metal and plastic barrels (Hermann 1991; Coulter et al. 1995; Gladstone et al. 2002). The objectives of the metal silo dissemination program were to improve the food security and livelihoods of poor rural households and to create employment and income for artisanal tinsmiths who produce the silos (SDC 2008b; Tefera et al. 2010).

There are several socio-economic studies that evaluate the Postcosecha Program which consider such aspects as food security, livelihoods and maize buying and selling, including price dynamics. According to the study by Raboud et al. (1984), metal silos reduce storage losses to less than 1%, compared to an average of 10% in conventional maize facilities. Considering the dissemination dynamics of the metal silos since 1983 (Table 1), SDC (2011) estimated that the accumulated grain saved from potential postharvest loss in the four considered countries accounted for up to 336,000 tons during the last 27 years. In 2009, the total storage capacity of metal silos in all four countries reached approximately 380,000 tons (SDC 2011); this means that about 34,000 tons (9%) were saved annually from potential storage losses, equivalent to the food requirement for almost 45,000 families.

For a group of farms ($N = 60$) in Honduras, Hermann (1991) reports that 20% of the surveyed farmers were metal silo users. Of the non-users, 83% indicated they would not have enough stored grains (maize, beans) to cover their food needs until the next harvest. The silo users said that they bought maize mainly during periods of low prices, whereas the non-users needed to buy mostly during the high price period from May to July. Coulter et al. (1995) report that due to the use of the metal silos, farmers stored more grain than previously (study conducted in El Salvador, Guatemala, Honduras and Nicaragua). Gladstone et al. (2002) found that among the farms surveyed in their study, 60% of the users still had some maize stored in the metal silo by the time of the new harvest compared to 29% of the non-users. All the above mentioned studies concluded that the use of metal silos has increased the food security of poor rural households.

Besides increasing food security, it is expected that farmers would gain flexibility in deciding when to sell the safely stored grain by taking advantage of seasonal price fluctuations caused by changes in supply and demand (Florkowski and Xi-Ling 1990). In Central America, prices are usually low during postharvest months (mid-August to February) when the supply is high, and peaks before the next harvest (from May until the beginning of August) when the grain available in the market becomes more scarce (Zappacosta 2005; Hernández 2008; Pérez et al. 2010). However, these recurring seasonal trends have also been distorted in Central America since 2007 due to the food price crisis (Pérez et al. 2010).

Hermann (1991) found that 67% and 40% of users and non-users of metal silos surveyed in Honduras sell maize. The users reported selling 66% of the stored maize before the new harvest during the high price period from May to August at an average price of 8.85 USD/quintal. The non-users stated they sold 50% shortly after harvest, i.e. from November to December, at an average price of 7.00 USD/quintal, but the rest was sold more evenly throughout the year. Additionally, the users reported selling 74% of the total maize to relatives or within the village, whereas the non-users were selling the bulk (87%) of their maize to intermediates or in the market. Coulter et al. (1995) found that the silo users have more freedom to sell maize when prices are higher and to sell more within the community, thereby reducing dependence on intermediates. In addition, Gladstone et al. (2002) reported that in their study, 67% of the women silo users reported having more opportunities to sell any desired quantity of the stored maize at their convenience to cover household costs.

The abovementioned socio-economic studies state that the use of the metal silo also has had a positive effect on the livelihoods of the families considering such aspects as food security, workload, hygiene and health. Hermann (1991) found that 50% of the silo users surveyed in Honduras indicated that they had a more balanced diet, i.e. the family consumed more rice, beans, meat, eggs, milk and milk products than prior to their use of the metal silo.

Another important aspect is related to a change in the postharvest workload of the family. The need to shell and remove the maize grain from the cob in order to fill the metal silo at one time involves considerable labor, which may be hired, or the use of mechanized equipment if available. However, the stored grain is ready for consumption and therefore there is no need for daily shelling and removal of grains from the cob, work which is mostly done by women when a household uses a traditional storage system. Consequently, 77% of the women in the surveyed households, who were users of silos, confirmed that their workload in postharvest operations had declined (Gladstone et al. 2002).

Finally, in the study by Hermann (1991) 93% of the silo-users indicated that their houses were cleaner than they were before silo adoption on account of fewer insect pests and rodents and therefore less spoilage. Better nutrition and hygiene also had a positive effect on health (especially that of children). Moreover, farmers indicated that they had fewer risks and health problems using aluminum phosphine in a hermetically sealed metal silo than when applying other storage pesticides in traditional storage systems (Gladstone et al. 2002)

All the above-mentioned studies were conducted during the implementation phase of the Postcosecha Program and thus provide an interim evaluation of the program's impact. In the present study we aim to give an ex-post assessment of the program's impact. In particular, the present study's objectives are: (i) to identify factors that significantly contributed to the adoption of the metal silo in the four countries; and (ii) to assess the impact of adoption of metal silos on the food security and well-being of rural households. In order to do this a household survey and an examination of existing secondary data on the number of disseminated silos was conducted in 2009. Based on the survey data, this paper presents an in-depth analysis of selected aspects pertaining to the effect of metal silo adoption on food security, well-being and grain-selling dynamics of staple grain producing households.

The remainder of the paper is organized as follows. Section 2 describes the methodology and the data employed to analyze factors relevant to the adoption of the technology and the effect of adoption on household food security and well-being. Section 3 presents the results and Section 4 draws conclusions and presents policy implications.

MATERIALS AND METHODS

Materials

The study utilizes data from a survey conducted with 800 non-users and 800 users of metal silos. Each of the four countries considered was represented by 200 users and 200 non-users. However, after calibrating the data, the initial sample size was reduced to 1,535 households.

The selection of the interviewed households was carried out by applying a random sampling procedure with multi-stage cluster sampling. The main selection criteria were the following: farms have (i) to be situated in one of the main maize-producing zones in each of the four countries; (ii) to belong to the farmer and be small or medium-sized maize farms with up to 15 manzanas (i.e. 10.5 hectares) (iii) to produce and store primarily maize grain (in metal silos or by other storage methods).

The structured interview with the surveyed households considered the following aspects[3]:

- basic individual characteristics such as, for example, household owner's gender, age, education level, and family size.
- production-related characteristics, e.g. area of owned and rented land, amounts of maize and beans production, number of hired workers, number of livestock, access to advisory services and credit.
- postharvest management characteristics, i.e. date of acquisition of the metal silo and size, storage and use of grain from different storage systems, beans and maize sales and prices.
- impact indicators: income and employment, investments, food security and livelihoods.

To assess the impact of adopting the metal silo, we focus on subsistence farms, i.e. small farms producing maize primarily for consumption by the farmer's family. We define subsistence farms by selecting farms for which maize production does not exceed maize consumption by a factor of 3.[4] As a result, the number of sample farms was further reduced to a total of 1,195 farms for this part of the analysis.

The same (reduced) dataset was used to analyze determinants of the household decision to invest in a metal silo.[5] However, in the adoption analysis, we selected only those user-households that invested in a metal silo for the first time between 2005 and 2009. This selection was dictated mainly by the availability of data.[6] Moreover, this time span refers to the period after 2003 when the SDC had stopped providing direct support to the Postcosecha Program in all four countries. An additional selection criterion for households used in the adoption analysis was that the farmers had paid at least a part of the metal silo market price. This criterion reduced the number of user-households to a total of 179: there were101 in Guatemala, 43 in Honduras, 21 in Nicaragua and 14 in El Salvador. The sample of 528 non-user-households consists of 141, 169, 92 and 126 households from Guatemala, Honduras, Nicaragua and El Salvador, respectively. Regarding the low number of the user-households for El Salvador and their small share compared to non-user households, we decided not to consider them further in the analysis.

Methods

Investment decision model

To explain a household's investment in metal silos, we employed a model which considers the investment decision as a two-stage process: first, the household has to decide to invest or not invest; second, if the decision is made to invest, it must decide how much to invest. To this end, the so-called double-hurdle model was employed (Cragg 1971; Aramyan et al. 2007).

According to the double-hurdle model, the households' investment decisions can be formulated as follows:

$$\mathbf{i}_j = \begin{cases} \mathbf{i}_j^* & if \quad \mathbf{i}_j^* \geq 0 \ and \ d_j = 1 \\ 0 & if \quad \mathbf{i}_j^* \leq 0 \ and \ d_j = 0 \end{cases} \tag{1}$$

where $\mathbf{i}_j$ is the observed level of investment (i.e. storage capacity of the metal silo(s) acquired by a household), $\mathbf{d}_j$ is a binary variable describing the decision to invest or not, and *j* is the household index and $\mathbf{i}^*_j$ is the latent value of the investment volume.

Cragg's model consists of two regressions: a binary choice model is estimated in the first step, while the second step involves the estimation of a truncated regression model, viz:

$$1^{st} \text{ step} : \mathbf{d}_j = \varphi' \mathbf{z}_j + \theta_j \tag{2}$$

$$2^{nd} \text{ step} : \mathbf{i}_j^* = \beta' \mathbf{x}_j + \varepsilon_j \tag{3}$$

where θ_j~N(0,1) and εj~N(0,σ2ε)εj~N(0,σε2). Vectors $\mathbf{z}_j$ and $\mathbf{x}_j$ are the vectors of explanatory variables in binominal and truncated regression models, respectively.

Accordingly, in our empirical analysis we employed two dependent variables: a binary variable signaling whether or not a particular household acquired a metal silo in the period 2005 to 2009, and a further variable which represents the capacity of the respective metal silo(s).[7] The vector of explanatory variables consists of different socio-economic characteristics of the sample households and is summarized in Table 8 (Appendix).

Modeling impact of metal silo adoption

The impact of using metal silos was investigated by focusing on three main aspects: a) food security; b) farmers' (and their families') well-being; and c) sales of maize. We employed linear regression models to assess how the metal silo adoption influenced food security and sales of maize.

Food security

To assess differences between silo users and non-users with regard to food security, farmers were asked how many months they had to buy (i.e. in addition to their own production) maize and beans. These questions covered the years 2008 and 2009. The average value of both years was used in the subsequent analyses. The investigated hypothesis is that silo users need to buy less staple grains from the market and can use their own production due to better storage capacity (Hermann 1991; Coulter et al. 1995; Gladstone et al. 2002). In a first step, empirical density functions of users and non-users are presented for each of the four countries (Guatemala, El Salvador, Honduras, Nicaragua). Because maize and bean production and consumption differ across these countries, we used a regression analysis to test if silo users in general needed to buy less maize and beans.[8] Since the dependent variable is censored (by 0 from the left and 12 from the right), we used a tobit model, in which the numbers of months farms needed to buy maize and beans ($Months_C$) are regressed against dummy variables for countries ($D_{Country}$, Guatemala was chosen as reference category), for silo non-use (the use of silos is the reference category), as well as interaction terms between both dummy variables, where β_0 is the regression intercept:

$$\mathrm{Months}_C = \beta_0 + \beta_1 D_{Non-User} + \beta_2 D_{Country} + \beta_3 D_{\mathrm{Non-User}} * D_{\mathrm{Country}}. \tag{4}$$

Livelihoods

To assess the impact of metal silos on non-economic factors (e.g. health, gender and education issues), as well as on factors that are difficult to quantify (e.g. income), farmers were asked how their situation changed from 2005 to 2009 with regard to the following variables: the family's food situation; the family's income situation; the workload of women; children's health situation, and the children's education situation. Thus, the questions covered aspects from various important fields, i.e. improvements in economic status, food security, gender and children's situation. The answer scale ranged from 1 (high improvement) to 5 (severe worsening), while 3 indicated no change. Category 6 was used if the interviewee indicated 'I don't know'. We tested the hypotheses that silo users faced better economic and social development. This was motivated by the fact that silo adoption allows adopters to generate more income and to reduce their workload, as well as that silo users are more resilient to certain shocks (e.g. price fluctuations or bad harvests). In a first step, cross tables and Pearson Chi-Square tests were used to identify whether significant differences between users and non-users existed. In a second step, regression analyses were also used to consider country-specific effects (and interaction terms) following the Tobit regression approach described in Equation 4.

Sales of production

We expected differences in the sale of stored grain and grain that is not stored (but sold immediately after harvest) with regard to the timing and location of sales as well as with regard to the distribution channel used. These differences were also expected to cause variation in the grain prices received for different storage technologies. The analyses presented in this section are focused on maize because it is the most important crop for the households interviewed.

To investigate whether the location of maize selling was affected by storage technology, the questionnaire included a question asking where maize was sold. The answer categories were as follows: on (one's own) farm; in the village; on the road; in the district town; others. Following the same structure, to whom the maize was sold was also asked. More specifically, the following answer categories were used: intermediates; retailers; super markets; farmers' organizations; direct sales to consumers; others. Furthermore, farmers indicated in which month they sold most of the maize and what price they received, on average. These questions were asked separately for the different categories of maize storage, i.e. for i) maize that is not stored; ii) maize stored in metal silos; and iii) maize stored in other storage systems.

For maize that is sold mainly directly after harvest, no difference between metal silo users and non-users was expected regarding the selling location and time, purchaser and price. In contrast, we expected differences between maize that is not stored and stored maize, because maize storage enables farmers to decide tactically where, when and at what price to sell their maize. Moreover, we expected that maize stored in metal silos could be kept longer than traditionally stored maize. Thus, the selling time as well as the price was expected to differ between these storage types. In order to test these hypotheses, we used cross tables and Pearson chi-square tests. In addition, group comparisons were conducted using the Mann–Whitney test. To test if maize prices from different storage systems lead to different prices, we used a regression analysis that also accounts for country-specific price differences:

$$Price = \beta_0 + \beta_1 D_{\text{StorageType}} + \beta_2 D_{\text{Country}}. \quad (5)$$

The dummy variable for the storage type ($D_{Storage\ Type}$) used maize that is not stored as a reference category, while Guatemala was used as a reference category for the country dummy ($D_{Country}$). Note that price data was only indicated by some farms, and interaction terms were thus not considered due to the lack of freedom in specific category combinations. All prices were given in local currencies and were converted into USD/qq in the results presented.

RESULTS

Investment decision analysis

Table 2 summarizes the estimation results of the double-hurdle model for three of the countries considered in this part of the analysis, i.e. Guatemala, Honduras and Nicaragua. We completed model estimations for the whole sub-sample, i.e. considering all three countries, and separately for Guatemala and Honduras. As the number of relevant user-households for Nicaragua was rather small, we did not estimate the double-hurdle model for it separately. Additionally, as several farm characteristics exhibited a substantial degree of correlation, we did not estimate the model employing all relevant farm characteristics, but only those which showed a low degree of correlation among each other and which scored significant parameter estimates.[9]

Table 2 Estimates of double hurdle model

Variables	All 3 countries	Guatemala	Honduras
1. step			
dummy Honduras	−1.03***	–	–
dummy Nicaragua	−1.15***	–	–
age	−0.02*	–	−0.11*
age^2	–	–	0.00
ownership	0.13**	–	0.21**
dummy training	0.62***	0.74**	–
maize self-sufficiency	0.19***	0.24***	0.20**
dummy altern. storage	−0.51**	−1.05***	–
dummy coffee	–	2.11**	–

constant	−4.24***	−2.95***	−1.23
Number of observations	566	241	212
R2	0.10	0.11	0.08
LR chi2	70.89***	37.50***	16.04***
2. step			
dummy Honduras	0.35***	–	–
dummy Nicaragua	0.12	–	–
age	–	–	0.01***
crop area	–	–	0.06**
dummy electricity	–	–	0.31***
dummy subsidy	–	0.15*	–
constant	3.13***	2.56***	2.15***
Number of observations	165	101	43
Wald chi2	12.09***	2.80*	20.58***

*, **, *** denote statistical significance at the 10%, 5% and 1% level, respectively.

Source: authors' estimates.

The first step model estimates (i.e. estimates of the logistic regression model) for three countries showed that the decision to invest varied significantly across countries.[10] In particular, the adoption of metal silos has been more extensive in Guatemala (the reference country) than in Honduras and Nicaragua in recent years. On the one hand, this might be related to dynamics of the metal-silo adoption in single countries. On the other hand, the adoption rate might be strongly influenced by governmental policies. Considering that more governmental efforts have been made to disseminate metal silos in Guatemala, our finding is indeed in line with the empirical evidence.

Furthermore, according to our estimates, significant differences in investment behavior existed regarding the age of the farm head, land ownership, completion of a training course, extent of maize self-sufficiency (calculated in months a household can cover its maize consumption by stored maize) and use of an alternative storage capacity such as metallic or plastic barrels, conical metal silos, etc. In particular, the estimation results showed that the probability of adoption declined with the age of the household head. This result is consistent with theoretical expectations (Rogers 2003) and findings of other empirical studies for both developing as well as developed countries (Barham et al. 2004; Ersado et al. 2004), suggesting that older people are more reserved regarding the introduction and acceptance of innovations due to declining cognitive and learning abilities. The possession of an alternative storage system also diminishes the probability of metal silo adoption. This result is very rational—a household's need for storage capacity declines if it already possesses an acceptable alternative storage system (e.g. metal or plastic barrels, conical silos).

Though we did not obtain a significant parameter estimate for the variable characterizing the household head's education,[11] the variable 'training' (access to training and advisory services for grain production) had a highly significant positive parameter estimate. The latter suggests that knowledge of production technology seems to spark the household's interest in metal silo acquisition. In our view, this result might be explained in two ways. First, training allows farmers to obtain new knowledge and thus become more aware of possibilities for more efficient utilization of their resources as well as farm organization. Second, training courses present an important communication channel for disseminating information about new technological solutions available on the market, and therefore play an important role in improving farmers' access to relevant markets and production factors.

A larger share of owned land in farm cropland also had a significant impact on the investment decision. On the one hand, farms that possessed a larger portion of their cropland were wealthier and thus might more easily be able to afford a metal silo than less wealthy farms.[12]On the other hand, they might also be more eager to invest in their farm in general, as they have less uncertainty regarding their land property rights, as well as a lower extent of agency costs due to potential information asymmetries between the land owner and tenant.

Finally, the probability of metal silo adoption increased with a household's maize self-sufficiency. This result is in line with the empirical evidence provided by other studies (e.g. Hermann 1991), in particular farms with lower levels of self-sufficiency often did not produce enough maize to fill a metal silo. Hence, metal silos present an effective instrument primarily for households with higher levels of self-sufficiency, while other policy instruments might be more effective for households with lower self-sufficiency levels. In particular, for the latter group of farms, instruments aimed at an increase of productivity through access to more advanced production technologies might trigger a more significant shift in food security and well-being.

The separate estimation of the logistic regression models for Guatemala and Honduras shows that different socio-economic characteristics determine the adoption decision in these two countries. Whereas in Guatemala the decision to acquire a metal silo was influenced significantly by the completion of a training course and the use of an alternative storage system, the main determinants of silo adoption in Honduras were age and the extent of land ownership. The only variable which obtained a significant parameter estimate for both countries was maize self-sufficiency.

The second-step estimates (i.e. the truncated regression model estimates) also suggested the presence of significant differences in investment behavior of farms in the different countries. Though none of the farm characteristics considered obtained a significant parameter estimate in the pooled truncated regression model, the overall significance of the model was high, which obviously pertains to the usage of the country dummies. In addition, the estimation results indicated that the investment extent is significantly higher in Honduras, i.e. the farms in this country invested in larger metal silos. This result presumably is related to the fact that farms in Honduras are on average larger than in Guatemala, and compared to farms in Nicaragua specialize more in maize that requires larger storage capacity than beans.

The extent of investment in Guatemala was found to vary only subject to the provision of governmental subsidy, i.e. farms that received a subsidy had a tendency to buy an additional silo. The estimates of the model for Honduras showed that the extent of adoption (i.e. the storage volume) depended significantly on the age of the household head. However, in contrast to the first-step model estimates, the parameter estimate for the age variable had a positive sign in the second-step model. Still, the results of the first- and second-step model estimates did not contradict each other. The first-step estimates showed that in general, adopters were younger than non-adopters. The second-step estimation results indicated that among the adopters, older farmers tended to acquire larger storage capacities. Older farmers usually had larger families than their younger counterparts, and thus required more storage capacity. Our results also showed that larger metal silos were usually requested by farms with a larger crop area. Access to electricity, which can be used as a proxy for infrastructure development, had a significant impact on investment volume; this suggests that farms with better access to markets, i.e. lower transaction costs, invested significantly more.

Impact assessment

Food security

Figure 1 shows the empirical density functions of months per year the surveyed farm-households had to buy maize from the market. All interviewed farmers produced maize and therefore had little or no need to buy additional maize. For all countries, we found that metal silo users needed to buy maize from the market in fewer months.

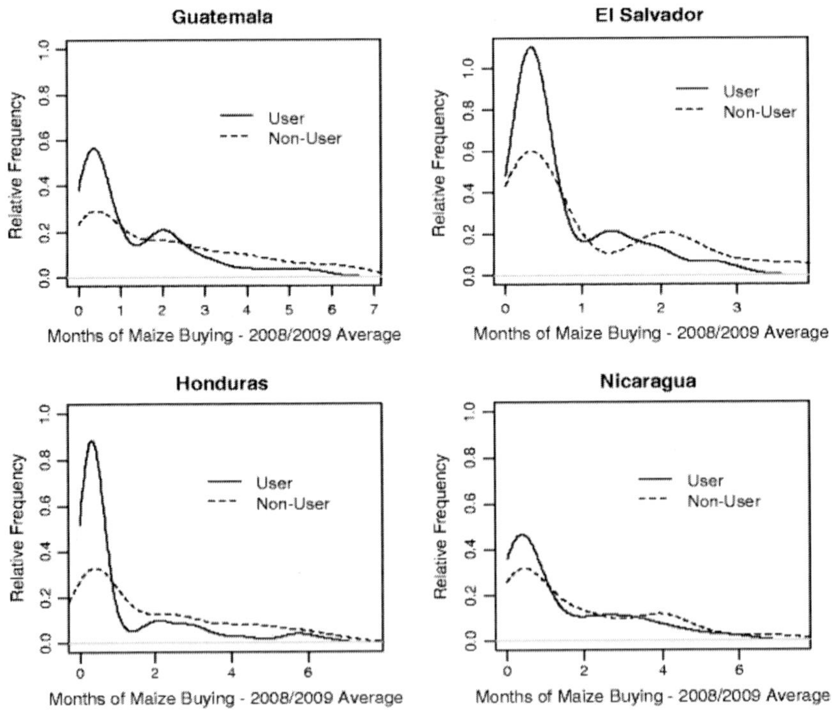

Figure. 1 Months of the year the surveyed farm-households had to buy maize from the market. *Source*: authors' estimates

The results for beans, which are presented in Fig. 2, contrast to our findings for maize. It shows that in all countries, two groups of producers exist that either produce almost enough for their own consumption, or buy most of their beans from the market. The second group of farmers is well represented in Guatemala and El Salvador, while farms in Nicaragua are rather more focused on their own bean production. The strongest differences between metal silo users and non-users are indicated for El Salvador, where users tend to rely much less on their own bean production and buy more beans from the market.

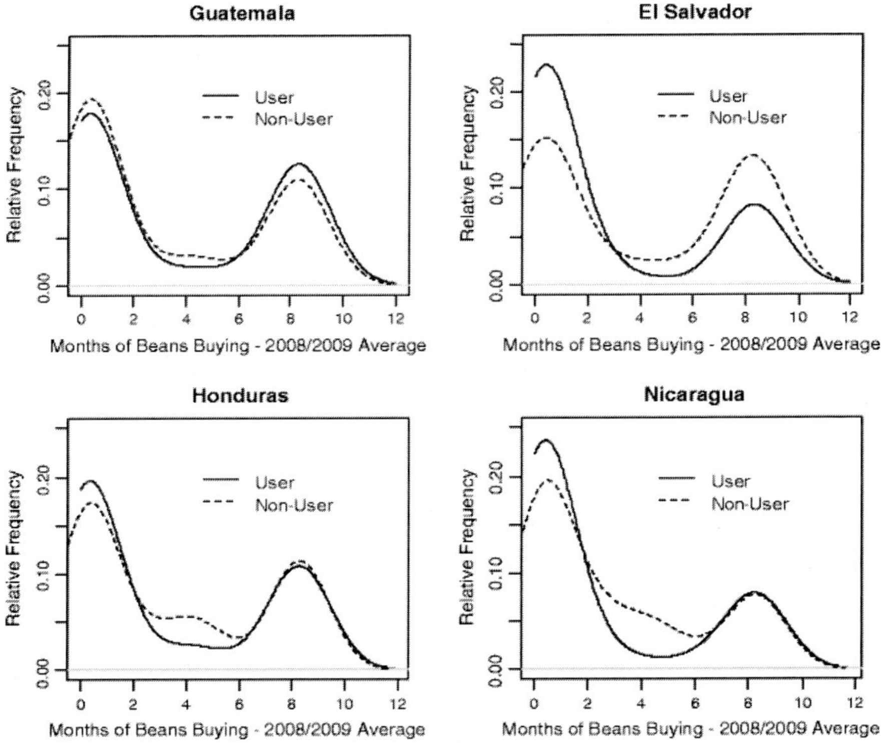

Figure. 2 Months of the year the surveyed farm-households had to buy beans from the market. *Source*: authors' estimates

The results of the Tobit model estimation presented in Table 3 show that non-users needed to buy maize in (significantly) more months than metal silo users. Country effects showed that households in El Salvador needed to buy (on average) less maize and beans (also for Nicaragua) than in Guatemala, which is the reference category in our analysis. For the number of months when beans were purchased, no general impact of silo use was found. However, the significant interaction effect of the dummy for silo non-use and the dummy for El Salvador shows that silo users in this country needed to buy beans in fewer months (compared with those in Guatemala). The latter result is furthermore underpinned by the empirical density function presented in Fig. 2, where the largest difference between metal silo users and non-users was indicated for El Salvador.[13]

Table 3 Tobit regression analysis on the determinants of number of months when additional food buying was necessary (average values for 2008/2009)

	Number of months when *maize* has to be bought	**Number of months when *beans* has to be bought**
Intercept	−0.23***	3.69 ***
Dummy Non-User (vs. User)	0.44 ***	−0.35 (n.s.)
Dummy El Salvador[a]	−0.42***	−1.16***
Dummy Honduras[a]	−0.27**	−0.43 (n.s.)
Dummy Nicaragua[a]	0.02 (n.s.)	−1.20 ***
Dummy Non-User x Dummy El Salvador[a]	−0.13 (n.s.)	1.90 ***
Dummy Non-User x Dummy Honduras[a]	0.14 (n.s.)	0.70 (n.s.)
Dummy Non-User x Dummy Nicaragua[a]	−0.17 (n.s.)	0.80 (n.s.)
Observations	1195	1195
Log-likelihood	−1534	−3212

[a]Country specific effects are evaluated against Guatemala as reference category.
** and *** denote significance at the 5% and 1% levels, respectively
Source: authors' estimates

For the interpretation of the results presented on the relationship between metal silo use and food security, it is important to take into consideration that metal silo users are usually characterized by a higher degree of self-sufficiency (cf. section 4.1). Thus, silo users already relied less on buying maize and beans before they acquired the metal silo. Therefore, the results presented were caused not only by the effect of the metal silo but also by higher general production levels of these farms.

Farmers' well-being

Cross tables for all answer categories (i.e. on developments of the family's food situation, the family's income, the workload of women, children's health and education) showed significant differences between metal silo users and non-users (data not presented). In order to test whether metal silo users tend to assess their situation more positively than non-users, regression analysis was used. Table 4 shows the results of the regression analysis on the assessment of these economic and social aspects of well-being. Note that answer scales ranged from 1 (high improvement) to 5 (severe worsening), while 3 indicated no change. Answers that indicated "I don't know" were not considered in the regression analysis.[14] The results showed that in all categories, households from Guatemala (reference category) had the smallest values, i.e. the best situation, because dummies for all other countries were significantly positive. More importantly, non-users had significantly higher (i.e. worse) responses compared to metal silo users. Thus, metal silo users assessed economic and social development more positively.

Table 4: Regression analysis on the assessment of economic and social aspects of family well-being

	Food Aspects	**Family Income**	**Women Workload**	**Children's Health**	**Children's Education**
Intercept	2.27 ***	2.46 ***	2.70 ***	2.41 ***	2.48 ***
Dummy Non-User (vs. User)	0.50 ***	0.41 ***	0.28 ***	0.28 ***	0.18 **
Dummy El Salvador[a]	0.46***	0.37 ***	0.33 ***	0.26 ***	0.20 **
Dummy Honduras[a]	0.25 ***	0.18 **	0.19 ***	0.28 ***	0.11 (n.s.)

Dummy Nicaragua[a]	0.78 ***	0.57 ***	0.23 ***	0.57 ***	0.22 ***
Dummy Non-User x Dummy El Salvador[a]	−0.42 ***	−0.25 (n.s.)	−0.30 ***	−0.19 **	0.02 (n.s.)
Dummy Non-User x Dummy Honduras[a]	−0.28 **	−0.14 **	−0.17 **	−0.18 **	−0.11 (n.s.)
Dummy Non-User x Dummy Nicaragua[a]	−0.41 ***	−0.34 ***	−0.26 ***	−0.33 ***	−0.23**
Degrees of Freedom	1176	1163	1080	1150	1099
Adjusted R2	0.11	0.07	0.03	0.06	0.02

[a]Country-specific effects are evaluated against Guatemala as reference category.

** and *** denote significance at the 5% and 1% levels, respectively

Source: authors' estimates.

We also asked farmers to indicate the main reason for positive developments in their food security and income situation in an open question. Farmers frequently mentioned increased off-farm employment as a contributing factor for improved food security and income. The reduction in workload for women is explained by the change in postharvest operation, which also requires a different division of labor. Due to the need for shelling, i.e. removing the kernels from the cob and drying all the grain at once for filling the silo, men are more actively engaged in these operations and sometimes use machines, thereby considerably reducing the workload of women. In addition, the removal of grain from the silo for daily consumption, mainly done by women, is more convenient compared to the traditional method of daily shelling.

Selling of maize

Selling of maize immediately after harvest was done mainly at the farm and in the village and, as expected, there was no significant difference in the numbers of respondents who used these two locations between users and non-users of metal silos (Table 5). More than 75% of all farmers who sold their maize immediately after harvest did so to intermediaries, while the rest was mainly sold directly to consumers (data not shown).

Table 5: Location of selling for maize that is not stored

	Users	Non-Users	Row Total
Farm	116	111	227
Village	87	87	174
Road	8	1	9
District town	10	8	18
Other	4	5	9
Column Total	225	212	437
Pearson's Chi-squared test statistic	5.50 (n.s.)		

[a]several households did not indicate sufficient details on maize selling and the total number of observations decreased from 1,195 to 437.

Source: authors' estimates.

In order to compare the selling location and purchaser of stored maize and maize that had not been stored, we focused our analysis on metal silo users.[15] Among the users, details of maize sold that was not stored were reported from 225 farms, while 123 observations were available for maize stored in metal silos. Details on selling maize stored in other systems were reported by 89 farms. Table 6 shows relative frequencies (in %) for the selling location, purchaser, and main month of selling. Stored maize was sold in the village (67%) mainly directly to consumers while non-stored maize was sold on the farm (52%) mainly to intermediaries (76%). With regard to the selling location and purchaser of maize stored in other systems, these ranged between maize that is not stored and maize stored in metal silos. It was mainly sold in the village or the district's town, but less often sold directly to the consumer compared to maize stored in metal silos.

Table 6 Analysis for silo users: stored maize vs. maize that is not stored—location and time of selling and purchaser

	Maize that is not stored ($N = 225$)	**Metal Silo Stored Maize ($N = 123$)**	**Maize stored in other system ($N = 89$)**
Selling Location			
Farm	52%	24%	24%
Village	39%	67%	57%
Road	4%	1%	1%
District town	4%	7%	11%
Other	2%	2%	7%
Pearson's Chi-squared test statistic	49.26***		
Purchaser			
Intermediaries	76%	41%	54%
Commercial house, Supermarket, and Farmers' Organization	0%	2%	5%
Direct to consumer	20%	50%	34%
Others	3%	7%	7%
Pearson's Chi-squared test statistic	56.12***		
Month of Selling			
August–October	12%	7%	14%

November–February	79%	20%	36%
March–July	9%	73%	50%
Pearson's Chi-squared test statistic	302.85***		

*** denotes significance at the 1% level

Source: authors' estimates.

The time of selling varied significantly: the main selling period for maize that is not stored was directly after harvest, 74% of the farmers indicating the period from November to January. In contrast, maize stored in metal silos was mainly sold from March until July, the most critical period before the new harvest, when selling prices were highest. Maize stored in other systems was usually sold after maize that is not stored but before maize that had been stored in metal silos

In order to compare the prices received for maize from different storage systems, we used observations from all 1,195 sample farms. Figure 3 shows box plots of price levels, where all values were converted to USD/qq; it shows that stored maize generates, in general, higher prices than maize that has not been stored (and sold immediately after harvest). Moreover, prices for maize stored in metal silos seem to be on average higher than for maize from other storage systems.

In order also to account for different price levels in the four countries considered, we conducted a regression analysis that used dummy variables for countries and storage system terms (see Equation 5). Maize price levels in El Salvador and Nicaragua were lower than in Honduras and Guatemala (the reference category of this analysis; Table 7). Moreover, if these country-specific effects are considered, maize stored in metal silos led to an average price markup of 1.85 USD/qq, while maize stored in other storage systems generates a price markup of 1.46 USD/qq compared to maize that has not been stored.

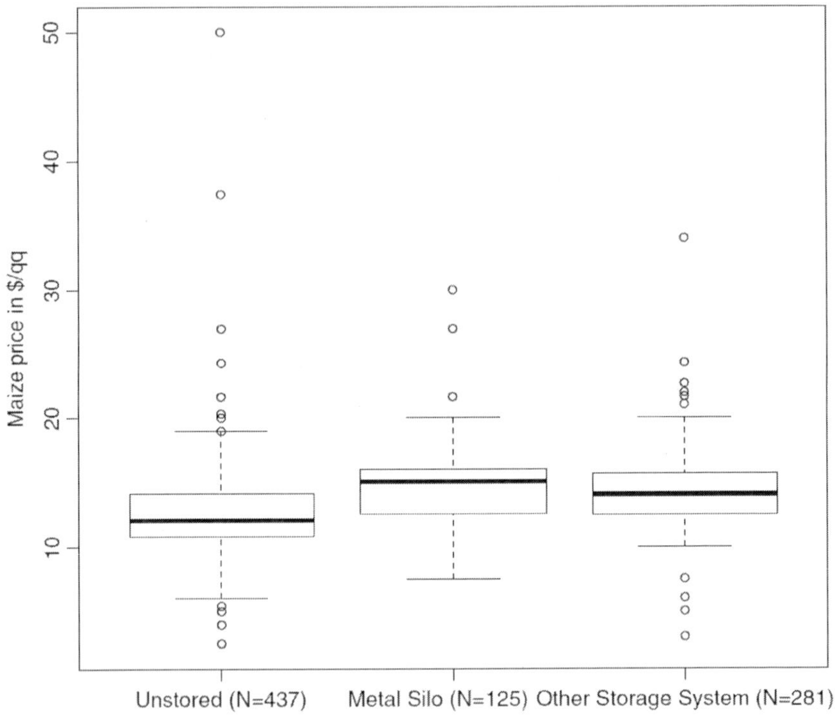

Figure. 3 Maize price by storage system. *Source*: authors' estimates

Table 7: Regression analysis: maize price for different storage systems

	Maize Price in USD/qq
Intercept	13.70 ***
Dummy Storage Metal Silo (vs. maize that is not stored)	1.85 ***
Dummy Other Storage System (vs. maize that is not stored)	1.46 ***
Dummy El Salvador[a]	−1.38 ***
Dummy Honduras[a]	0.54 (n.s.)
Dummy Nicaragua[a]	−2.75 ***

Degrees of Freedom	837
Adjusted R2	0.14

*** denotes significance at the 1% level

Source: authors' estimates.

[a]Country-specific effects are evaluated against Guatemala as the reference category.

DISCUSSION

Postharvest losses in yield are a major factor that negatively affects household food security in rural areas of Central America. From 1983 to 2003, SDC implemented the Postcosecha Program, which aimed at reducing smallholder farmers' postharvest losses by supporting the manufacture and use of metal silos. This study evaluated determinants of the adoption and impact of the use of metal silos on the food security and well-being of rural households in El Salvador, Guatemala, Honduras and Nicaragua from 2005 to 2009.

Our results showed that both the demand for metal silos and the impact of their adoption was different across the four countries. The highest demand, as well as the highest impact of the adoption of metal silos from 2005–2009 was observable in Guatemala, suggesting that this country has found an effective policy for metal silo dissemination. In contrast to the other three countries, where the government might spasmodically disseminate metal silos free of charge, e.g. prior to an election, the government of Guatemala provides regular support in the form of a subsidy to smallholder farmers interested in acquiring metal silos. Additionally, farmers from Guatemala more often reported that they had access to extension services in the form of training in subjects relevant to grain production.

Furthermore, our results indicated that the desire for household self-sufficiency in maize was an important factor in explaining farmers' demand for metal silos. These would only be fully effective if the farmer had the potential to produce enough maize to be self-sufficient, alternative strategies for improving productivity might be more appropriate for subsistence farms with lower levels of potential self-sufficiency. Besides, the adoption of metal silos was influenced by a group of socio-economic characteristics of the households which included age of the head of the household, land ownership, access to extension services and quality of basic infrastructure.

We found that, compared to the non-adopters, the adopter households experienced a significantly higher improvement in their well-being from 2005 to 2009. For instance, as adopters had to buy less food and had improved the marketing of their harvests, they had greater financial freedom. This resulted in better health and education of their children and is likely to have direct spillovers (and multiplier effects) on future generations, ensuring sustainable long-term improvements. Furthermore, silos reduced gender inequalities by reducing women's workload owing to the elimination of the daily shelling. Therefore, the promotion of metal silos seems to provide a path for sustainable social and economic development, which should be considered when evaluating policy. Households were also asked to assess the development of their food security in this period; metal silo users indicated much better development than non-users.

Our results on maize-selling patterns showed that metal silo users are much more flexible regarding when and where they sell their harvest. Primarily, this provides higher economic returns for the users. However, this also has an indirect effect on other households. Because the supply of food from local producers is not limited to harvest periods, price peaks on local markets are expected to decline. Thus, this storage technique is expected to contribute to less variable prices and more affordable food for poor households.

The results of the analysis allow two important policy implications to be derived. First, more targeted policies are required. This aspect primarily concerns the design of policy instruments considering different household clusters. In particular, policy design should differentiate between the needs of and effective instruments for farms performing under and over the subsistence level. Additionally, as our results suggest that factors determining the adoption of an innovative technology might vary from country to country, when developing policy instruments more attention should be paid to regional specifics and should incorporate a careful examination of the main needs and limitations to development in each country. The example of the subsidy model in Guatemala targeting poor rural households is interesting in this respect. Second, regarding the relatively low educational level of farmers, more efforts should be carried out to improve farmers' professional skills, as well as their access to and awareness of innovative technological solutions.

FOOTNOTES

1 Manzana is the local land measurement unit. One manzana is equal to approx. 0.7 hectare.

2 1 quintal (qq.) = 45.45 kg

3 The complete questionnaire is available upon request.

4 This ratio was calculated based on the farm maize production and consumption data for 2008, which was more representative than 2009 considering weather conditions for grain production. We employed a relatively high upper limit of this indicator to consider that in bad harvest years maize production might drop substantially and thus seriously affect farm food security.

5 We excluded from the sample farms with higher levels of self-sufficiency (i.e. maize production exceeds family consumption by three times), as the focus of our analysis is on small subsistence households. Additionally, considering that the adoption of metal silos might have been influenced for commercially-oriented farms by factors different from those for subsistence farms, a joint estimation of the model for these two groups of farms might have caused biased estimates.

6 The farm responses refer to the situation in 2009, 2008, 2005 and also the year prior to metal silo acquisition, which is individual for each farm-adopter. Both periods, from 2009 to 2010 and from 2008 to 2010, were too short to form a sufficiently large sub-sample of farm-adopters. Accordingly, the period from 2005 to 2009 was the only available option for obtaining a sufficiently large sub-sample and also for using the data corresponding to the same reference period for all farms.

7 In the truncated regression, we used the Box-Cox transformation of the dependent variable.

8 We included a control question regarding how many months farmers' own production of maize and beans was sufficient for family food provision. This variable confirmed the presented results.

9 For example, several farm characteristics representing farm size, e.g. crop land, number of employees, etc., had a rather high correlation with the household maize self-sufficiency indicator.

10 This result might, however, be related to the sample composition in this part of the analysis, in particular the share of user-households in the total number of households were substantially lower for Honduras and Nicaragua than for Guatemala. If this composition of the sub-samples corresponded well to the real situation, then our finding is correct.

Conversely, if the sample composition did not represent the actual stratification of the farmers, then this estimation result was biased due to inappropriate sampling.

11 As can be seen from Table 8 in the Appendix, educational background does not vary sufficiently across household heads; most household heads are alphabetized or have completed primary school. This low variation in educational background can explain insignificant parameter estimates for this variable.

12 Also, farms with a higher level of land ownership might more easily gain access to credit for buying a silo, as they can use their own land as collateral.

13 We also included a control question in the survey on how many months households could cover their consumption from stored grain (maize and beans) production. These results confirmed the presented results: metal silo users indicated a higher degree of self-sufficiency.

14 Furthermore, missing values were generated if the question did not apply to certain families that were interviewed (e.g. they had no children).

15 This restriction on adopters is necessary to ensure that we compare farms at the same level of consumption and with the same selling patterns.

Open Access

APPENDIX

Table 8 Descriptive statistics of variables employed in the double hurdle model

Variable	Description	165 metal silo users				401 non-users			
		Mean	S.D.	Min	Max	Mean	S.D.	Min	Max
silosizesum	farm total metal silo storage	15.82	7.62	0	68	0.00	0.00	0	0

	capacity, quintals								
age	household head age, years	44.09	14.5 3	18	86	45.13	14.0 4	18	82
education	educational background (1 - none, 5 - professional education and higher)	2.43	0.98	1	5	2.45	1.01	1	5
dummy training	1, if the household head completed a course in a grain production subject; 0 otherwise	0.32	0.47	0	1	0.15	0.36	0	1
family size	number of family members	5.27	2.46	1	13	5.14	2.26	1	14
dummy off-farm job	1, if one of the children left family to find an employment ; 0 otherwise	0.22	0.42	0	1	0.26	0.44	0	1
crop area	farm crop area, manzanas	1.57	1.28	0.0 8	10	1.74	1.40	0.0 6	13
ownership	share of own land in farm crop area	1.35	1.81	0	10	1.33	1.75	0	20
dummy workers	1, if the farm employs permanent or temporal workers; 0 otherwise	0.49	0.50	0	1	0.49	0.50	0	1
dummy livestock	1, if the farm possesses livestock; 0 otherwise	0.30	0.46	0	1	0.24	0.43	0	1
cattle	number of cattle	0.58	1.30	0	6	0.74	3.34	0	50

pigs	number of pigs	0.16	0.46	0	2	0.31	1.15	0	15
dummy coffee	1, if the farm has a coffee plantation; 0 otherwise	0.07	0.25	0	1	0.05	0.22	0	1
maize self-sufficiency	number of months covering the family demand in maize from stored maize.	10.92	2.09	3	12	9.64	3.13	0	12
bean self-sufficiency	number of months covering the family demand in beans from stored beans	5.04	5.83	0	12	5.25	5.78	0	12
dummy altern. storage	1, if the farm possesses an alternative storage capacity; 0 otherwise	0.18	0.39	0	1	0.22	0.42	0	1
dummy subsidy	1, if the farm obtained a subsidy when purchasing a metal silo; 0 otherwise	0.51	0.50	0	1	0.00	0.00	0	0
dummy electricity	1, if the farm has access to electricity; 0 otherwise	0.70	0.46	0	1	0.50	0.50	0	1

REFERENCES

1. Abeleira, N., Pérez, O. C., Ferrer, J. G., Sánchez, E. G., & Figueroa, H. H. (2008). Postharvest losses of common bean seed in a traditional storage system. *Agricultura Técnica en México, 34*(001), 91–100.

2. Aramyan, L., Oude Lansink, A., & Verstegen, J. (2007). Factors underlying the investment decision in energy-saving systems in Dutch horticulture. *Agricultural Systems, 94*, 520–527.
3. Barham, B. L., Foltz, J. D., Jackson-Smith, D., & Moon, S. (2004). The Dynamics of Agricultural Biotechnology Adoption: Lessons from series rBST Use in Wisconsin, 1994–2001. *American Journal of Agricultural Economics, 86*, 61–72.
4. Baumeister, E. (2010). *Pequennos productores de granos basicos en Ameica Central.* FAO-RUTA, pp. 38
5. Bravo, J. (2009). *Metal silos and food security.* Lessons learned from a successful Central American Post-Harvest Program, pp. 270
6. Coulter, J., Brüssel, J., & Wright, M. (1995). *Programa Postscosecha en Centroamerica: Evalucion de impacto y de la sostenibilidad.* SDC, pp. 56
7. Cragg, J. G. (1971). Some statistical models for limited dependent variables with application to the demand for durable goods. *Econometrica*: 829–844
8. Ersado, L., Amacher, G., & Alwang, J. (2004). Productivity and land enhancing technologies in northern ethiopia: health, public investments, and sequential adoption. *American Journal of Agricultural Economics, 86*, 321–331.
9. Florkowski, J., & Xi-Lling, J. (1990). Simulating the impact of pecan storage technology on farm price and grower's income. *Southern Journal of Agricultural Economics, 22*(2), 217–222.
10. Gladstone, S., Astuias, L., & Hruska, J. (2002). *Estudio de adopción y de impactos de tecnologias Postscosecha.* COSUDE, pp. 82
11. Hermann, H. (1991). *Seguridad alimentaria.: comparación de impactos socio-eoconomicas en la tenencia del silo métalico versu sistema tradicional de almacenamiento.* SDC, pp. 41
12. Hernández, M. (2008). *Analisis de la evolucion de los precios de maiz, arroz trigo y de sus productos derivados en Guatemala: un studio de los impactos de los precios en el consume de estos cereals en la coyuntura actual.* Infore final de investigacion de coyunctura. Universidad de San Carlos de Guatemala, pp. 33
13. Pérez, F., Barrios, J., & Pavón, K. (2010). *Responses to high world food prices.* Nitlapan/IDS/ODI/DFID
14. Raboud, G., Narvaez, M., & Sieber, J. (1984). *Evaluacion de peridas post-produccion de maiz a nivel de pequeños y medianos productores en Honduras.* COSUDE, pp. 18
15. Rogers, E. M. (2003). *Diffusion of Innovations* (5th ed.). New York: Free Press.
16. SDC (2008a). Metal Silos. Manual for manufacturing metal silos for grain storage. 2nd edn
17. SDC (2008b). Latin Brief. Central America: Fighting poverty with silos and job creation. www.postharvest.ch, October 07, 2009
18. SDC (2011). Five year ex-post impact study Postcosecha Program in Central America. Report prepared on behalf of the Swiss Agency for Development and Cooperation. BernGoogle Scholar
19. Tefera, T., Kanampiu, F., Groote, H. D., Hellin, J., Mugo, S., Kimenju, S., et al. (2010). The metal silo: An effective grain storage technology for reducing post-harvest insect and pathogen losses in maize while improving smallholder farmers' food security in developing countries. Crop Protection, in press, 1–6
20. World Bank (2008). World Development Indicators. http://data. worldbank. org/indicator
21. Zappacosta, M. (2005). *Honduras: Market profile for emergency food security assessments.* World Food Program, United Nations

CHAPTER 16

Farmers' Knowledge and Perception of Postharvest Physiological Deterioration in Cassava Storage Roots in Ghana

R. Prempeh[1], J. A. Manu-Aduening[1], B. O. Asante[1*], I. K. Asante[2], S. K. Offei[2] and E. Y. Danquah[2]

[1]CSIR-Crops Research Institute, P.O. Box 3785, Kumasi, Ghana

ABSTRACT

Background

In spite of the essential role of cassava in ensuring household food security and employment for most rural farm households, postharvest physiological deterioration (PPD) of the roots which is serious abiotic stress in cassava renders the roots unmarketable, thereby reducing the economic value of the crop. This paper investigates farmers' knowledge and perception of PPD in the storage roots of cassava among smallholders in Ghana.

Methods

A participatory rural appraisal and a formal survey involving 137 farmers across four agro-ecological zones were conducted using focus group discussion and semi-structured questionnaires, respectively. With the appropriate perception indices, farmers' perceptions of each PPD attribute were assessed by the use of a 5-point Likert scale. The Kendall's coefficient of concordance was used further to rank identified cassava production constraints.

Results

The result shows that a significant number of farmers were aware of PPD and the corresponding changes in the roots. It was also confirmed that PPD commences from 2 to 3 days after harvest. Farmers perceived PPD to be caused by pathogens, sunlight, mechanical injuries during harvesting and atmospheric air. The results further indicated that farmers have in-depth knowledge of PPD in cassava. The findings suggest the importance of PPD which is ranked as one of the major constraints in cassava production with about 40% agreement among the farmers. Storing of cassava roots in polythene and jute sacks were found to delay PPD for a few days.

Conclusion

There is the need to concert cassava research and development efforts to tap into, and investigate further the adaptability and sustainability of these storage practices and methods to minimize both PPD and deterioration of cassava roots during storage.

Keywords: Cassava Constraints Perceptions Postharvest physiological deterioration Storage roots Ghana

BACKGROUND

Cassava (*Manihot esculenta* Crantz) is the sixth most important food crop in the world (Food and Agriculture Organization [1]). It is the third most important source of carbohydrates in Africa [2] and the second most important staple crop in Ghana [3]. Cassava serves as food for over 800 million people worldwide [4], providing about 500 calories daily for more than 70 million people [5]. It is estimated that 250 million people in sub-Saharan Africa (SSA) derive half of their daily calories from cassava [1]. The roots are used for animal feed, industrial starch production and income generation for many small-scale farmers [6]. Cassava roots and leaves are available all year round [7] making it an important food security crop, even in drought-prone areas [5]. Cassava occupies a key position in Ghana's agricultural economy contributing 22% of the agricultural gross domestic product [8]. It is the most important crop in terms of area cropped and total production. Cassava is produced by over 70% of Ghanaian farmers and consumed by more than 80% of the population [9], indicating its importance in food security.

In spite of the importance of the crop in the Ghanaian economy, its production is faced by a number of challenges. The key among these challenges is its short shelf life termed postharvest physiological deterioration (PPD). PPD is an oxidative reaction that starts immediately after harvesting when the root is detached from the mother plant [10, 11]. It starts from the central vascular bundles of the root, spreading to the adjacent storage parenchyma, and subsequently, the stored starch undergoes structural changes [12]. The roots deteriorate 24–48 h after harvest and subsequently changes colour. Visible signs of PPD are vascular streaking with blue or black discolouration, rendering the roots unpalatable and unmarketable [11]. PPD is a serious challenge confronting farmers, processors and consumers. To delay PPD, storage roots have to be consumed or processed immediately after harvesting. However, this is only practical on smaller scales but not on commercial levels. Participatory rural appraisal has been employed successfully in crop improvement. It recognizes the importance of farmer indigenous knowledge and skills to understand the target area and identifies production constraints, which result in the breeding of demand driven crop varieties that meet the needs of farmers. This paper evaluates farmers' knowledge and perception on PPD as a major constraint in cassava production. Besides PPD tolerance, this paper identifies and ranks other characteristics that farmers prefer in a new cassava cultivar. Such information is vital for guiding the development of improved demand driven cassava cultivars for increased productivity and incomes of farmers.

METHODS

Study areas

The study was conducted in the Rain Forest, Semi-deciduous Rain Forest, Forest Savanna Transition and Coastal Savanna agro-ecological zones in Ghana (Table 1). The selected communities were Nkawie and Seidi in the Atwima Nwabiagya District of the Ashanti region, Ayigbe, Amangoase and Akrofrom in the Wenchi District of the Brong Ahafo region. Other communities were Ohawu in the Akatsi South District of the Volta region and Abodobi and Yeyaso in the Fanteakwa District of the Eastern region. These communities were selected because cassava farming and processing are the major occupation of the inhabitants. In addition, cassava is one of the major staples in these communities and plays an important role in ensuring household food security.

Table 1 List of agro-ecological zones, districts and communities the study was conducted

Agro-ecological zone	District	Community
Rain Forest	Fanteakwa	Yeyaso
		Abodobi
Semi-deciduous Rain Forest	Atwima Nwabiagya	Nkawie
		Seidi
Forest Savanna Transition	Wenchi	Ayigbe
		Amangoase
Coastal Savanna	Akatsi South	Ohawu

These agro-ecological zones are characterized by bimodal rainfall pattern ranging between 800 and 2500 mm per annum with an average of about 950 mm per annum. The major season begins in March and ends in July for the Rain Forest, Semi-deciduous and Coastal Savanna and April to July for the Forest Savanna Transition agro-ecological zone. The minor rainy season is from September to October for the Forest Savanna Transition and Coastal Savanna while that for the Rain Forest and the Semi-deciduous zones begin in September and ends in November.

Sampling procedure

Multi stage sampling procedure was employed to obtain the sample respondents for this study. One region was purposefully selected from each of the four agro-ecological zones. One district was randomly selected from a list of major cassava-growing districts in the region. From each of these districts, two communities were sampled with the exception of Akatsi South. Participatory rural appraisal (PRA) was conducted using focus group discussion (FGD) and a carefully constructed checklist. This was followed by a formal survey in which data were collected through interviews using semi-structured questionnaires. The FGD was used to elicit information on constraints in cassava production and marketing, perceptions of PPD, existing methods used to delay PPD and preferences for improved varieties with delayed PPD among the farmers at the community level. In each community, the FGD involved two main focus groups (male and female focus groups) after which they were merged into a single group for synthesis and confirmation of the issues discussed at the group level. After the FGD, a minimum of 15 farmers per community were randomly selected for the semi-structured interviews. Farmers were interviewed individually to solicit their views on the different topics. Overall, 137 farmers were interviewed.

Data analysis

Data collected were coded and analysed using Statistical Package for Social Sciences (SPSS). Descriptive statistics were used to summarize the data. Farmers' perceptions of each PPD attribute were assessed by the use of a 5-point Likert scale (Likert 1932). The following perception indices; 2 = strongly agree: 1 = agree: 0 = indifferent: −1 = disagree: −2 = strongly disagree were used. The mean perception for each PPD attribute was measured using Eq. 1.

$$\bar{x} = \frac{\sum n_i - x_i}{N} \tag{1}$$

where n_i = number of individuals who chose the *i*th response, x_i = the *i*th response, N = the total number of respondents.

For the Likert scale, a positive mean denotes agreement, whereas negative implies disagreement. To evaluate the importance of cassava production constraints among the farmers, the Kendall's coefficient of concordance (W) described by Mattson [13] was used to rank the constraints. A lower mean rank indicates the importance of the constraint and vice versa. The Kendall's W was computed as shown in Eq. 2.

$$W = \frac{12\sum \overline{R}_i^2 - 3N(N-1)^2}{N(N-1)} \tag{2}$$

where W = Kendall's value, N = total sample size, R = mean of the rank.

The Kendall's coefficient of concordance (W) is a measure of the extent of agreement or disagreement among farmers of the rankings obtained. The value of W is positive and ranges from zero to one where one denotes perfect agreement among farmers of the rankings and zero denotes maximum disagreement. This approach was further applied to identify and examine farmer preferred varietal characteristics in a new cassava cultivar.

RESULTS AND DISCUSSIONS

Characteristics of farmers

Distribution of the cassava farmers and their characteristics are presented in Tables 2, 3, respectively. Over 50% the sampled farmers were males with an average age of 45 years. The total number of farmers constituted of about 35% from the Ashanti region, 32% from Brong Ahafo region, 24% from the Eastern region and 8% from the Volta region (Table 2).

Table 2 Distribution of cassava farmers by community and gender

Community	Males	Females	Total	Percentage (%)
Nkawie	14	9	23	17
Seidi	20	5	25	18
Ayigbe	10	18	28	20
Amangoase	7	10	17	12
Yeyaso	14	8	22	16
Abodobi	8	3	11	8.5
Ohawu	7	4	11	8.5
Total	80 (58%)	57 (42%)	137	100

The farmers cultivated an average of 1 ha of cassava and travelled an average of 26 km from their farm to the market. Mixed cropping was practised by over 80% of the farmers and the cassava is largely intercropped with other food crops such as maize. Over 80% of respondents conserve their planting materials before planting.

Cassava production constraints

Seven major cassava production constraints were identified and ranked by farmers (Table 4). Overall, inadequate capital was ranked as the most important constraint and this was consistent across all the four agro-ecologies. This was followed by inadequate processing centres as this was also consistent across all the agro-ecologies.

Table 3 Characteristics of sampled farmers

Variable	Percent/mean (SD)
Age of farmers (years)	45.13 (12.87)
Farm sizes (hectares)	1.0 (0.7)
Distance from farm to market (km)	25.7 (22.7)
Cropping system	
Monocropping (%)	19.6
Mixed cropping (%)	80.4
Major cassava intercrops	
Plantation/tree crops (%)	8.8
Food crops (%)	73.7
Vegetable crops (%)	17.5
Conservation of plant material (%)	84.8

Table 4 Cassava production constraints across agro-ecological zones

Production constraints	Coastal		Forest		Semi-deciduous		Transition		Overall	
	Mean rank	Rank	Mean rank	Rank	Mean rank	Rank	Mean rank	Rank	Mean rank	Rank
Inadequate capital	1.91	1	1.94	1	1.94	1	1.85	1	1.91	1
Inadequate processing centres/buyers	2.27	2	2.12	2	1.94	1	2.34	2	2.14	2
High cost of transportation	4.82	5	4.47	3	4.52	4	4.73	4	4.31	3
Deterioration of roots after harvest	4.27	3	4.53	4	4.48	3	3.93	3	4.58	4
High cost of labour (in the dry season)	5.09	6	4.65	5	4.69	5	4.82	5	4.74	5
Pests and diseases	4.55	4	4.79	6	5.04	6	4.84	6	4.89	6
High cost of agrochemicals	5.09	6	5.5	7	5.4	7	5.49	7	5.43	7
W ($p < 0.000$)	0.384		0.4137		0.447		0.411		0.416	

As fresh cassava roots cannot be stored for long due to the high moisture content and had to be processed immediately after harvest in order to increase the shelf life, deterioration of cassava roots after harvest was indicated as the next most important production constraint by the farmers.

Other constraints identified were high cost of labour (due to the long cropping cycle and harvesting in the dry season), pests and diseases and high cost of agrochemicals. At the agro-ecological zone level, however, pest and diseases were of priority to the farmers in the Coastal Savanna as they ranked it as the fourth most important constraint. The extent of agreement among the farmers of the rankings as indicated by the Kendall's '*W*' suggests that overall more than 40% of the farmers are in agreement with the rankings. However, the highest agreement was recorded by farmers in the Semi-deciduous zone (44.7%) and the least by farmers in the Coastal Savanna (38.4%) agro-ecological zone.

Farmers' indigenous knowledge on postharvest physiological deterioration

Majority of the farmers (95.3%) were aware of PPD and its associated changes in the roots (i.e. colour change to brown, blue–black or black and subsequently the roots becoming dry, hard and dark) (Table 5). PPD was perceived to generally commence between 2 and 3 days after harvest. The main causes of PPD as perceived by the farmers were pathogens, sunlight, mechanical injuries during harvesting and atmospheric air. A number of methods were identified for storing cassava roots after harvest in order to reduce deterioration. However, it was common among the farmers that no PPD-tolerant cultivars exist, but rather, some cultivars were perceived to deteriorate faster than others.

Table 5 Farmers' knowledge on PPD in cassava

Attribute	Response by farmers
Symptoms of PPD	Roots change colour to brown or blue black and becomes dry, hard and dark
Number of days for cassava roots to deteriorate after harvest	3 days
Causes of PPD	Air, mechanical injuries, sunlight and heat
Methods for cassava root storage	In water or polythene bags, sacks, pits, under shade
Availability of PPD-tolerant cultivars	No PPD-tolerant cassava cultivars available

Farmers' perception of PPD in cassava storage roots

For better understanding of the dynamics among farmers regarding their perceptions of PPD, the general perception among the farmers was evaluated using a 5-point Likert scale.

The results of farmers' perception of PPD are presented in Table 6. Perceived challenges with deteriorated roots were difficult to pound, reduced root quality and reduced market value. Also, PPD was perceived to start between 2 and 3 days after harvest and it was believed to be caused largely by mechanical injuries during harvesting.

Table 6 General perception on PPD among sampled farmers

No.	Perception attribute	Strongly agree (2)	Agree (1)	Indifferent (0)	Disagree (−1)	Strongly disagree (−2)	Mean score
1	Deteriorated roots taste better	35	23	5	54	20	−0.01
2	Deteriorated roots are difficult to pound	64	69	1	3	0	1.42
3	Deteriorated roots affect root quality	30	105	0	2	0	1.19
4	Roots deteriorate 2–3 days after harvest	46	88	0	2	1	1.28
5	Mechanical injuries cause PPD	69	55	1	11	1	1.31
6	Micro-organisms cause PPD	7	42	26	22	39	−0.32
7	PPD reduce market value of roots	36	95	0	5	1	1.17
8	Deteriorated roots are easy to process	9	16	9	75	27	−0.69
9	Deteriorated roots can be used to prepare many dishes	3	36	2	61	35	−0.65
10	PPD can be prevented	9	34	27	39	24	−0.26

These perception attributes had mean score values between 1.17 and 1.42, implying that largely farmers were in agreement with the attributes represented by these indices. All the other indices had negative mean score values (−0.01 to −0.69) indicating their disagreement with those perception attributes. Farmers were somehow indifferent in some attributes with the fact that deteriorated roots taste better and also that PPD was caused by micro-organisms and thus can be prevented. Conversely, the farmers disagreed on the perception that PPD roots were easy to process and PPD roots can be used to prepare many dishes.

Bulk harvesting and methods for cassava root storage

The methods used by farmers to store cassava roots after harvest are presented in Fig. 1. Farmers harvest cassava roots in bulk only when there is availability of buyers; otherwise, harvesting is mostly done in bits for contingency purposes such as for sale and home consumption. Over 68% of farmers store the roots after harvest while the remaining farmers do not store but sell or consume immediately after harvest. However, among the farmers who store their roots, about 34% store in sacks; about 27% in polythene; about 26% in water; about 8% in pits and a little below 6% store the roots under shade.

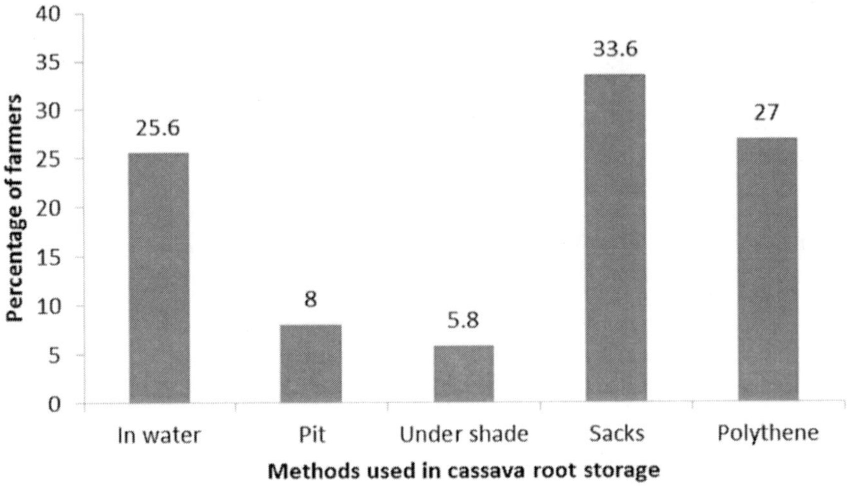

Figure. 1 Methods of storing cassava roots after harvesting

Ranking of farmer preferred traits

On the whole, high yield was identified as the most preferred varietal trait and this was similar across the different agro-ecological zones (Table 7). This was followed by early maturity, high dry matter, delayed PPD, poundability, large roots, resistance to pests and diseases and peel colour (in order of importance).

Table 7 Ranking of farmer preferred cassava traits

Desirable trait	Coastal		Forest		Semi-deciduous		Transition		Overall	
	Mean rank	Rank	Mean rank	Rank	Mean rank	Rank	Mean rank	Rank	Mean rank	Rank
High yield	1.64	1	1.53	1	1.52	1	1.5	1	1.52	1
Early maturity	3	2	2.71	2	2.77	2	2.64	2	2.7	2
High dry matter	4.36	5	4.47	5	4.65	5	4.43	5	3.59	3
Delayed PPD	3.82	4	4	4	3.4	3	3.91	4	3.76	4
Poundability	3.27	3	3.41	3	3.6	4	3.64	3	4.5	5
Larger roots	5.98	6	5.98	7	6.2	7	5.76	6	5.97	6
Resistance to pests and diseases	6	7	5.91	6	5.98	6	6.02	7	6.72	7
Peel colour	7.27	8	7.26	8	7.29	8	7.18	8	7.25	8
Kendal's W, $p < 0.000$	0.641		0.666		0.694		0.663		0.671	

More than half of the farmers were in agreement with the rankings (67.1%); however, the highest agreement was recorded by farmers in the Semi-deciduous Forest (69.4%) while farmers in the Coastal Savanna recorded the least (64.1%).

DISCUSSION

Major constraints ranked by the famers were inadequate capital to expand their businesses, poor bargaining ability resulting in low pricing as well as the physiological deterioration of roots 2–3 days after harvest. These findings concert with that of [14], under the collaborative study of cassava in Africa project, in which two-thirds of Ghanaian farmers identified postharvest losses as a major risk factor in cassava production. This makes postharvest physiological deterioration a serious issue for farmers and other stakeholders involved in the cassava value chain. Other constraints identified were inadequate processing centres and high cost of transportation from farm gate to processing centre/market. This could be due to poor accessibility to farmers' field, poor road conditions and inappropriate means of transport to carry the bulky and highly perishable roots. This agrees with report by [15,17]. Thus, identifying ways of reducing the deterioration of cassava roots has a high potential of increasing the incomes of farmers.

The results further indicate that farmers have in-depth knowledge of postharvest physiological deterioration in cassava. This could be explained by their experience acquired over the years and indigenous knowledge in cassava cultivation. The discolouration of the roots and the number of days it takes before PPD sets in as perceived by farmers confirm earlier studies [12, 18 19]. This implies that PPD has been known among farmers for a while and has been a constraint in production, marketing and processing of cassava. Farmers' indigenous knowledge on PPD makes them to believe that PPD cannot be prevented or controlled because it is a natural process. It is also evident from the results that there are no PPD-tolerant cultivars although some cultivars deteriorate faster than others; thus, developing cultivars that are tolerant to the PPD is very essential. This will not only erase the perception of no PPD-tolerant cultivars but also reduce PPD in harvested roots and increase the income of farmers and other stakeholders along the cassava value chain.

One of the ways of reducing PPD in cassava roots is through the storage of roots using appropriate methods. Several researchers have designed and evaluated cassava storage methods that have the potential of extending PPD in cassava roots [19,-21] such as refrigeration at 4 °C and coating with paraffin or waxing. However, farmers have not adopted any of these cassava storage methods but have rather improvised their own means of storage. This may be due to the fact that these methods have not been well adapted to their local conditions and may also be cost-ineffective. In Ghana, most cassava farmers are poor and live in rural areas, and thus may not be able to afford refrigerators for storing the roots. Subsequently, even in situations where farmers are able to obtain refrigerators, they may not be able to afford the associated high cost of electricity. This makes such a storage method non-viable and cost-ineffective especially for storing large quantities of roots for the markets.

The findings suggest that storage of cassava roots in water is not an appropriate method while storage in polythene and jute sacks could delay PPD for a few days. Hence, these methods could be modified and adapted well into farmers conditions to be more effective and sustainable for minimizing both PPD and root rot during storage. The key reasons why farmers use their own storage methods could be attributed to the fact that their methods are less expensive and user-friendly. Although farmers perceived that PPD could not be solved, their preference for an improved cassava cultivar with delayed PPD was substantial as delayed PPD was ranked after high yield, early maturity and high dry matter. The rankings confirm the assertion that high yield is a very important attribute that farmers always desire in improved cultivars. The long maturity of the crop with its associated high cost of labour was also identified as another important constraint; hence, they prefer early maturing cultivars to late maturing cultivars that they currently cultivate. Since about 50% of the total production of cassava is consumed as *fufu*, the roots should also have good cooking qualities particularly for pounding. The peel colour was also of interest to the farmers because they perceived pink peel delays PPD more than the cream and white peels.

The high market value of cassava cultivars with such preferred traits could possibly explain the selection of those traits. This finding is in parallel with that of [22] where these traits were identified as farmer preferred traits in improved cassava variety. Adequate considerations should therefore be given to these varietal attributes in the development of improved cassava cultivars with delayed PPD. This would ensure that stakeholders obtain the best from the cassava roots and subsequently increase their incomes and improve their livelihoods.

CONCLUSION AND RECOMMENDATION

The study examined farmers' knowledge and perception of PPD in the storage of cassava roots among smallholders in Ghana. Major cassava production constraints identified were inadequate capital, low pricing, physiological deterioration, inadequate processing centres, pests and diseases, high cost of labour and transportation. Due to the pivotal role played by processing in managing postharvest losses in cassava, establishment of more processing centres is vital to the development of a more effective and efficient cassava value chain in Ghana.

Farmers prefer cassava cultivars with good and stable yield, early maturing, high dry matter, delayed PPD, larger roots, poundability, resistance to pests and diseases and pink peel colour. Combination of these cultivar traits would facilitate their adoption by farmers and other stakeholders along the cassava value chain. Polythene and jute sacks were found to be the best storage technique for the cassava roots.

Farmers were aware of and had some indigenous knowledge about PPD. Research and development need to direct efforts at investigating this indigenous knowledge in relation to the available storage methods to develop sustainable cost-effective storage techniques capable of minimizing PPD during storage. The findings revealed that PPD is a major problem that needs to be addressed. It was evident that there are no PPD-tolerant cassava cultivars available to the farmers. However, some cultivars were found to deteriorate faster than others. This suggests that genetic diversity exist in farmers materials for improvement on delayed PPD. Determining the extent of diversity among cassava germplasm from farmers collections, research institutes and genetic resource centres could help in the identification of new sources of tolerance for PPD as well as other important economic traits. This would improve the storability of cassava roots and enhance cassava value chain in Ghana.

ABBREVIATIONS

PPD: postharvest physiological deterioration

FAO: Food and Agriculture Organization

SSA: sub-Saharan Africa

MoFA: Ministry of Food and Agriculture

FGD: focus group discussion

PRA: participatory rural appraisal

SPSS: Statistical Package for Social Sciences

NRI: Natural Resources Institute

GCP: Generation Challenge Program

WACCI: West African Centre for Crop Improvement

ACKNOWLEDGEMENTS

The authors are grateful to the Generation Challenge Program (GCP) for providing funds for the PhD study of the first author from which this paper was developed as well as the West African Centre for Crop Improvement (WACCI).

REFERENCES

1. Food and Agriculture Organisation (FAO). Facts and figures. Rome, Italy www.faostat.org; 2008
2. Owolade OF, Dixon AGO, Adeoti AYA. Diallel analysis of cassava genotypes to anthracnose disease. World J Agric Sci. 2006;2:98–104
3. Adjei-Nsiah S. Yield and nitrogen accumulation in five cassava varieties and their subsequent effects on soil chemical properties in the forest/savanna transitional agro-ecological zone of Ghana. J Soil Sci Environ Manag. 2010;1:15–20
4. Nassar NMA, Elkholy H, Eltantawy A. Cassava productivity worldwide: an overview. Ceres. 2002;248:369–81
5. Chavez AL, Sanchez T, Jaramillo G, Bediya JMI, Echeverry J, Bolanos EA, Ceballos H, Iglesias CA. Variation of quality traits in cassava roots evaluated in landraces and improved clones. Euphytica. 2005;143:125–33
6. Kawano K. Thirty years of cassava breeding for productivity: biological and social factors for success. Crop Sci. 2003;43:1325–35
7. Ntawuruhunga P, Ssemakula G, Ojulong H, Bua A, Ragama P, Kanobe C, Whyte J. Evaluation of advanced cassava genotypes in Uganda. Afr Crop Sci. 2006;14:17–25
8. Ministry of Food and Agriculture, (MoFA). Agriculture in Ghana: facts and figures. Produced by Statistics, Research, and Information Directorate: Crops Research Institute; 2011
9. Parkes EY, Allotey DFK, Lotsu E, Akuffo EA. Yield Performance of five cassava genotypes under different fertilizer rates. Int J Agric Sci. 2012;2(5):173–7
10. Beeching JR, Dodge AD, Moore KG, Phillips HM, Wenham JE. Physiological deterioration in cassava: possibilities for control. Trop Sci. 1994;34:335–43
11. Reilly K, Yuanhuani H, Thome J, Beeching J. Isolation and characterization of cassava catalase expressed during postharvest physiological deterioration. Biochim Biophys Acta. 2001;15:317–23
12. Plumbley RA, Rickard JE. Postharvest deterioration in cassava. J Trop Sci. 1991;31:295–303
13. Mattson DE. Statistics: difficult concepts, understandable explanations. Chicago: Bolchazy-Carducci; 1986. p. 281–3
14. Natural Resources Institute, (NRI). Collaborative study of cassava in Africa (COSCA). Phase 1: processing component, COSCA Working Paper No. 7. Chatham, UK, Natural Resources Institute; 1992
15. Nweke FI, Haleegoah J, Dixon AGO, Ajobo BO, Ugwu BO, Al-Hassan R. Cassava production in Ghana: a function of market demand and farmer access to improved production and processing technologies. Collaborative Study of Cassava in Africa (COSCA) Working Paper No. 21; 1999
16. Manu-Aduening JA, Lamboll RI, Ampong Mensah G, Lamptey JN, Moses E, Dankyi AA, Gibson RW. Development of superior cassava cultivars in Ghana by farmers and scientists: the process adopted outcomes and contributions and changed roles of different stakeholders. Euphytica. 2006;150:47–61

17. Fakoya EO, Banmeke TOA, Ashimolowo OR, Fapojuwo OE. Cassava production and poverty eradication among crop farmers in Ondo State, Nigeria. OIDA Int J Sustain Dev. 2010;2(05):67–72
18. Wenham, JE. Postharvest deterioration of cassava. A biotechnology perspective. FAO Plant Production and Protection Paper 130.NRI/FAO. Rome. pp 90; 1995
19. Wheatley CC. Conservation of cassava roots in polyethylene bags. CIAT Study Guide 04SC-07.06.Cali, Colombia, CIAT. 1989
20. Rickard JE, Coursey DG. Cassava storage. Part 1: storage of fresh cassava roots. Trop Sci. 1981;23:1–32
21. Balagopalan, C. (2000) Integrated technologies for value addition and postharvest management in tropical tuber crops. Thiruvanathapuram, India. p. 46
22. Manu-Aduening JA, Lamboll R, Dankyi A, Gibson R. Cassava diversity in Ghanaian farming systems. Euphytica. 2005;144:331–40

CHAPTER 17

Preliminary Observations on a Hot - Water Equipment for Disinfesting Maize Grain from Larger Grain Borer, Prostephanus Truncatus (Horn) and Greater Rice Weevil, Sitophilus Zeamais (Motschulsky) (Coleoptera: Bostrichidae): A Tool for Rural Farmers

Moses Zimba

Faculty of Science, Department of Biological Science, University of Zimbabwe, P. O. Box MP167 Mount Pleasant Harare, Zimbabwe.

ABSTRACT

The present study describes and evaluates a new tool for disinfesting maize grain from weevils using heat from boiling water. In developing countries, rural farmers lose huge amounts of grain due to weevils. Cost of insecticides and resistance of grain weevils to current insecticides are major threats. The grain losses affect the nutritional status and general quality of life of rural communities. Most grain is infested by weevils while in the field before harvest, therefore disinfestation of grain before storage is crucial. The equipment was made from 0.3 mm galvanized sheets and consisted of an inner pot for holding grain, outer pot for boiling water and a two litre metal cup. The most common maize variety grown in Zimbabwe, infested predominantly with Prostephanus truncatus and Sitophilus zeamais was used to evaluate the equipment. Empty maize cobs (after removing the grain) were used as source of heat. The equipment achieved 100% mortality of both species of weevils at 48°C (when grain became hot to the palm) after approximately 30 min of maize exposure to boiling water. There was no loss in grain viability as there was no significant ($\chi2 = 1.4$, $df = 1$, $p > 0.05$) difference in germination between treatment and control seeds. The equipment has a number of advantages. There are no health hazards which are normally associated with direct chemical treatment, especially when farmers apply higher doses of insecticide than recommended in an effort to prevent grain losses. Farmers will use empty cobs as source of heat. The equipment can be made locally by the rural farmers. The equipment has the potential to disinfest other small grains.

Keywords: Boiling water, maize weevils, heat, economic losses, germination, empty cobs, rural farmers.

INTRODUCTION

Insect pests cause heavy economic losses to stored grains throughout the world and their impacts are more devastating in poor countries (Boxall, 2002; FAO, 2016) where these losses are of serious nutritional and economic importance (Tefera, 2012). The insect pests which cause damage to stored grains are beetles (Coleoptera) and moths (Lepidoptera) with beetles being more distructive to stored grain in comparison to moths (Upadhyay and Ahmad, 2011).

In rural Africa, smallholder farmers' losses caused by the weevils *Sitophilus zeamais* on maize have been worsened by the introduction in the late 1970s of a neotropical pest, *Prostephanus truncatus* onto the continent (Hill et al., 2002). According to Mvumi et al. (2006), the pest complex of stored maize in Zimbabwe like in most countries in sub-Saharan Africa comprises of the cosmopolitan maize weevil, *S. zeamais*, Angoumois grain moth *Sitotroga cerealella*, and the less cosmopolitan larger grain borer (LGB), *P. truncatus*, among others. These pests attack a range of stored products, from small grains to maize, and even cassava. Some of these pests start infesting cereals in the field before harvest (Sori and Ayana, 2012). In the absence of stored crops, the natural vegetation is an important reservoir of *P. truncatus*, where it is capable of breeding and surviving in the dead wood of some tree hosts, some of which are actually used in the construction of grain stores (Nang'ayo et al., 2002). Of the above-mentioned stored product pests, *P. truncatus* is the most destructive in Zimbabwe, followed by the maize weevil. In some parts of Africa, losses as high as 50% have been reported on stored maize due to *P. truncatus* and *S. zeamais* in recent decades (FAO, 2016). Unlike the maize weevil that is an indigenous pest in Zimbabwe, LGB was first recorded in the country around 2006 and only declared as present by the government in 2010 and by that time the pest had spread to many parts of the country (Nyagwaya et al., 2010).

Although insecticidal control is still the most widely used (Benhalima et al., 2004); other strategies whose efficacy has been studied include use of plant products, inert dusts. There is often a misconception that other insecticides, especially botanicals, are safer than synthetic products (Odeyemi et al., 2008). Several leguminous plants widely used as broad spectrum insecticides are extremely toxic to aquatic life and show some level of toxicity to mammals. Charleston et al. (2005), reported that neem and Chinaberry plant extracts effectively controlled diamondback moth with no negative effect. As a requirement, detailed information on the mammalian toxicity and residual effect of botanicals should be provided before their use as insecticides (Odeyemi et al., 2008). Integrated pest management (IPM) which is essentially a holistic approach to pest control aimed at optimising the use of two or more methods for the management of pests is strongly encouraged (Dent, 2000; Odeyemi et al., 2008).

The trade in illegal and counterfeit public health pesticides is also a major concern in the region, which requires adequate legislation and capacity for enforcement as well as intersectoral and cross-border collaboration. In 2001, FAO/WHO estimated that around 30% of pesticides marketed in developing countries, with an estimated value of US$ 900 million annually, did not meet internationally accepted quality standards (FAO/WHO, 2001).

Control efforts over the years have come to rely solely on the use of synthetic insecticides in the form of fumigants or as grain protectants. However, the use of some of the traditional registered grain protectants has been reported to be ineffective on many occasions against these pests. This has led to farmers using higher doses of insecticide than recommended in a bid to try and curb storage losses caused by stored product pests. The high doses have in turn created a selection pressure against the pests leading to development of insect strains that are resistant to the insecticides. Giga and Mazarura (1990) reported that natural populations of *S. zeamais* showed malathion resistance of up to six-fold compared to a susceptible laboratory strain.

Despite the existence of stringent LGB-specific phytosanitary measures in international trade the bostrichid is now established in nearly all of sub-Saharan Africa including Zimbabwe (Kasambala and Chinwada, 2011). The favourable climatic conditions and poor storage systems in Africa often favours growth and development of these pests, resulting in considerable crop losses. For example, losses as high as 40% are reported on stored maize due to *P. truncatus* and *S. zeamais* and in Africa, where subsistence grain production supports the population, such grain losses may be substantial. With increasing pressure to avoid the use and minimise the choice of organophosphorus insecticides or fumigants on potential foodstuffs due to concerns about residues and human health, there is a need to find equivalently rapid non-chemical means of disinfesting grain, particularly for organic use (FAO/WHO, 2001). Fumigants have a limitation in that they fail to sterilize eggs of grain weevils (Das et al., 2013).

Temperature treatment of stored grains has been described as the best physical method which successfully kills several life stages of insects at a time (Upadhyay and Ahmad, 2011). According to Fields and White (2002) heat treatment for insect pests has been used for many years under different methods and situations. Heat has also been used extensively for fruit disinfestation, using both forced air and hot water (Hansen et al., 1992). Vapour heat has been demonstrated as effective for disinfestation of cut flowers and foliage (Hansen et al., 1992). Many approaches to heat disinfestations of stored products have been developed in the past (Beckett et al., 2007). The use of radio frequency heating, both with and without additional heating (Janhang et al., 2005), has been evaluated for the treatment of insects in rice and nuts.

Different insect species have been reported to have different heat tolerance ranges. A study of desert *Drosophila* species (Stratman and Markow, 1998) showed greater tolerance of high temperature by some species, but it was not consistent across all species and even the most tolerant conformed to the general rule of being within 45 to 51°C. Some studies have shown that insect mortality in grain can be achieved at lower grain temperatures but with longer exposure time which is also influenced by the moisture content of the grain (Beckett et al., 1998; Yan et al., 2014; Lu and Zhang, 2016). Beckett et al. (1998) studied the effects of both grain temperature and moisture content on the mortality of *Rhyzopertha dominica* and *Sitophilus oryzae* L. and concluded that the effects on insect mortality of grain moisture content within the range suitable for stored grain are small compared to the temperature effects. It was therefore the purpose of this study to develop and evaluate a simple equipment for use by rural communities in disinfecting stored grain products.

MATERIALS AND METHODS

Description of equipment

The Zimba grain weevil killer was produced at a cost of USD 60.00. The equipment makes use of boiling water to kill weevils in grain. It consists of an outer pot for boiling the water, an inner pot for holding the grain, a lid, a two litre metal cup and a square stand (Figure 1). Figure 2 shows a picture of the equipment. The pots are made of galvanized metal, 0.3 mm thick. The outer pot has a diameter of 50 cm and a height of 40 cm and has two handles at the rim. The inner pot has a diameter of 48 cm and a height of 50 cm and has a rim which is folded to form a lip which prevents steam from entering inside it (Figure 1). The inner pot has two handles with the lower part of the handles situated at 39 cm from the bottom so that when the inner pot is placed inside the outer pot, the lower parts of the handles rest on the rim of the outer pot in such a way that the inner pot is suspended 2 cm above that of the outer pot (Figure 1). The equipment uses 2 L of water resulting in a depth of 2 cm when poured into the outer pot. The function of the space between the bases of the two pots is to allow free movement of boiling water and steam. The lid has a fitting rim 1 cm wide and a protruding verandah 2 cm wide to aid in preventing entry of steam into the grain. The stand has sides of 52 cm, height 20 cm and a rim 1 cm wide for holding the outer pot in place. A plank, 10 cm long, 5 cm wide and 2 cm thick is used to stir the grain.

Evaluation of the equipment

The evaluation was conducted in January 2016. The most common maize variety in Zimbabwe which was in storage for approximately five months, with a visible natural infestion of predominantly *P. truncatus* (Horn) and *S. zeamais* (Motschulsky) (Coleoptera: Bostrichidae), the most common maize weevils in the country, was obtained from a communal farmer, 30 km north of Harare city. Sampling of maize was done following method in Giga and Mazarura (1990). The pre-trial weevil infestation level was determined by sampling about 2 kg of maize from each bag. The sample was then thoroughly mixed and a sub-sample of 250 g taken for insect identification. The sampling was done using a 2-metre bulk sampling probe which drew grain from four different levels vertically. The weevils were identified at the Department of Biological Sciences, University of Zimbabwe.

Maize from the three bags with a total weight of approximately 180 kg was poured onto a cement floor and thoroughly mixed using a shovel. Maize weighing 20 kg was poured into the inner pot (Figure 3). The pot was closed, carefully transferre to the outer pot (Figure 4) and inserted into the outer pot containing two litres of boiling water (Figure 5). The water was maintained at boiling point to the end of experiment. Maize cobs (with grain removed) and maize stalks were used as source of heat. The maize was stirred after every 5 min so that all the grain would come in contact with the bottom and lower sides of the inner pot which were the hottest parts and to allow homogeneous distribution of heat throughout the grain (Figure 6). In the first experiment, temperature was measured every five minutes to establish the relationship of temperature and time when the maize was exposed to heat from water which was constantly at boiling point. A Habor Digital Thermometer-Long Probe with Instant Read was used to measure the temperature. In the second experiment, temperature was measured at the beginning, at maximum insect activity and when activity was totally lost following guidelines from the first experiment. According to Chapman (1982) insects show a sharp rise in activity at the upper end of the preferred temperature range, and at still higher temperature, the activity is followed by an inability to move, a condition known as heat stupor and is followed by death of the insect.

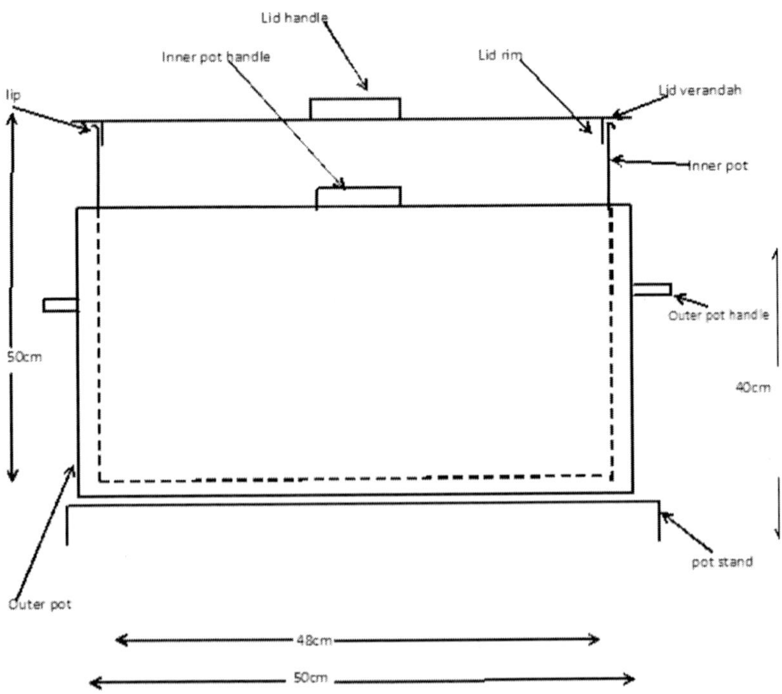

Figure 1. Diagram and specifications of a Zimba grain weevil disinfesting equipment.

Figure 2. Picture of a Zimba grain weevil disinfesting equipment.

Figure 3. Transferring maize grain into the inner pot.

Figure 4. Transferring the inner pot to the outer pot with boiling water.

Figure 5. Insertion of the inner pot containing maize grain into the outer pot with boiling water.

Figure 6. Stirring the maize grain every five minutes using a wooden plank 15×15×0.5 cm to ensure that all grain gets in contact with the bottom and sides of the inner pot and to achieve homogeneous distribution of heat in the grain.

At five minutes intervals, approximately 500 g of grain was taken out of the inner pot and carefully inspected for insect activity. At the same time, 20 damaged kernels were examined and insects dislodged using dissecting needles to determine status of the insects.. After treatment, the maize was exposed to open air for 20 min to allow cooling. The experiment was replicated three times.

Grain germination

The effect of heat on germination viability of the maize was evaluated by planting 100 uninfested seeds from each replicate treatment and untreated control. The seeds were planted in paper cups containing soil, placed in a greenhouse and watered daily until germination.

Data analysis

Descriptive statistics were used to analyse data. The analysis was focused on mortality of insects and maize seed viability after exposure to heat form boiling water. Effect of heat on insect mortality was expressed as percentages of kernels containing dead insects at various temperature points. The χ^2 test was used to measure the difference in germination viability between heat treated and control grain.

RESULTS

Pre-trial maize damage assessment showed that more than half the experimental grain was infested. Results of three replicates from the first experiment showed no variation on effect of exposure to constantly boiling water on maize temperature and insect activity (Table 1). A sudden increase in insect mobility was observed at between 43 and 45°C, followed by attempts by the insects to escape In the second experiment, examination of 20 infested kernels from each replicate showed that the Zimba grain weevil killer achieved 100% mortality of adult *P. truncutus* and *S. zeamais* in maize grain following total loss of insect activity at ±48°C (when grain was just hot to the palm) (Table 2), achieved at 25 min post exposure to heat (Table 1). Maximum insect activity was observed when grain temperature was between 43 and 45°C (Table 1) attained between 22 and 25 minutes of grain exposure to heat (Table 1). Activity response to heat observed for both *P. truncatus* and *S. zeamais* was similar.

Table 1. Insect activity according to period of maize exposure to constantly boiling water and temperature.

Time post exposure to heat (min)	Treatments			
	Control	Replicate 1	Replicate 2	Replicate 3
5	Nil (21°C)	Nil	Nil	Nil
10	Nil (21 °C)	Nil	Nil	Nil
15	Nil (21 °C)	Nil	Nil	Nil
20-22	Nil (21°C)	Start of activity (43-45°C)	Start of activity (43-45°C)	Start of activity (43-45°C)
22-25	Nil (21 °C)	End of activity (±48°C)	End of activity (±48)	End of activity (±48°C)

At 48°C grain was just hot to the palm.

Table 2. Insect mortality in infested kernels according to temperature recorded before exposure (A), at maximum insect activity (B) and at total loss of activity (C).

Parameter	Treatments											
	Control			Replicate 1			Replicate 2			Replicate 3		
	A	B	C	A	B	C	A	B	C	A	B	C
Temperature (°C)	21	21	21	21	±43	±48	21	±43	±48	21	±43	±48
Kernels examined	20	20	20	20	20	20	20	20	20	20	20	20
Number of kernels with dead insects	0	0	0	0	2	20	0	1	20	0	4	20
% kernels with dead insects	0	0	0	0	10	100	0	5	100	0	20	100

Table 3. Germination viability of maize seeds exposed to between 48°C in the Zimba grain weevil killer.

Parameter	Control			Treatment		
	Replicate 1	Replicate 2	Replicate 3	Replicate 1	Replicate 2	Replicate 3
Number of seeds used	100	100	100	100	100	100
Number germinated	100	96	100	97	95	100
Proportion germinated	1.0^{a}	0.96^{a}	1.0^{a}	0.97^{a}	0.95^{a}	1.0^{a}

Values in the same row with a different superscript letter are significantly different at $p<0.05$.

DISCUSSION

The present study described the Zimba grain weevil killer and the experiments conducted to evaluate the weevil killing efficacy of the equipment and assess germination viability of the heat treated grain. Many studies have been conducted in the past to evaluate heat treatment to disinfest grain using different types of equipment (Beckett et al., 2007). The equipment achieved 100% mortality of adult *P. truncutus* and *S. zeamais* in maize grain when the maize was exposed to a maximum temperature of 48°C (when grain was just hot to the palm) using heat from boiling water. The lethal temperature of 48°C observed in the present study is within the range reported in earlier studies (Beckett et al.. 2007). The heat treated grain showed germination viability of between 95 and 100% which was not significantly different from the control seeds.

Insect activity and mortality in the maize grain were monitored upto 48°C, a temperature at which there was total loss of adult insect activity. Immature stages were not monitored as they were reported to be more susceptible to heat than the adults (Chapman, 1982). Maximum insect activity was observed at a temperature of 43 to 45°C, reached after 20 min of grain exposure to heat from water kept constantly at boiling point. However, Yan et al. (2014) reported that heat tolerance in Sitophilus could vary with species. Prolonged exposure to a temperature of 43°C was reported to be lethal to *S. zeamais* (Lu and Zhang, 2016) and similarly *S. oryzae* exposed to a temperature range of 44 to 50°C for different time periods showed variations in mortality (Yan et al., 2014). Insect activity was totally lost when the grain temperature reached ±48°C, resulting in 100% insect mortality, observed at 25 min post exposure. These results support previous findings (Beckett et al., 1998; Evans, 1987) where stored grain pests *R. dominica* and *S. oryzae* were reported to have mortality thresholds of 42 to 48°C and 45 to 53°C respectively.

According to Fields and White (2002) heat treatment for insect pests is not new. In this study, heat from boiling water was exploited using a novel equipment to kill maize grain weevils. Extensive use of heat for fruit disinfestation, using both forced air and hot water has been reported (Hansen et al., 1992). Vapour heat has also been demonstrated as effective for disinfestation of cut flowers and foliage (Hansen et al., 1992). The use of radio frequency heating, both with and without additional heating (Janhang et al., 2005), has been evaluated for the treatment of insects in rice and nuts. In the present study, the killing equipment and the method of heat application and utilization to effect insect kill of the maize grain weevils could be new.

Different insect species have been reported to have different heat tolerance ranges. A study of desert *Drosophila* species (Stratman and Markow, 1998) showed greater tolerance of high temperature by some species, but was not consistent across all species and even the most tolerant conformed to the general rule of being within the 45 to 51°C. The loss of *S. zeamais* and *P. truncatus* activity observed at 48°C in the present study agrees with earlier reports of a 45 to 51°C heat tolerance range in many insect species.

Some studies have shown that insect mortality in grain can be achieved at lower grain temperatures but with longer exposure time which is also influenced by the moisture content of the grain (Beckett et al., 1998). The present study did not take into account the moisture content of the grain. However, the grain used in this study had been in storage for five months a period generally considered adequate for drying by local communities.. Beckett et al. (1998) studied the effects of both grain temperature and moisture content on the mortality of *R. dominica* and *S. oryzae* L. and concluded that the effects on insect mortality of grain moisture content within the range suitable for stored grain are small compared to the temperature effects. This tool is ideal for rural communities because firewood, maize cobs or dry maize stalks can be used as source of heat. The tool also uses very little water.

In the present study, maize grain was exposed to a maximum temperature of 48°C attained between 25 and 30 min of exposure to heat from boiling water. The viability results showed no significant differences in germination between treated and control grain. According to Akman (2009), prolonged dry heat treatment of 40°C for 72 h and 60°C for 24 h was reported to cause lower seed germination. The maximum exposure period of grain to heat was 30 min. It is possible that the high germination viability observed in the current study was due to the short exposure of grain to heat. As also reported earlier the extent of the loss of grain viability depends on the initial quality of the grain, the temperature and moisture content of the grain and the time of exposure. The maize grain used in this study was assured of good quality as it was obtained from a seasoned communal farmer and had been in storage for five months.

CONCLUSIONS

The current new tool disinfects maize grain by killing existing insects in a similar way as fumigants. The tool has the advantage that it can be used against insect pests which have become resistant to insecticides and there are no health hazards normally associated with direct chemical treatment The equipment has the potential for use in disinfesting stored grain insect pests infesting other grains such as sorghum, cow peas, beans and millet. The tool requires simple instructions. The tool costs USD60 and can be used over a number of years. Maize cobs (after removing the grains) and maize stalks can be used as a source of heat.

REFERENCES

1. Akman Z (2009). Comparison of High Temperature Tolerance in Maize, Rice and Sorghum Seeds by Plant Growth Regulators. J. Anim. Vet. Adv. 8:358-361.
2. Beckett SJ, Fields PG, Subramanyam BH (2007). Disinfestation of stored products ans associated structures using heat. In: Tang J, Mitcham E, Wang S,Lurie S (Eds), Heat treatments for postharvest pest control: Theory and practice. CAB International 2007, Cromwell Press, Trowbridge, United Kingdom pp. 182-237. http://www.grains.kstate. edu/spirel/docs/publications/chapters/heat-rev.pdf
3. Beckett SJ, Morton R, Darby JA (1998). The mortality of *Rhyzopertha dominica* (F.) (Coleoptera:Bostrychidae) and *Sitophilus oryzae* (L.) (Coleoptera: Curculionidae) at moderate temperatures. J. Stored Prod. Res. 34:363-376.

4. Benhalima H, Chaudhry MQ, Mills KA, Price NR (2004). Phosphine resistance in stored-product insects collected from various grain storage facilities in Morocco. J. Stored Prod. Res. 40:241-249.
5. Boxall RA (2002). Damage and losses caused by larger grain borer Prostephanus truncatus. Integrated Pest Manag. Rev. 7:105-121.
6. Charleston KF, Kfir R, Dickie M (2005). Impact of botanical pesticides derived from *Melia azedarach* and *Azadirachtaa indica* on the biology of two parasitoid species of the diamondback moth. Biol. Control 33:131-142.
7. Chapman RF (1982). The Insects: Structure and Function (Eds Chapman RF) Hodder and Stoughton, Kent, England. 919pp. http://trove.nla.gov.au/work/10976380
8. Das I, Kumar G, Shah NG (2013). Microwave Heating as an Alternative Quarantine Method for Disinfestation of Stored Food Grains. Int. J. Food Sci. Article ID 926468, 13 p.
9. Dent D (2000). Insect Pest Management, 2nd edition. CABI Publishing, Wallingford, 410 p. DOI: 10.1046/j.1439-0418.2001.0538a.x
10. Evans DE (1987). Some biological and physical restraints to the use of heat and cold for disinfesting and preserving stored products. In Donnahaye E, Navarro S (Eds). Proceedings of 4th International Working Conference on Stored Product Protection, Tel Aviv, Israel, September 1986. pp. 149-164.
11. FAO (2016) Food losses and waste: a challenge to sustainable development. http://www.fao.org/save-food/news-andmultimedia/news/news-details/en/c/429182/
12. FAO/WHO (2001). Amount of poor-quality pesticides sold in developing countries alarmingly high. http://www.fao.org/WAICENT/OIS/PRESS_NE/PRESSENG /2001 /pren0105.htm
13. Fields PG, White NDG (2002). Alternatives to methyl bromide treatments for stored product and quarantine insects. Annu. Rev. Entomol. 47:1-39.
14. Giga DP, Mazarura UM (1990). Malathion resistance in *Sitophilus zeamais* (Motsch.) in Zimbabwe. Int. J. Pest Manage. 36:3-320.
15. Hansen JD, Hara AH, Tenbrink VL (1992). Vapour heat: a potential treatment to disinfest tropical cut flowers and foliage. HortScience 27(2):139-143.
16. Hill MG, Borgemeister C, Nansen C (2002). Ecological studies on the larger grain borer *Prostephanus truncatus* (Horn) (Coleoptera: Bostrichidae) and their importance for integrated pest management. Integrated Pest Manag. Rev. 7:201-221.
17. Janhang P, Krihigamas N, Lucke W (2005). Using radio frequency heat treatment to control the insect *Rhyzopertha dominica* during storage in rice seed. Conference on International Agricultural Research for Development. Stuttgart-Hohenheim October, 2005 pp. 11-13.
18. Kasambala T, Chinwada P (2011). Modelling the occurrence of Prostephanus truncates (Coleoptera: Bostrichidae) in Southern Malawi. J. Stored Prod. Postharvest Res. 2(4):72-78.
19. Lü J, Zhang H (2016). The Effect of Acclimation to Sublethal Temperature on Subsequent Susceptibility of Sitophilus zeamais Mostchulsky (Coleoptera: Curculionidae) to High Temperatures. doi:10.1371/journal.pone.0159400.
20. Mvumi BM, Stathers TE, Kaparadza V, Masiiwa P, Jowah P, Riwa W (2006). Comparative insecticidal efficacy of five of five raw diatomaceous earths against three tropical stored grain Coleopteran pests: Sitophilus zeamais, Rhyzopertha dominica and Tribolium castaneum. 9th International Working Conference on Stored Product Protection, Campinas, Brazil. pp. 868-876.
21. Nang'ayo FLO, Hill MG, Wright DJ (2002). Potential hosts of Prostephanus truncates (Coleoptera: Bostrichidae) among native and agro-forestry trees in Kenya. B. Entomol. Res. 92:499-506.
22. Nyagwaya LDM, Mvumi BM, Saunyama IGM (2010). Occurrence and Distribution of Prostephanus truncatus (Horn) (Coleoptera: Bostrichidae) in Zimbabwe. Int. J. Trop. Insect Sci. 30 (4):221-231.

23. Odeyemi OO, Masika P, Afolayan AJ (2008). A Review of the use of phytochemical for insect pest control. Afr. Plant Prot. 14:1-7.
24. Sori W, Ayana A (2012). Storage pests of maize and their status in Jimma Zone. Afr. J. Agric. Res. 7:4056-4060.
25. Stratman R, Markow TA (1998). Resistance to thermal stress in desert Drosophila. Funct. Ecol. 12(6):965-970.
26. Tefera T (2012). Post-harvest losses in African maize in the face of increasing food shortage. Food Secur. 4:267-277.
27. Upadhyay RK, Ahmad S (2011). Management Strategies for Control of Stored Grain Insect Pests in Farmer Stores and Public Ware Houses. World J. Agric. Sci. 7(5):527-549.
28. Yan R, Huang Z, Zhu H, Johnson JA, Wang S (2014). Thermal death kinetics of adult *Sitophilus oryzae* and effects of heating rate on thermotolerance. J. Stored Prod. Res. 59:231-236.

CHAPTER 18

Management of Bean bruchids *(Accanthoscelides obtectus* Say.) (Coleoptera: Bruchidae) using botanical oils in Western Amhara, Ethiopia

Mihret Alemayehu[1] and Emana Getu[2*]

[1]*Plant Science Department, College of Agriculture and Natural Resource, Debre Markose University, P. O. Box 269, Addis Ababa, Ethiopia.*

[2]*Zoological Sciences Department, College of Natural and Computational Sciences, Addis Ababa University, P. O. Box 1176, Addis Ababa, Ethiopia.*

ABSTRACT

Faba bean, Vicia faba L. is an important legume crop that substitute animal proteins especially in developing countries. Bean bruchids or bean weevil, Accanthoscelides obtectus Say is a major insect pest of the faba bean grains mainly in the store. An experiment was conducted in Debremarkose University to screen effective botanical oils for the management of bean bruchids on faba bean grains. The tested oils were from seeds of Noug (Guizotia abyssinica L. f.), Rapeseed (Brassica napus L.), Lantana (Lantana camara L.), Pepper (Capsicum annum L.), Tephrosia (Tephrosia vogelii Hook), Groundnut (Arachis hypogaea L.) and Castor (Ricinus communis L.). The untreated check, Acetone (solvent used to extract oils from the botanicals) and Ethiothion 5% dust (a standard check) were used for comparison. One day after treatment application, the dead bruchid in

Ethiothion 5% dust, noug oil and rapeseed oil treated grains were 100, 90.85±0.63 and 78.35±0.78%, respectively indicating the fast knock down effect of the treatments. Four days after treatment, 100% of the parent bruchids were killed in all oil treated grains. The mortality of the parent bruchids four days after treatment application was 10.85+0.6% in acetone treated grains and 11.65±0.5% in untreated grains. Statistically lower number of eggs and progeny (adults) were recorded from Ethiothion 5% dust and oil treated grains than the acetone treated and untreated grains. Bean grain losses due to bruchids were 18.3±0.4% in acetone treated grains and 18.4±0.44% in the untreated check, while 0 to 5.2±0.27% losses were recorded in other treatments. Ethiothion 5% dust and the oils inhibited emergence of bruchids from 71.7±2.56 to 100±0.53%. Lower germination percentages were recorded from acetone (67.3±4.23%) treated grains and the untreated grains (66.7±3.02%). From the results of this experiment, it can be concluded that the Ethiothion 5% dust and all the tested oils can effectively control bean bruchid and recommended for the management of the same.

Keywords: Bean bruchid/weevil, botanical oil, faba bean grain (Vicia faba L.), mortality, survival, progeny.

INTRODUCTION

The faba bean, Vicia faba L. is one of the earliest domesticated legumes after chickpea and field pea which was vastly distributed throughout the world (Kanaji, 2007). It is an important crop used as a source of organic protein for peoples who cannot afford feeding expensive animal protein on regular basis (El-Tokhy and Kasem, 2012). However, bean storage at small scale subsistence farming level is limited due to bruchid infestations that result in heavy losses (Mushobozy et al., 2009). The common bean weevil, Acanthoscelides obtectus (Say) and the Mexican bean weevil, Zabrotes subfasciatus (Boheman) are the common insect pests in a wider area where faba bean is growing (Mushobozy et al., 2009; Taponjou et al., 2002). Their attack results in yield reduction, poor grain quality, loss of seed viability and unsuitable taste for human consumption (Prakash et al., 2008).

Farmers are using synthetic insecticides to keep faba bean grains free of insect infestation and to reduce damage during storage (Taponjou et al., 2002). They use insecticides without having appropriate knowledge on its application and drawbacks like development of resistance, environmental pollution and poisoning of fauna and flora (Prakash et al., 2008; Taponjou et al., 2002). To date, attentions were diverted to look for effective, biodegradable, simple to apply and homemade control options such as oil extracts of botanical origin for the management of bruchids (Olotuah, 2013). Oils of many plant species are known to have one or more insecticidal properties in terms of pest control such as fumigants, repellents, toxicant and oviposition deterrent among others (Mushobozy et al., 2009; Olotuah, 2013).

In Ethiopia, specifically in Amhara region small scale farmers are facing problems of storing their bean grains due to damage by A. obtectus. The faba bean grains are prone to damage by A. obtectus starting from field at the maturity of the crop and extended to storage where heavy infestation occurs. In the region, the availability of plants with insecticidal value is tremendous. Hence, the objective of this study was to screen effective botanical oils for the management of A. obtectus on the faba bean grains.

MATERIALS AND METHODS

Description of the study area

The study was conducted in Eastern Gojjam zone, Debre Markose University located 300 and 265 km away from Addis Ababa and Bahir Dar, respectively. The average elevation of the study area was 2400 masl. The rainfall pattern was uni-modal with a mean annual rainfall of 1500 mm. Daily average temperature was 24°C. The major crops grown are Teff, wheat, maize, faba bean, chickpea and rough pea.

Rearing of A. obtectus

Cultures of A. obtectus were established at Debremarkose University to obtain the same age group and required numbers of adult bruchids for the experiment. The bruchids used for rearing were collected from local farmer faba bean stores and reared in three plastic pots having 5 L capacity each. Each pot was half filled with 3 kg grains to serve as food for the bruchids. Newly harvested faba bean grains with no bruchid eggs were collected from local market. The grains were carefully examined by hand lens for the presence of bruchid eggs. The selected grains were washed with potable water to avoid any obscure sources of infestation and frozen at a temperature of -4±1ºC for three weeks to protect fungal development and to ensure uniform moisture content of the grains (Ileke and Olotuah, 2012). Grains having 14% moisture contents were used as a substrate for A. obtectus rearing (Kanaji, 2007).

About 100 unsexed adults of A. obtectus were added to each jar assuming that 50% of the adults are females and the rest are males as the sex ratio for the series of sex ratio experiments was 1:1 (Ileke and Olotuah, 2012). The jars were covered with muslin cloths held in place by rubber bands. Rearing was conducted at 28±1.5ºC and 65±5% RH (Ileke and Olotuah, 2012; Kanaji, 2007). The temperature used for rearing was adjusted by electric power, while the follow up record was taken by Thermo hygro and a thermometer. Frequent inspection of the culture for emergence of the progeny was carried out daily starting from twenty two days after parent bruchids introduction (Bhardwaj and Verma, 2012). The newly emerged one day old adult F_1 progenies were used for the experiment (Bhardwaj and Verma, 2012).

Treatment preparation

The botanical seeds were collected from the localities nearby Debre Markos University. Oils were extracted from various plant species using acetone. The plant species used for the experiment were seed of Noug (Guizotia abyssinica L.f.), rapeseed (Brassica napus L.), Lantana (Lantana camara L. (Sensu lato)), pepper (Capsicum annum L.), Tephrosia (Tephrosia vogeliiHook.), Groundnut (Arachis hypogaea L.) and Castor oil (Ricinus communis L.) (Table 1). Untreated check, Ethiothion 5% dust (standard check) and Acetone (a solvent to extract the oils) were used in the experiment for comparison.

Table 1. Botanical oils evaluated for their efficacy against *A. obtectus* on faba bean grains.

Common name	Scientific name	Plant parts used	Dosage
Noug oil	*Guizotia abyssinica*	Seed	4 ml/kg
Castor oil	*Ricinus communis*	Seed	4 ml/kg
Groundnut	*Arachis hypogaea*	Seed	3 ml/kg
Pepper	*Capsicum annum*	Seed	3 ml/kg
Tephrosia	*Tephrosia vogelii*	Seed	3 ml/kg
Lantana	*Lantana camara*	Seed	4 ml/kg
Rapeseed	*Brassica napus*	Seed	4 ml/kg
Ethiothion 5% dust		-	50 g/Qt
Acetone	Dimethyl ketone	-	4 ml/kg
Untreated check	-	-	-

The seeds were dried under shade and grinded with pestle and mortar to obtain fine dusts. The dusts were further grinded until it passed through 0.5 mm perforated sieves (Ileke and Olotuah, 2012). The dusts were dissolved in acetone in a two liters plastic pot to separate oils from fine dusts. Each mixture was stayed for one month and shaken daily for 15 min (Bhardwaj and Verma, 2012). The liquid that contains oil and acetone was float on the upper surface of the containers. Each container was pierced with scissor at the site of layer formation between dust residues and the floated liquids. The liquids that float on the surface of the containers were come out through the hole prepared between the layers. Each liquid was collected and filtered again using muslin cloths. The oils and solvents were exposed to air for two days to evaporate the solvent (Acetone). Subsequently, the oils collected from each botanical were stored separately in clean vials at a temperature of 4°C in a refrigerator until use (Ileke and Olotuah, 2012).

A concentration of 3 to 4 ml/kg oils were added in one litter plastic pots containing 500 g of the faba bean grains. The pots containing faba bean grains, the botanical oils, Ethiothion 5% dust and acetone were vigorously shaken and rolled for one minute to coat the treated grains uniformly (Ileke and Olotuah, 2012).

Effects of botanical oils on parent adult bruchid fertility and mortality

The treated bean grains in each pot were infested with 24 hour old 20 unsexed adult of A. obtectus (Ileke and Olotuah, 2012). The pots were covered with muslin cloth to allow sufficient aeration, to protect escape of A. obtectus and entrance of other pests. The experimental design was a Completely Randomized Design (CRD) in three replications. Parent adult A. obtectusmortality and the number of eggs laid per 20 grains were recorded from each pot 1, 2, 4, 8 and 10 days after treatment application. In each subsequent count, dead bruchids were counted and removed from each pot. Ten days after treatment application all dead and alive bruchids were counted and removed from the pots.

Emergence of adult progeny of A. obtectus

Bruchid progenies inspection and continuous follow up started from 22 days after treatment application. Subsequent counts of dead and alive adult bruchids were done. The bruchids were separated from bean grains with the help of three mm sieve made from wire mesh (Swella and Mushobozy, 2007). Data on grains with and without eggs per 20 randomly selected grains were taken for three months with ten days interval.

Percentage protection

Protective efficacy of the botanical oils was calculated based on the emerged progenies using Taponjou et al. (2002) and El-Ghar et al. (1987) procedures. Percentage reduction to adult emergence or inhibition rate (% IR) was calculated as:

$$\% \, IR = \frac{(Cn - Tn)}{Cn} X100$$

Where: C_n = the number of newly emerged insects in the untreated pot, and T_n = the number of newly emerged insects in the treated pot.

Seed weight loss determination

On the 90th day of treatment application, 1000 grains were randomly taken from each treatment. The selected grains were separated and categorized into damaged grains with exit holes and undamaged grains. Grains with and without exit holes were counted and weighed separately and the obtained data were used to calculate the percentage weight loss. Percentage weight losses were determined by the count and weigh method applied by Gwinner et al. (1996):

$$Percent\ weight\ loss = \frac{(W\mu \times Nd) - (Wd \times N\mu)}{W\mu \times (Nd + N\mu)} X100$$

Where: Wμ = weight of undamaged grains; Nμ = number of undamaged grains; Wd = weight of damaged grains; Nd = number of damaged grains.

Germination test

The effect of botanical oils on faba bean grain germination was done three months after treatment application. Faba bean grains from the untreated check, acetone and the standard check were used for comparison. Fifty grains were randomly selected from each treatment. Grains from each treatment were treated separately with sodium hypochlorite (Clorox 10%) for one minute to eliminate fungal contamination. Grains treated with sodium hypochlorite were washed by water for one minute to avoid physical damage. The seeds were placed on moist filter paper in petri dish for seven days. The number of sprouted seeds was counted seven days after incubation. Subsequently, germination percentage was determined using the following formula (Gwinner et al., 1996):

$$Germination(\%) = \frac{Total\ No.of\ bean\ grains\ sprouted}{Total\ No.of\ bean\ grains\ added\ to\ the\ petridish} X100$$

Statistical analysis

Analysis of variance (ANOVA) was done according to Gomez and Gomez (1984) procedures. All obtained data were transformed using square-root, logarithmic and arc sign transformations before the analysis. Tukey's studentized range test (HSD) at p=0.05 was used to separate significant means. SAS (SAS 9.2) and MS Excel 07 soft wares were used for these analyses. Results were reported using back transformed values.

RESULTS AND DISCUSSION

Number of eggs laid by adult parent female bruchid on treated and untreated faba bean grains

The number of eggs laid by parent bruchidson treated (the oils, the Ethiothion 5% dust and the Acetone) and untreated grains are shown in Table 2. Parent bruchids were started mating with in the first days of an experiment due to their eggs were recorded on day one after treatment application. Significantly ($p<0.05$) higher number of bruchid eggs were laid on untreated check (7.33±0.53) and acetone (7.5±0.68) treated faba bean grains compared to the botanical oils (0.33±0.33 to 3.17±0.48) and Ethiothion 5% dust treated grains which did not yield any bruchid egg one day after treatment. The current result was similar to that of Ibrahim (2012) who reported sesame, olive and sunflower oils effectively decreased the egg laying capacity of C. maculates on chickpea grain starting from treatment application up to the end of the experiment. Similar results were reported by Abdulahi (2011) who tested four levels of Balantie aegyptiaca leaf extract and found that the botanical treatments were an efficient ovipositional deterrent compared to the untreated check. Tabu et al. (2012) conducted similar experiment that bruchids inhibited oviposition on chickpea grains treated by seed powders of Azadiracta indica at 2% (w/w) andChenopodium ambrosiodes at 4% (w/w) starting from day one of treatment application up to three months.

Table 2. *Mean(±se) number of eggs laid by adult parent bruchids during the first ten days of treatment application.

Year	Treatment	Dose	**Parent bruchid eggs after treatment application				
			1st day	2nd day	4th day	8th day	10th day
	Noug	4 ml/kg	0.33 ± 0.33^{c}	0.67 ± 0.33^{de}	0 ± 0^{d}	0 ± 0^{b}	0 ± 0^{b}
	Rapeseed	3 ml/kg	0.50 ± 0.35^{c}	1.33 ± 0.33^{de}	0 ± 0^{d}	0 ± 0^{b}	0 ± 0^{b}
	Groundnut	3 ml/kg	0.50 ± 0.35^{c}	2.00 ± 0.58^{d}	0 ± 0^{d}	0 ± 0^{b}	0 ± 0^{b}
	Castor	4 ml/kg	2.17 ± 0.75^{b}	4.33 ± 0.78^{c}	4.83 ± 0.95^{c}	0 ± 0^{b}	0 ± 0^{b}
2013-2014	Lantana	4 ml/kg	2.83 ± 0.63^{b}	5.17 ± 0.63^{bc}	6.67 ± 2.67^{bc}	0 ± 0^{b}	0 ± 0^{b}
	Tephrosia	3 ml/kg	2.67 ± 0.33^{b}	6.17 ± 0.75^{b}	7.17 ± 0.75^{bc}	0 ± 0^{b}	0 ± 0^{b}
	pepper	3 ml/kg	3.17 ± 0.48^{b}	5.13 ± 0.26^{bc}	10.67 ± 0.91^{b}	0 ± 0^{b}	0 ± 0^{b}
	Ethiothion 5% dust	50 g/Qt	0 ± 0^{c}	0 ± 0^{e}	0 ± 0^{d}	0 ± 0^{b}	0 ± 0^{b}
	Acetone	4 ml/kg	7.50 ± 0.68^{a}	11.67 ± 0.67^{a}	16.33 ± 0.89^{a}	19.33 ± 0.67^{a}	20.67 ± 1.05^{a}
	Untreated check		7.33 ± 0.53^{a}	11.83 ± 0.79^{a}	16.67 ± 1.13^{a}	18.33 ± 1.2^{a}	20.83 ± 1.27^{a}

Means followed by the same letter (s) within a column are not significantly different at 5%.Tukey's studentized range test HSD.
*Data from both years pooled together; **Means presented in the table are square root back transfomed values.

Two days after treatment the number of eggs laid by bruchids on noug oil and rapeseed oil treated the faba bean grains were 0.67±0.3 and 1.33±0.33, respectively. Significantly ($p<0.05$) higher number of bruchid eggs were laid on pepper (5.13±0.26), lantana (5.17±0.63) and tephrosia (6.17±0.75) than the rest of the oil treatments. The highest number of bruchid eggs were recorded on untreated grains (11.83±0.79) and acetone (11.67±0.67) treated faba bean grains. Similar results were reported by Uddinll and Sanusi (2013) such that no bruchid eggs were recorded on cowpea grains treated with groundnut and palm kernel oils starting from two days after treatment application. Yahaya et al. (2013) reported similar result on chickpea grains treated with groundnut or palm oil at the rate of 2.5 to 3 ml/kg which significantly reduced C. maculates oviposition.

There were no extra bruchid eggs after four days of treatment application than the two days of experiment on Ethiothion 5% dust, noug oil, rapeseed oil and groundnut oil treated grains. Based on Okonkwo and Okoye (2008) finding, essential oils of D. tripetela and P. guineense achieved 100% mortality of C. maculates and S. zeamais in 24 h after treatment application as a result no extra eggs were recorded till death of the parent insects. Statistically higher numbers of bruchid eggs were recorded on oils of castor (4.83±0.95), lantana (6.67±2.6), tephrosia (7.17±0.75) and pepper (10.67±0.91) treated faba bean grains four days after treatment. The numbers of bruchid eggs were significantly higher than the rest of the treatments in untreated (16.67±1.13) and acetone (16.33±0.89) treated grains four days after treatment (Table 2). Yahaya et al. (2013) indicated that, significantly lower number of C. maculates eggs were recorded on cowpea grains treated with botanical oils (0-4.5+1) compared to 62.87+0.2 eggs laid in the control plot.

Eight days after treatment, no extra eggs of bruchids were recorded in Ethiothion 5% dust, and oil treated faba bean grains. However, the numbers of bruchid eggs laid in acetone treated and untreated grains were recorded to 19.33±0.67 and 18.33±1.2 eggs, respectively. The number of A. obtectus eggs laid in acetone (20.67±1.05) treated and the untreated (20.83±1.27) faba bean grains were slightly increased ten days after treatment (Table 2). Ibrahim (2012) remarked that sesame (98.49%), olive (96.54%) and sunflower (95.37%) oils at the rate of 7.5 ml/kg (v/w) were effectively deterring bruchids oviposition on chickpea grain. Comparable result was reported by Shukla et al. (2007) that Murraya koenigii L. and Eupatorium canabinum L. dusts deter bruchid oviposition by 90.62 and 86.46%, respectively. Sangeeta and Apte (2015) reported that green gram grains treated with botanical oils were 100% free from parent Callosobruchus maculates infestation.

Effects of botanical oils on the survival of adult parent bruchids

The oils killed parent bruchids started from day one after treatment application. Parent bruchids died in Ethiothion 5% dust, noug oil and rapeseed oil were 100, 90.85±0.63 and 78.4±0.78%, respectively within the first day of experiment. 28±0.52 to 70.85±0.85% the parent bruchids were died in groundnut, lantana, tephrosia and pepper oil treated grains. On the contrary, 100% of the parent bruchids were survived in acetone and untreated check treated grains day one after treatment application (Table 3). Olotuah (2013) conducted similar experiment to control C. maculates (F.) that the oils were killed 100% of the parent bruchids with in six minutes while100% of the parents survived in the untreated check. Chickpea grains treated with dusts of A. indica at 2% (w/w) were killed parent bruchids 80, 91.67 and 93 67% after one, two and four days respectively (Taba et al., 2012). Tadele et al. (2014) also confirmed that botanical dusts were completely eradicated soldiers and worker termites a day after treatment. Okonkwo and Okoye (2008) evaluated the efficacy of D. tripetela dusts at 1.5 to 2.5 g/kg to manage S.zeamais on maize grains that 100% of the parent weevils were died at ($P < 0.05$) within 24 h of treatment.

Table 3. *Dead adult parent Bruchids in the first ten days of treatment application.

Year	Treatment	Rate	**Mortality of bruchids (%) after treatment application					
			1st day	2nd day	4th day	8th day	10th day	Live bruchids after 10 days
	Noug	4 ml/kg	90.85±0.63^{ab}	100±0^{a}	0±0^{c}	0±0^{b}	0±0^{b}	0±0^{b}
	Rapeseed	3 ml/kg	78.35±0.78^{bc}	100±0^{a}	0±0^{c}	0±0^{b}	0±0^{b}	0±0^{b}
	Groundnut	3 ml/kg	70.85±0.85^{c}	98.35±1.27^{a}	100±0^{a}	0±0^{b}	0±0^{b}	0±0^{b}
	Castor	4 ml/kg	52.5±1.0^{d}	90±0.98^{b}	100 ±0^{a}	0±0^{b}	0±0^{b}	0±0^{b}
2013-	Lantana	4 ml/kg	37.5±0.97^{e}	75±1.06^{c}	100±0^{a}	0±0^{b}	0±0^{b}	0±0^{b}
2014	Tephrosia	3 ml/kg	28.35±0.52^{e}	70±0.67^{c}	100 ±0^{a}	0±0^{b}	0±0^{b}	0±0^{b}
	pepper	3 ml/kg	30±0.81^{e}	73.35±0.88^{c}	100 ±0^{a}	0±0^{b}	0±0^{b}	0±0^{b}
	Ethiothion 5% dust	50 g/Qt	100±0^{a}	0±0^{e}	0±0^{c}	0±0^{b}	0±0^{b}	0±0^{b}
	Acetone	4 ml/kg	0±0^{f}	3.35±0.53^{d}	10.85±0.6^{b}	50±0.53^{a}	86.65±0.41^{a}	13.35±0.63^{a}
	Untreated check		0±0^{f}	3.35±0.56^{d}	11.65±0.5^{b}	55±1.49^{a}	85.85 ±0.35^{a}	14.15±0.26^{a}

Means followed by the same letter (s) within a column are not significantly different at 5%, Tukey's studentized range test HSD); *Data from both years pooled together, **Means presented in the table are square root back transformed values.

Treatments like Ethiothion 5% dust, noug, rapeseed and groundnut oils were statistically shown similar results that 98.35±1.27% to 100% of the parent bruchids were died after two days of treatment application. The rest of the botanical oils were shown superior efficacy (70±0.67 to 90±0.98%) to kill parent bruchids than acetone treated and untreated check faba bean grains that 96.65% of the parents were continued to exist. Four days after treatment 100% of the parent bruchids were died in Ethiothion 5% dust and oil treated faba bean grains. In acetone and the untreated check 88.65 and 88.35% of the parent bruchids respectively were alive to lay their eggs (Table 3). Similar result was reported by Bhardwaj and Verma (2012) and

Tabu et al. (2012) management of beetles by botanical dusts, edible oils and inert materials were appeared to be more effective compared with control after four days of treatment. According to Yohannes et al. (2013) 85 to 100% parent bruchids were eradicated by botanical dusts within seven days after treatment application. C. ambrosiodes leaf dusts at 4% (w/w) killed the parent bruchids 88.34 and 91.67% after two and four days of treatment application (Taba et al., 2012).

The dead bruchids in acetone and the untreated check after eight days of experiment were 50±0.53 and 55%±1.49, respectively. The numbers of dead parent bruchids were increased in acetone and the untreated check with time due to natural death was expected. As a result, the dead parent bruchids were 85.85±0.35% in untreated check and 86.65±0.41% in acetone ten days after treatment application. 14.15±0.26 and 13.35±0.63% of the parent bruchids in the untreated check and acetone treated grains respectively were continued and alive after ten days of the experiment (Table 3). Waktole (2014) was reported similar finding that a dust prepared from Chenopodium spp. was achieved 66.67% mortality of S. zeamais parents at par with Acetellic 2% dust (70.39%). Sangeeta and Apte (2015) report was also confirmed that C. maculatus were not survived in any of the green gram grains treated with botanical oils. Shukla et al. (2007) report indicated that parent bruchids were effectively controlled by M. koenigii and E. canabinum dusts of leaves.

Effects of botanical oils on adult progeny emergence and the capacity of adult progeny in egg laying

Eggs laid by bruchids on bean grains were vary due to various efficacy of treatments. Numbers of eggs laid in noug, rapeseed and groundnut oils were statistically lower (0±0) at par with Ethiothion 5% dust treated grains. Yahaya et al. (2013) reported that palm, groundnut and coconut oils treated chickpea grains at the rate of 4 ml/kg completely inhibited C. chinensis infestation for three months of storage periods. Sangeeta and Apte (2015) indicated that green gram grains treated with botanical oils were free from bruchid eggs one year after treatment application while100% of the grains were infested with the insect eggs within the same period at the same conditions of experiment.

Significantly lower numbers of bruchid eggs were laid on castor, lantana, tephrosia and pepper oil (37.67±3.16 to 98±8.19 eggs) treated grains compared to 498.3±41.5 eggs in untreated check and 493.3±41.1 eggs in acetone treated grains. The numbers of eggs laid in acetone and untreated check were fivefold that of pepper oil shows better numbers of eggs laid than other oil treatments (Table 4). A comparable result was reported by

Abdulahi (2011) that the eggs laid by bruchids on cowpea grains treated with actellic 2% and botanical dusts were statistically lower (88.6±4.6 to 203.6+ 5.9) than acetone (798.6±27.4) and untreated check (794.3+15). Shukla et al. (2007) report was in line with the current finding that bruchids were laid lower numbers of eggs on 2% (w/w) leaves dusts of M. koenigii (17.67±5.5) and E. canabinum (22.34±4.5) treated chickpea grains compared to control (127.67±8.05).

Year	Treatment	Dose	Numbers of bruchids and their eggs**		
			Bruchid eggs	Emerged progenies	Live bruchids
2013-2014	Noug	4 ml/kg	$0±0^{f}$	$0±0^{f}$	$0±0^{d}$
	Rapeseed	3 ml/kg	$0±0^{f}$	$0±0^{f}$	$0±0^{d}$
	Groundnut	3 ml/kg	$0±0^{f}$	$0.67±0^{f}$	$0±0^{d}$
	Castor	4 ml/kg	$37.67±3.16^{e}$	$22.7±2.69^{e}$	$4.3±0.49^{cd}$
	Lantana	4 ml/kg	$61.67±5.15^{d}$	$41.7±4.91^{d}$	$11.7±1.36^{bc}$
	Tephrosia	3 ml/kg	$72.67±6.05^{c}$	$58±6.86^{c}$	$14.7±1.75^{b}$
	pepper	3 ml/kg	$98.00±8.19^{b}$	$77.7±9.19^{b}$	$21.7±2.56^{b}$
	Ethiothion 5% dust	50 g/Qt	$0±0^{f}$	$0±0^{f}$	$0±0^{d}$
	Acetone	4 ml/kg	$493.3±41.1^{a}$	$465.3±54.83^{a}$	$350.7±41.31^{a}$
	Untreated check		$498.3±41.5^{a}$	$462.3±54.48^{a}$	$360.3±42.45^{a}$
	Noug	4 ml/kg	$0±0^{f}$	$0±0^{f}$	$0±0^{d}$
	Rapeseed	3 ml/kg	$0±0^{f}$	$0±0^{f}$	$0±0^{d}$
	Groundnut	3 ml/kg	$0±0^{f}$	$0±0^{f}$	$0±0^{d}$
	Castor	4 ml/kg	$39.3±4.62^{e}$	$28.7±3.38^{d}$	$9.3±1.11^{c}$
	Lantana	4 ml/kg	$63.7±7.49^{d}$	$52.3±6.24^{c}$	$15.3±1.82^{bc}$
	Tephrosia	3 ml/kg	$73.7±8.67^{c}$	$62±7.29^{bc}$	$15.3±1.79^{bc}$
	pepper	3 ml/kg	$95.3±11.22^{b}$	$71.7±8.43^{b}$	$17.7±2.07^{b}$
	Ethiothion 5% dust	50 g/Qt	$0±0^{f}$	$0±0^{f}$	$0±0^{d}$
	Acetone	4 ml/kg	$486.3±57.3^{a}$	$449.7±54.2^{a}$	$363.3±42.84^{a}$
	Untreated check	-	$496±57.3^{a}$	$448±52.4^{a}$	$353.3±42.23^{a}$

Means followed by the same letter (s) within the column are not significantly different at 5%, Tukey's studentized range test (HSD)
Means presented in the table are logarithmic (**) back transformed values.

There was no bruchid progeny emerged in oil treatments like noug, Rapeseed, Groundnut and Ethiothion 5% dust treated grains (Table 4). Okonkwo and Okoye (2008) were reported the same result that there was no F_1 S. zeamais progeny emerged on maize grains treated with leaves dusts of D. tripetela and Piper guineense till three months of experiment. According to Okonkwo and Okoye (2008) seed oils of D. tripetela and P. guineense treated maize and Cowpea grains were completely suppressed the emergence of F_1 progenies and given protection up to 4 months of storage.

The maximum numbers of bruchid progenies emerged from castor, lantana, tephrosia and pepper oils were 28.8±2.69, 52.3±4.91, 62±6.86 and 77.7±9.19, respectively. Statistically higher numbers of bruchid progenies were recorded in acetone (465.3±54.83) and the untreated check (462.3±54.48) (Table 4). Abdulahi (2011) was reported related result on numbers of bruchids emerged from actellic 2% dust and botanical dust treated cowpea grains were ranged from 9.3±1.2 to 40.3±7.9 compared to acetone (698±24.46) and control (672.3±19.6). The similar result was reported by Regmil and Dhoj (2011) that numbers of bruchids emerged in oil and dust treated chickpea grain were significantly lower (0 to 2292±204.6) compared to control (4713±204.6). Tabu et al. (2012) reported similar results on the numbers of progenies emerged from A. indica at 2% (w/w) and C. ambrosiodes at 4% (w/w) were 42 and 24.33 respectively compared to 320 bruchids were recorded from untreated check.

A comparable result was reported by Waktole (2014) that S. zeamais progenies emerged from maize grains treated with Chenopodium leaves dusts at 10% (w/w) and control were 20 and 80% of the recorded eggs respectively. Based on Ibrahim (2012) experiment sesame seed oils treated chickpea grain at 5 and 7.5 ml/kg were decreased adult emergence by 96.03 and 96.22% respectively compared to the untreated check. A similar result was reported by Yahaya et al. (2013), e.g., applications of groundnut or palm oil at the rate of 2.5 to 3 ml/kg on cowpea grains were significantly suppressed C. maculates progeny emergence.

Effect of botanical oils, Ethiothion 5% dust and acetone treatments on the germination and grain weight loss by bean bruchid

The oil treatments and Ethithion 5% dust were statistically shown lower percentage weight loss than untreated check and acetone treated bean grains. There was no weight loss recorded in Ethiothion 5% dust, noug, rapeseed and groundnut oil treated grains. Higher weigh losses were recorded from tephrosia oil (3.2+0.51%) and pepper oil (5.2±0.27%) treated bean grains than the rest of the oil treatments, however, the loss recorded from untreated check was 3.5 to 18.4 folds exceeded than the oil treatments and the standard check (Table 5). Regmil and Dhoj (2011) reported comparable results on chickpea grains treated with oils and botanical dusts were no loss compared to 38.44%+9.43 chickpea grain weight loss was recorded in the control treatments. Loth et al. (2007) were recorded significantly lower bean seed weight loss (0 o 8.54%) in Ngongwe, pyrethrum flower and garlic powder treated grains than 14.64% in the untreated check after three months of experiment.

Higher bean grain germination was recorded from Ethiothion 5% dust (98.7±1.05%), noug (98.7±1.33%), rapeseed (97±1.73%) and groundnut oil (96±2.13%) treated grains. All the treatments were statistically shown better bean grains germination than acetone and untreated check after three months of the experiment (Table 5). Treatments like oils and Ethiothion 5% dust inhibited emergence of bruchids from 71.7±2.56 to 100% compared to 0±0% in acetone and untreated check. Mushobozy et al. (2009) were conducted similar research on common bean to manage Zabrotes sabfasciatus that the oils treatments were shown superior

Table 5. Weight loss and germination percentage of bean grains and inhibition rates of botanical oils on Bruchids, 2013-2014.

Treatment	Rate	Weight loss	Germination percentage	IR
Noug	4 ml/kg	$0±0^{c}$	$98.6±1.33^{a}$	$100±0^{a}$
Rapeseed	3 ml/kg	$0±0^{c}$	$97±1.73^{ab}$	$99.3±0.51^{a}$
Groundnut	3 ml/kg	$0±0.1^{c}$	$96.2±2.13^{ab}$	$97.3±1.35^{a}$
Castor	4 ml/kg	$0.4±0.39^{c}$	$93.8±1.65^{bc}$	$92.2±1.55^{ab}$
Lantana	4 ml/kg	$2.9±0.43^{c}$	$84.3±1.81^{d}$	$78.9±3.64^{bc}$
Tephrosia	3 ml/kg	$3.2±0.51^{bc}$	$87.2±1.84^{dc}$	$81.4±2.47^{bc}$
pepper	3 ml/kg	$5.2±0.27^{b}$	$79.3±1.75^{d}$	$71.7±2.56^{c}$
Ethiothion 5% dust	50 g/Qt	$0.71±0^{c}$	$98.7±1.05^{a}$	$100±0.0^{a}$
Accetone	4 ml/kg	$18.3±0.4^{a}$	$67.3±4.23^{e}$	$0±0^{d}$
Untreated check	-	$18.4±0.44^{a}$	$66.7±3.02^{e}$	$0±0^{d}$

Means followed by the same letter (s) within the column are not significantly different at 5% Tukey's studentized range test HSD)
Data from both years are pooled together; Means presented in the table are square root back transformed values.

efficacy to inhibit emergence of bruchids and better germination (87.5 to 95%) ability of grains. As Taba et al. (2012) reported that on A. indica at 2% (w/w) and C. ambrosiodes at 4% (w/w) seed dust treated chickpea grains were effectively germinated to 100 and 98.33% respectively. Ibrahim (2012) was added that 80 to 93% of the chickpea grains treated with botanical oils were effectively germinated after the 90 days of experiment.

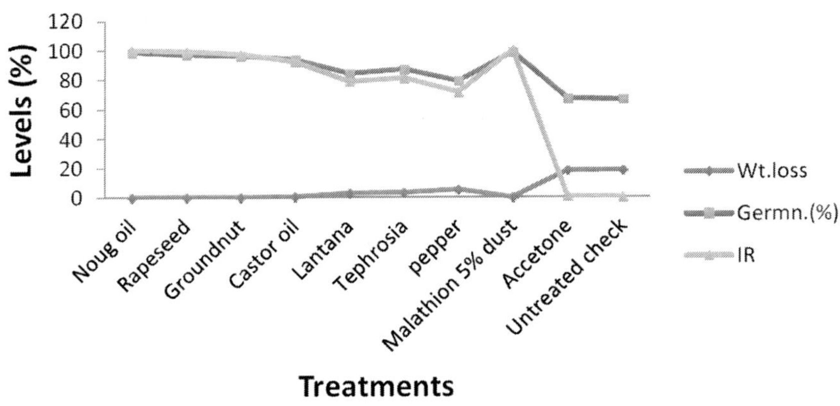

Figure 1. Effectiveness of botanical oils on faba bean grain weight loss and germination due to bruchids.

Higher bean grain germinations were achieved form treatments of better inhibition rate on bruchids emergence and their correlation was strongly positive (Figure 1). Okonkwo and Okoye (2008) conducted similar experiment on essential oils of D. tripetela and P. guineense to control C. maculates and S. zeamais and their finding was indicated that 100% of the parent insects mortality were observed within the first 24 h as well as the insects were inhibited to lay their eggs and emerge to adult stages. Tabau et al. (2012) also indicated that A. indica at 2% (w/w) and C. ambrosiodes at 4% (w/w) seed dust treated chickpea grains were inhibited emergence of bruchids 86.85 and 92.36% respectively. In general oils and Ethiothion 5% dust were shown lower faba bean grain weight loss, better germination percentage and better seedling performance even after germination than acetone treated and untreated faba bean grains.

CONCLUSION

Ethiothion 5% dust and oils were eradicated 30±0.81 to 100% parent bruchids day one after treatment application while 100% of adult bruchids were survived in acetone and untreated check treated grains. The parent bruchids were totally eradicated in Ethiothion 5% dust and oil treated faba bean grains four days after treatment application. The numbers of dead bruchids in acetone and the untreated check treated grains were only 10.85±0.6 and 11.65±0.5%, respectively after four days of experiment. Ten days after treatment application in acetone (13.35±0.63%) treated grains and untreated check (14.15±0.26%) of the parent bruchids were survived. The numbers of eggs laid by parent bruchids during the first ten days of experiment were significantly lower in Ethiothion 5% dust (0±0) and oil (0.67 to 10.67±0.91) treated bean grains. However, very significantly higher numbers of bruchid eggs were laid in acetone (20.67±1.05) and the untreated check (20.83±1.27) and the eggs were continued to lay till ten days of experiment.

Bean grains treated with Ethiothion 5% dust and oils were shown lower numbers of eggs laid by bruchids and lower numbers of progeny emerged adults than acetone and untreated check (Table 4). Statistically lower percentage of bean grain weight loss, no adult bruchids emergence and better germination percentage of faba bean grains were recorded from Ethiothion 5% dust, noug, rapeseed and groundnut oil treated grains. Higher weight losses of faba bean due to bruchids and lower germination percentages of the grains as well as freely emergence of bruchids were recorded from acetone treated and the untreated check grains.

Based on the results recorded, the botanical oils tested were effective to control bruchids from day one to 90 days of experiment. The botanical oils were effectively controlled parent bruchids and their progenies. The numbers of eggs laid by parent bruchids and their progenies were effectively inhibited by the oils evaluated up to the whole durations of the experiment. As a result bean grains treated with the oils were effectively sprouted (79.3±1.75 to 98.7±1.05%).

It is needed to assess efficacy, biodegradability, chemistry of the evaluated botanical oils and their effect on bruchids biology on faba bean and other pulse grains at different agro-ecological zones started from field to storage conditions to be used as one component of Integrated Pest Management (IPM) to manage bruchids and other storage insect pests. The farmers can be used to manage bruchids of based on the evaluated botanical oils due to their availability and ease of extraction without adversely affecting the health of farmers, other consumers and the environment.

REFERENCES

1. Abdulahi N (2011). Studies on the efficacy of leaf extract of Balanites aegyptiaca on the oviposition and survival of immature stages (larvae and pupae) of Callosobruchus maculates (F.) on treated cowpea seeds. Bavero J. Pure Appl. Sci. 4:40-43.
2. Bhardwaj A, Verma SC (2012). Evaluation of vegetable oils against pulse beetle, Callosobruchus chinensis (Coleoptera: Bruchidae) infesting pea seeds (Pisium sativum L.). J. Pest Manage. Hortic. Ecosyst. 18:46-53.
3. El-Ghar G, El-Shelken AEG, Gesa EEL (1987). Effectiveness of some plant extracts as surface protectant of cowpea seed against the pulse beetle, Calosobruchus chinensis. (L.). Phytoparasitica 17:109-113.
4. El-Tokhy F, Kasem A (2012). Studies on fertilizer requirements on faba bean. A Thesis submitted to Cairo University Department of Agronomy, Faculty of Agriculture, in Partial Fulfillment of the Requirement for the Degree of Master of Science. pp. 1-13.
5. Gomez KA, Gomez AA (1984). Statistical Procedures in Agricultural Research. Johnwiley and Sons. 2nd Edn. New York. P. 680.
6. Gwinner JR, Harnisch J, Muck O (1996). Manual of the Prevention of Post-harvest Grain Losses. Deutsche Gesellschaft fur Tecnische Zusammenarbeit (GTZ) GmbH. P 338.
7. Ibrahim YM (2012). Efficacy of some plant oils against stored-product pest cowpea weevil, Callosobruchus maculates (Coleoptera: Bruchidae) on chickpea seeds. Persian Gulf Crop Prot. 1(1):4-11.
8. Ileke KD, Olotuah OF (2012). Bioacttivity of Anacardium occidentale (L.) and Allium sativum (L.) powder and oil extracts against cowpea Bruchid Callosobruchus maculates (Fab.) (Coleoptera; Chrysomelidae). Int. J. Biol. 4:96-103.
9. Kanaji GAD (2007). A study of bruchid resistance and its inheritance in Malawian dry bean germplasm. A Thesis submitted in fulfillment of the requirements of the degree of Doctor of philosophy (PhD) in plant breeding. Submitted to Kwazulu University, Natal Republic of South Africa.
10. Loth SM, Elice NL, Shazia OWM, Robert NM (2007). Effectiveness of local botanicals as protctants of stored beans (Phaseolus vulgaris L.) against bean bruchid (Zabrotes subfasciatus Boh) (genar: Zabrotes. Family: Bruchidae). J. Entomol. 4:210-217.
11. Mushobozy DMK, Nganilevanu G, Ruheza S, Swella GB (2009). Plant oils as common bean (Phaseolus vulgaris L.) seed protectants against infestations by Mexican bean weevil Zabrotes subfasciatus (Boh.). J. Plant Prot. Res. 49:35-40.
12. Okonkwo OU, Okoye WI (2008). The efficacy of four seed powders and the essential oils as protectants of cowpea and maize grains against infestation by Callosobruchus

maculates (Fabricus) (Coleoptera: Bruchidae) and Sitophilus zeamais (Motschulsky) (Coleoptera: Curculionidae) in Nigeria. Int. J. Pest Manage. 42(3):143-146.

13. Olotuah OF (2013). Comparative use of botanical oil extracts in pest management. Agric. Biol. J. North Am. 4(4):419-421.
14. Prakash A, Rao J, Nandagopal V (2008). Futures of Botanical pesticides in rice, wheat, pulses and vegetable pest management. J. Biopesticides 1:154-169.
15. Regmil H, Dhoj Y (2011). Eco-friendly management of pulse beetle. J. Agric. Environ. 12:81-90.
16. Sangeeta G, Apte SD (2015). SEffects of botanicals (oils) on green gram (Vigna radiate L. Wilczek) against Callosobruchus maculates (Fab). Eur. J. Exp. Biol. 5(8):31-33.
17. Shukla R, Srivastava B, Kumar R, Dubey NK (2007). Potentials of some botanical powders in reducing infestation of chichpea by Callosobruchus chinensis L. (Coleoptera Bruchidae). J. Agric. Technol. 3(1):11-19.
18. Swella GB, Mushobozy DMK (2007). Evaluation of the efficacy of protectants against cowpea bruchid Callosobruchus maculates (F.) on Cowpea seeds (Vigna unguiculata (L.) Walp.). J. Plant Prot. Sci. 43:68-72.
19. Tabu D, Selvaraj T, Singh SK, Mulugeta N (2012). Management of Adzuki bean beetle (Callosobruchus chinensis L.) using some botanicals, inert materials and edible oils in stored chickpea. J. Agric. Technol. 8(3):881-902.
20. Tadele S, Habtamu A, Mulugeta N (2014). Effects of some botanicals against termite, macroterms Spp (Isoptera: Termitidae) under laboratory conditions. Int. J. Sustain. Agric. Res. 1:52-57.
21. Taponjou LA, Adler C, Bouda H, Fontan DA (2002). Efficacy of powder and essential oil from Chenopodium ambrostodes leaves as post-harvest grain protectants against six- stored product beetles. J. Stored Prod. Res. 38:395-402.
22. Uddinll RO, Sanusi SA (2013). Efficacy of olive oil, Soybean oil and palm kernel oil in the control of Callosobruchus maculates (F.) in stored cowpea (Vigna unguculata L. Walp). J. Agro-res. 13:67-72.
23. Waktole S (2014). Effects of selected botanicals and local seed storage practices on maize insect pests and health of maize seed in Jimma Zone. Singapore J. Sci. Res. 4:19-28.
24. Yahaya MM, Bandiya HM, Yahaya MA (2013). Efficacy of selected seed oils against Fecundity of Callosobruchus maculates (F.) (Coleoptera: Bruchidae). Int. J. Sci. Technol. 8:513-321.
25. Yohannes E, Setu B, Tigist S, Samuel S (2013). Evaluation of different botanicals for the management of cowpea Bruchid (Callosobruchus maculates) in North-Eastern Ethiopia. Arch. Phytopathol. Plant Prot. 46:1331-1337.

CHAPTER 19

Post-Harvest Losses and Handling Practices of Durable and Perishable Crops Produced in Relation with Food Security of Households in Ethiopia: Secondary Data Analysis

Abadi Gebre Mezgebe[1]*, Zemenu Kerie Terefe[1], Tafese Bosha[1], Tilku Desalegn Muchie[1] and Yosef Teklegiorgis[2]

[1]*School of Nutrition, Food Science and Technology, College of Agriculture, Hawassa University, P. O. Box 05, Hawassa, Ethiopia.*

[2]*School of Animal and Range Sciences, College of Agriculture, Hawassa University, P. O. Box 05, Hawassa, Ethiopia.*

ABSTRACT

This secondary data analysis was conducted to assess post-harvest losses and handling practices. The postharvest section of the Living Standard Measurement Survey (LSMS) data set of the Ethiopian Rural Socio-economic Survey (ERSS) was used as a data source. The analysis result showed losses of 153.29 kg of cereal crops (7.7% households), 120.16 kg of pulses (5.5% households), 320 kg of oilseeds (4.1% households), 102.19 kg of fruit crops (11% households), 181.86 kg of vegetable (8.2% households), 105.56 kg of root crops (5% households), and 556.13 kg of cash crops (8.2% households). Higher number of households (30%) lost 30% of fruits,

10% of vegetables (21.62% household), 50 and 60% of root and cash crops (26.5% households), respectively. The major cause for cereals and oilseeds loss was rodents/pests, others for pulses and diseases for the perishable crops. Additionally, 84% of the households stored cereal, 63.9% pulses, 80.9% oilseeds, 7.6% fruits, 16.6% vegetables, 25.7% roots and 29% cash crops. The main storage method of durable crops was bags in house and sacks for perishable crops. Furthermore, 91% of the households protected cereal crops, 59.97% pulses, 74.8% oilseeds, 78.4% fruits, 81.7% vegetables, 71% roots and 78.5% cash crops. The major technique used for protection of cereal, pulse, vegetable and cash crops was elevation, while other techniques were used for fruits and root crops. About 50.4% of households did not protect oilseeds. Postharvest losses varied among crop types and handling practices. Reduction of losses could contribute to food and nutritional security; hence attention should be given towards improving postharvest handling practices.

Keywords: Durable crops, perishable crops, postharvest losses, handling practices, food security.

INTRODUCTION

Achieving food security continues to be a challenge as it is affected by a complexity of factors (Nicholas et al., 2006; FAO, 2009; Dercon and Krishnan, 2015). Increasing the food availability is therefore not only increasing the productivity in agriculture, there is also a need to lower the losses (Kader, 2005; Parfitt et al., 2010; Victor, 2014). Food losses after harvest until the food reach the consumer are significant (FAO and World Bank, 2010; FAO, 2014). A large amount of food and products are not reaching the consumer particularly due to post-harvest losses (Kader, 2005; FAO, 2009; FAO and World Bank, 2010) during harvesting, handling, transporting, storage, processing, packaging and distribution. Handling and processing of the food are of high importance in order to ensure food-safety (Kader, 2003). Post-harvest loss leads to an inadequate food intake and it could be manifested by seed loss, monetary loss, food loss and loss of reputation which in turn affect marketing (Gross et al., 2000; FAO and World Bank, 2010). Post-harvest losses can be caused by mechanical damage and injury, physiological processes, poor handling, lack of processing, inadequate packaging, poor logistics and sub-optimal storage conditions (Chakraverty et al., 2003). According to FAO (2014) post-harvest losses in developing countries can range from 15 to 50%.

Quantitative and qualitative losses could occur in crops (Kidane et al., 2006; FAO and World Bank, 2010). Qualitative losses, such as loss in edibility, nutritional quality, caloric value, and consumer acceptability of the products, are much more difficult to assess than quantitative losses (Kader, 2005; Adeoye, 2009; Buyukbay et al., 2011). Post-harvest losses reduce the availability of food crops and income that could be generated by selling these products, thus in terms of quantity are linked to food security (FAO and World Bank, 2010; Jayne et al., 2010; Hodges et al., 2011). The qualitative losses are related directly to nutritional and they are more complicated to measure. However the quantitative losses are of greater importance to measure in developing countries (Humble and Reneby, 2014).

Food grains (cereals, pulses and oil crops) constitute the major sources of food intake in Ethiopia (Abebe, 2000; CSA, 2006, 2012). Some preliminary estimates show that post-harvest losses could range between 5 and 26% of production, which is enormous given the state of food insecurity of the country (Abebe, 2000). It is important to recognize that post-harvest management practices and capacities (not only just production and marketing) are consequential for many reasons including attaining high level of food and nutritional security. As described by FAO (2014), Felleke (2004), Kader (2005), Nega and Semeon (2006) and Abebe and Bekele (2011) post-harvest losses studies are generally rare in Ethiopia; and even these few available are related to aspects of engineering structures and not with aspects of food and nutrition security.

Post-harvest losses may vary greatly among commodities, production, areas, seasons, handling and management practices. Nearly 20% of losses of fresh fruits and vegetables are due to product deterioration. This may include the excess of perishable products that are discarded and the food that is wasted, not consumed by the purchaser) (Kader, 2005; Victor, 2014). Some estimates would suggest the magnitude of post-harvest losses in Ethiopia to be tremendous; for example depending on the type of post-harvest handling method, the losses could range between 5 and 19% for maize, 6 and 26% for millet, 6 and 23% for wheat, and 5 and 20% for teff (Dereje, 2000). The international congress of Save Food (FAO, 2011) reported that the percent weight of food losses in the post-harvest chain for different economic status countries and it has shown estimates of food losses in low-income sub-Saharan African countries including Ethiopia (Table 1).

Mrema and Rolle (2002), Jayne et al. (2010) and Hodges et al. (2011) indicated an evolution of priorities within the post-harvest sector of developing countries from a primarily technical focus geared towards the reduction of losses, to a greater holistic approach designed to link farm activities to processing, marketing, and distribution. However, the major constraints continue to be high post-harvest losses, poor marketing systems, weak research and development capacity, and inadequacies in policies, infrastructure, and information exchange. Overcoming the socio-economic constraints is essential to achieving the goal of reducing the post-harvest losses of food (Kader, 2005; Abebe and Bekele, 2006).

Table 1. Percentage of food losses in low-income sub-Saharan African countries.

Food type	Agricultural production (%)	Post-harvest handling practices (%)	Processing and packaging (%)	Distribution (%)	Consumption (%)	Total loss (%)
Cereals	6	8	3.5	2	1	20.5
Roots and tubers	14	18	15	5	2	54
Oilseeds and pulses	12	8	8	2	1	31
Fruit and vegetable	10	9	25	17	5	66
Meat	15	0.7	5	7	2	29.7
Fish and seafood	5.7	6	9	15	2	37.7
Milk	6	11	0.1	10	0.1	27.2

Source: International congress of Save Food (2011), with modification on the total loss (total was calculated through summing up of all the losses along the chain).

Reduction in food losses are sometimes considered as the 'third dimension' to the world food supply equation, that is, in addition to increase in food production and population. In Ethiopia, grain storage at farm level is carried out by means of grain pits, underground holes, and sacks. Grain losses due to poor storage by farmers are reported to range between 11 and 19% (Abebe, 2000; Abebe and Bekele, 2006). Grain traders also store grains in warehouses with small but varying capacities; moreover, these are characterized by poor ventilation and dirt floors (Kader, 2005). Hence, poor storage naturally leads to high grain damages from pests and moisture. Inadequacy of storage, combined with the vulnerability of crops to damage, makes traders unwilling to store stocks beyond the minimum turnover period. Similarly, pre-damage treatment (aeration, application of pesticides, etc) is seldom practiced by most farmers, and post-damage treatment is limited to aeration.

In Ethiopia, consumers buy unprocessed crops actually unchanged since processing is not undertaken either by producers or middlemen. Processing is essentially undertaken as and when grains are prepared for consumption rather than as an important economic activity (Abebe and Bekele, 2006). The aim of this study was to assess post-harvest and handling practices of crops produced in Ethiopia in relation with food security of households.

METHODOLOGY

The Living Standard Measurement Survey (LSMS) data set of the Ethiopian Rural Socio-economic Survey (ERSS) of the year 2012/2013 from the agriculture postharvest section was used as the data source for this secondary analysis. The data were organized to answer the specific research objectives. All the households that produces crops (Durable crops: cereals, pulses, oilseeds and perishable crops: fruits, vegetables, root and cash crops) were considered in the analysis. Crops were categorized as cereals (barley, maize, millet, oats, rice, sorghum, teff and wheat), pulses (chickpeas, haricot beans, horse beans lentils, field peas and soya beans), oilseeds (linseed, ground nuts, nueg, rape seed, sesame and sunflower), fruits (banana, lemon, mango, orange, papaya, pineapple and avocado), vegetables (red pepper, cabbage, carrot, cauliflower, garlic, kale, lettuce, onion, green pepper, pumpkins and tomatoes), root crops (potato, sweet potato, *Godere* and *Enset*) and cash crops (coffee, cotton, *Gesho* and sugar cane).

The specific questions from the LSMS - ERSS - Agriculture postharvest data considered for this research were: For durable crops: (1) Were there losses of crop? (2) Was any portion of the crop lost postharvest due to rotting, insects, and rodents? (3) Did you have any of this harvest stored now? (4) Did you protect the stored crop? (5) What is the reason of the losses of the stored crops? (6) What is the method of storage? (6) What are the techniques used for protection of stored crops? (8) The unit of losses of the crop? For perishable crops: (1) Were there losses of fruit/crop? (2) Was any portion of the fruit/crop lost postharvest due to rotting, insects, and rodents? (3) Did you have any of this harvest stored now? (4) Did you do something to protect the stored crop? (5) What is the reason of the losses of the stored crops? (6) What is the method of storage? (7) What are the techniques used for protection of stored crops? (8) Out of 10 fruits how many was lost?

The data in the original source was collected from different regions of Ethiopia which had produced the crops listed. Analysis was conducted in SPSS version 16.0. Descriptive statistical methods (relative frequency, mean, range, percentage) were used to report the results of the analysis. Postharvest losses, causes of losses, availability of stored crop, storage method, and protection of stored crops and techniques of protection were reported.

RESULTS AND DISCUSSION

Postharvest losses and causes

As shown in Table 2, concerning the cereal producers (5639 households), 7.7% reported to have crops losses with mean of 153.29 kg. About 5.2% of the pulse producers (from the total of 1612 households) responded to have losses with average of 120.16 kg. From the oil seed producers (860 households), 4.1% responded to have losses with mean of 207.16 kg. The percentages of losses of the crops were not calculated as the data for production was not collected in the original data source. However, these findings are within the range reported by Dereje (2000), Abebe (2000), Abebe and Bekele, (2006) and Kader (2005). It is known that food grains (cereals, pulses and oil crops) constitute the major source of food in Ethiopia, accounting for 82 and 70% of total calorie intake and food expenditure, respectively (Abebe, 2000; CSA, 2006, 2012). Cereals alone provide about 70% of the average Ethiopian's calorie intake (Howard et al., 1995). Hence efforts should be done to reduce the losses.

In relation to the fruits producers (1007 households, 11% responded to have losses of the crops with average of 102.19 kg, and about the vegetable producers, 8.2% of a total of 1913 households, reported to have 181.86 kg of losses. About 5% of the root crops producers (out of the total of 1398 households) lost around 105.56 kg. Regarding the cash crop producers, 8.2% reported losses of 556.46 kg) (Table 2).

Table 2. Respondents (%) and crop loss (kg) reported by sampled households in Ethiopia.

Crop type	No. of HH	% respondents reported loss	Average crop loss (kg)
Cereals	5639	7.7	153.29
Pulses	1612	5.2	120.16
Oilseeds	860	4.1	207.16
Fruits	1007	11.0	102.19
Vegetables	1913	8.2	181.86
Root Crops	1398	5.0	105.56
Cash crops	1249	8.2	556.13

HH= Households producing the crops.

The percentage of losses of perishable crops show that higher number of household (30%) lost their fruit as much as 30%, and 21.62% of the household lost 10% vegetables. With regard to root crops and cash crops higher number of households (26.5%) reported about 50 and 60% of losses, respectively (Table 3). The percent of reported losses (about 20%) of fruits, vegetables, root crops and cash crops was reported by corresponding respondents of 17.9, 16.2, 25 and 25%, respectively. The results are in line with reports of Admassu (2005), Save Food (2011), Abebe and Bekele (2006), and Humble and Reneby (2014). According to FAO (2014) post-harvest losses in developing countries can range from 15 up to 50%. Horticultural crops are perishable products and they are more prone to greater losses than for non-perishable crops (Parfitt et al., 2010).

Table 3. Percentage of losses reported by sampled household (perishable crops).

Loss reported (%)	Percentage of respondents with respective crop type (%)			
	Fruits (n=106)	Vegetables (n=148)	Root crop (n=68)	Cash crops (n=68)
10	16.00	21.62	14.70	14.70
20	17.90	16.20	25.00	25.00
30	30.00	8.78	11.00	2.90
40	9.40	4.73	7.40	11.00
50	24.50	2.70	26.50	7.40
60	2.80	5.40	2.90	26.50
70	1.90	2.03	1.50	2.90
80	5.70	2.70	1.50	1.50
90	0.90	12.20	1.50	1.50
100	0.94	2.03	0.00	1.50

n = number of households sampled.

As shown in Table 4, the main cause of the losses for cereal crops was reported to be rodents/pests (41.5%). Rotting and insects were also reported to be the cause of the losses of cereals with 14.9 and 14.7%, respectively. With regard to pulse crops the main reason for the losses was reported to be rodents/pests (22.8%) and rotting (21.2%). insects were also the cause for the losses of pulses (13.4%). The oil seed losses were mainly due to rodents/insects (41.2%). Insects were also reported to be the cause of the losses of the oil seeds with 13.2%. Comparing with all the crops, the losses due to flood and theft were minor. It has been addressed that pests, rodents, rotting/insects, flood and theft could cause the losses of grains. More detailed causes of postharvest losses were reported by Boxall (2001), Kader (2005), Magan and Aldred (2007), Parfitt et al. (2010), Tefera et al. (2011, 2012), Hodges et al. (2011), and Humble and Hundie (2014).

Table 4. Percentage of main causes for crop loss reported (durable crops).

Crop type	Total no. HH	Rotting	Insects	Rodents/pests	Flood	Theft	Other specify
Cereals	834	14.9	14.7	41.5	4.6	1.4	22.9
Pulses	179	21.2	13.4	21.8	2.8	3.9	36.9
Oilseeds	68	8.8	13.2	41.2	5.9	2.9	27.9

HH= Households producing the crops.

The main causes of the losses of fruits (Table 5) were reported to be primarily diseases (30.1%), birds (21.9%), insects (16.4%) and thefts (11.5%). The losses due to animals seem to be low as compared to the others. Vegetables losses were mainly reported primarily due to diseases (44.8%). Next to the diseases, the main causes for the losses of vegetables were insects (19.4%) and Animals (11%). The losses of vegetables due to thefts and birds were less. As secondary causes of the losses of due to birds (33.6%), animals (18.2%) and thefts (11.8%) were reported for fruits and animals (36.1%), birds (25.2%) and insects (8.4%) for vegetables. Moreover some detail of the main causes of losses of fruits and vegetables after harvest have been studied by Kader (2002), Kader (2005), Wu (2010), Victor (2014) and Devkota et al. (2014).

The main cause of root crops loss was reported to be diseases (36.5%) primarily (Table 5). Insects and animals were also main causes. Thefts and birds were reported less likely to cause the losses. The main reason for losses of cash crop was reported to be primarily due to diseases (51.8%). Insects and thefts were also reported to be the main causes. Birds and animals were less likely reported to cause losses of cash crops. With regard to the secondary main causes of the losses of root crops, animals (33.3%) and birds (17.4%) were reported. As for the cash crops, animals (25.2%) and thefts (19.4%) were reported. With differences in the relative frequency, root crops and cash crops seemed to have similar causes of with fruits and vegetables. Moreover, the possible main causes were addressed in the reports of Adda et al. (2002), Chakraverty et al. (2003), Kader (2002) and Rembold et al. (2011).

Table 5. Percentage of 1st and 2nd causes of the crop losses of the perishable crops.

Crop type	Causes priority	No. of HH	Birds	Animals	Insects	Diseases	Thefts	Other specify
Fruits	1st	183	21.9	6.0	16.4	30.1	11.5	14.2
	2nd	110	33.6	18.2	4.5	0.0	11.8	31.8
Vegetables	1st	355	2.8	11.0	19.4	44.8	2.0	20.0
	2nd	155	25.2	36.1	8.4	1.3	4.5	24.5
Root crops	1st	137	0.7	10.9	16.1	36.5	5.8	29.9
	2nd	69	17.4	33.3	4.3	1.4	7.2	36.2
Cash crops	1st	282	3.5	3.5	16.0	51.8	9.9	15.2
	2nd	103	5.8	25.2	1.9	0.0	19.4	47.6

HH= Households producing the crops.

Storage and methods used for storage

As shown in Table 6, about 84% of the producers stored cereals, 63.9% pulses, 80.9% oilseeds, 7.6% fruits, 16.6% vegetables, 25.7% root crops and 29% cash crops. This result is probably due to the nature of the commodities, since the perishable one scan not stay long in the storage.

Table 6. Relative frequencies of households (HH) who had crops in storage.

Crop type	No. of HH	% Positively responded
Cereals	5562	84.0
Pulses	1587	63.9
Oilseeds	881	80.9
Fruits	1008	7.6
Vegetables	1913	16.6
Root Crops	1403	25.7
Cash Crops	1252	29.0

HH= Households producing the crops.

With regard to the storage methods used, the bags in house was the mainly used for cereals, pulses and oil seeds storage (Table 7), while the producers were using mainly sacks for fruits, vegetables, root crops and cash crops (Table 8). There were no producers that used unprotected pile for storing the perishable crops. The storage methods used by the producers is in agreement with the study done by Abebe and Bekele (2006); and the possible options that could be considered for choosing these methods is the cost as indicated by McFarlane (1998).

Table 7. Percentage of storage methods used by the households for durable crops.

Crop type	No. HH	Unprotected pile	Heaped in house	Bags in house	Metallic silo	Other specify
Cereal	4574	8.6	7.7	48.9	0.4	34.4
Pulses	997	5.0	8.9	61.6	0.2	24.3
Oilseeds	685	1.6	4.2	68.8	0.1	25.3

HH= Households producing the crops.

Table 8. Percentage of storage methods used by households for perishable crops.

Crop type	No. HH	Locally made traditional structure	Improved locally made structure	Modern store	Sacks	Other
Fruits	76	1.3	2.6	0.0	94.7	1.3
Vegetables	318	7.9	0.6	0.3	90.3	0.9
Root Crops	359	12.3	0.0	0.0	86.4	1.4
Cash Crops	359	3.9	0.3	0.3	95.0	0.6

HH= Households producing the crops.

It is well known that perishable crops should be kept in cold places with proper temperature, and humidity.

Selection of storage depends on the volume of production and the operation and establishment costs (David and David, 1998). The common storage structure used by most of Ethiopian farmers are the traditional ones with poor construction that expose the stored grains to different conditions and deterioration agents and in which the appropriate management of all the influencing factors has not been dealt in detail (Kemeru, 2007; Befikadu, 2014).

Handling and methods used for protection

As indicated in Table 9, about 91% of the producers protected cereals, 59.97% pulses, 74.8% oilseeds, 78.4% fruits, 81.7% vegetables, 71% root crops and 78.5% cash crops. According to the results in Table 10, elevation was the method used mainly for protection of cereals, pulses and oilseeds. There was higher percent of respondents that did nothing for protecting. Spraying was used by some of households. The protection method for fruits was not stated in the list and this indicates the respondents were using a different method. About 10% of the households did nothing to protect the fruits. Vegetables were mainly protected by elevating followed by smoking. Still there were producers that did not protect the vegetables. Protection method for root crops was not stated in the list and this indicates that the respondents were using a different method. There were also about 13.1% of households that did nothing to protect the root crops. As shown in the Table 9, the cash crops were protected mainly by elevation.

Table 9 Relative frequency of households who protect crops in storage.

Crop type	No. of HH	% Positively Responded
Cereals	100	91.00
Pulses	1589	59.97
Oilseeds	897	74.80
Fruits	74	78.4
Vegetables	317	81.7
Root Crops	359	71.0
Cash Crops	362	78.5

HH= Households producing the crops.

Table 10. Percentage of protection method of crops used by the Households.

Crop type	No. of HH	Spraying	Smoking	Hired guard	Did nothing	Elevation	Other specify
Cereal	4517	18.9	2.7	0.2	33.6	41.4	3.3
Pulses	967	15.0	3.7	0.4	34.7	42.1	4.0
Oilseeds	671	3.3	0.9	0.0	50.4	38.9	6.6
Fruits	58	0.0	1.7	0.0	10.3	8.6	79.3
Vegetables	259	0.8	15.8	0.4	13.1	59.5	10.4
Root Crops	254	0.0	1.2	1.2	13.4	13.8	70.5
Cash Crops	283	0.0	1.1	0.0	4.9	87.6	6.4

HH= Households producing the crops.

Crops that have been put into the storage after harvest without proper management and handling can rapidly be deteriorated, becoming a worthless mass (David and David, 1998; Befikadu, 2014).

CONCLUSION

Most of the households reported losses of both durable and perishable crops. These crops are source of income and calorie intake, hence reducing the losses could contribute to food security of the households. The main causes of durable crops losses were reported to be rodents, pests, rotting and insects. It has been observed that there were producers that didn't protect stored crops. Protection of the crops during storage helps to reduce the losses thereby contributing to food security. Moreover diseases, birds and animals were the major reasons for the loss of perishable crops. The loss/ spoilage by diseases, birds and animals could be prevented by appropriate postharvest handling. Regarding the storage most of the households used bags in house to store durable crops and sacks to store perishable crops. However, there were very few of the households using improved and modern local, for storing the crops. The majority of the households stated that the elevating was the main used method of protection. In addition to the increased income, reducing postharvest losses could increase the amount of food available for consumption by farmers, rural and urban consumers. The final consumer can also be benefited by the reducing prices as a result of the increased availability of foods. Improve postharvest handling and technology can create some job opportunities in long term. Therefore, reducing losses by postharvest handling practices will generally improve the livelihoods if it is done appropriately. Postharvest has been relatively little studied in relation to food security and nutrition. Considering the potential of postharvest for large impacts in food and nutrition security, this study recommends a primary research on the relationship of postharvest to food and nutrition security.

ACKNOWLEDGEMENTS

The authors would like to appreciate and thank the institutions which contributed to the secondary data analysis. This research work is made possible by the support of the American people through the United States Agency for International Development (USAID) under Agreement No.AID-663-A-11-00017. The contents of the document are sole responsibility of the researchers and do not necessarily reflect views of USAID or the United States Government.

REFERENCES

1. Abebe HG, Bekele H (2006). Farmers' Post-Harvest Grain Management Choices under Liquidity Constraints and Impending Risks: Implications for Achieving Food Security Objectives in Ethiopia. Poster paper presentation, International Association of Agricultural Economists Conference, Gold Coast, Australia.
2. Abebe HG (2000). Development Strategies and the Ethiopian Peasantry: Supply Response and Rural Differentiation. Shaker Publishing: Maastricht
3. Adda C, Borgemeister C, Biliwa A, Meikle WG, Markham RH, Poehling HM (2002). Integrated pest management in post-harvest maize: a case study from the Republic of Togo (West Africa). Agric. Ecosyst. Environ. 93:305-321.
4. Adeoye IB (2009) Economic Analysis of Tomato Losses in Ibadan Metropolis, Oyo State, Nigeria. Afr. J. Basic Appl. Sci. 1(5-6):87-92.
5. Admassu S (2005). Post-harvest sector challenges and opportunities in Ethiopia. Ethiopian Agricultural Research Organization, Addis Ababa, Ethiopia
6. Befikadu D (2014). Factors affecting quality of grain stored in Ethiopian traditional storage structures and opportunities for improvement. International J. Sci. Basic Appl. Res. 18(1):235-257.
7. Boxall RA (2001). Post-harvest losses to insects – a world review. Int. Biodeterior. Biodegradation 48:137-152.
8. Buyukbay EO, Uzunoz M, Bal HSG (2011). Post-harvest losses in tomato and fresh bean production in Tokat province of Turkey. Sci. Res. Essays 6(7):1656-1666
9. Chakraverty A, Mujumdar AS, Vijaya Raghavan GS, Ramaswamy HS (2003). Handbook of Postharvest Technology. Marcell Dekker, Inc., New York, USA.
10. CSA (2012). Ethiopia Demographic and Health Survey. Central Statistical Agency: Addis Ababa, Ethiopia.
11. CSA (2006). Ethiopia Demographic and Health Survey. Central Statistical Agency and ORC Macro: Addis Ababa, Ethiopia.
12. David PS, David DJ (1998). Management - The key to maintaining stored grain quality. Available at: http://lancaster.unl.edu/ag/crops/inservice/manage-i98.html
13. Dercon S, Krishnan P (2015). Vulnerability, Seasonality and Poverty in Ethiopia. J. Dev. Stud. 36(6):25-53.
14. Dereje A (2000). The Utilization and Conditions of Postharvest Concept in Ethiopia. In EARO. Postharvest Food Preparation and Storage, and Research and Extension on Agricultural by-products Utilization Technologies. Amharic Version.
15. Devkota AR, Dhakal DD, Gautam DM, Dutta JP (2014). Assessment of fruit and vegetable losses at major wholesale markets in Nepal. Int. J. Appl. Sci. Biotechnol. 2(4):559-562.
16. FAO (2009). How to Feed the World in 2050. Rome: FAO. Available at:

17. FAO W (2014). IFAD (2012) The State of Food Insecurity in the World 2012: Economic growth is necessary but not sufficient to accelerate reduction of hunger and malnutrition. FAO, Rome, Italy.
18. FAO (2011). Global food losses and food wastage: study conducted for International congress of Save Food. Rome, Italy.
19. FAO/World Bank Work (2010). Food and Agricultural Organization of the United Nations. Reducing post-harvest losses in grain supply chains in Africa: Lessons learned and practical guidelines. FAO/World Bank Work. FAO Headquarters 18–19 March 2010, Rome, Italy
20. Gross R, Schoeneberger H, Pfeifer H, Joachim AH (2000). The Four Dimensions of Food and Nutrition Security: Definitions and Concepts. Available at:
21. Hodges RJ, Buzby JC, Bennett B (2011). Foresight project on global food and farming futures: Postharvest losses and waste in developed and less developed countries: opportunities to improve resource use. J. Agric. Sci. 149:37-45.
22. Humble S, Reneby A (2014). Post-harvest losses in fruit supply chains – A case study of mango and avocado in Ethiopia. Karin Hakelius, Swedish University of Agricultural Sciences, Department of Economics. Available at:
23. Jayne TS, Mather D, Mghenyi E (2010). Principal Challenges Confronting Smallholder Agriculture in Sub-Saharan Africa. World Dev. 38(10):1384-1398.
24. Jones M, Alexander C, Lowenberg-DeBoer J (2014). A simple methodology for measuring profitability of on-farm storage pest management in developing countries. J. Stored Prod. Res. 58:67-76.
25. Kader AA (2002). Postharvest technology of horticultural crops, third edition. Univ. California Agric. Nat. Resour. 3311:535.
26. Kader AA (2003). A perspective on postharvest horticulture (1978–2003). HortScience 38(5):1004-1008.
27. Kader AA (2005). Increasing food availability by reducing postharvest losses of fresh produce. In V International Postharvest Symposium 682:2169-2176.
28. Kemeru D (2007). Farm Level Survey on Farm Tools and Implements in Jimma and Iluababor Zones. Agricultural mechanization for productivity and sustainable use of the natural resource. Proceedings of the first agricultural mechanization post-harvest and food science research completed research forum, 5-7 June 2007, Addis Ababa, Ethiopia.
29. Kidane H, Alemu ZG, Kundhlande G (2010). Causes of household food insecurity in Koredegaga peasant association, Oromiya zone, Ethiopia. Agrekon 44(4):543-560
30. Kimatu NJ, McConchie R, Xie X, Nguluu SN (2012). The Significant Role of Post-Harvest Management in Farm Management, Aflatoxin Mitigation and Food Security in Sub-Saharan Africa. Greener J. Agric. Sci. 2(6):279-288.
31. Magan N, Aldred D (2007). Post-harvest control strategies: Minimizing mycotoxins in the food chain. Int. J. Food Microbiol. 119:131-139.
32. McFarlane JA (1988). Storage methods in relation to post-harvest losses in cereals. Insect Sci. Appl. 9(6):747-754. Parfitt J, Barthel M, Macnaughton S (2010). Review Food waste within food supply chains: quantification and potential for change to 2050. Philos. Trans. Royal Soc. B: Biol. Sci. 365(1554):3065-3081.
33. Nicholas H, Nisar M, James D (2006). A review of emergency food security assessment in Ethiopia. Himanitarian Policy Group Report, a study commissioned by and prepared for the World Food Programme, Rome
34. Parfitt J, Barthel M, Macnaughton S (2010). Review Food waste within food supply chains: quantification and potential for change to 2050. Philos. Trans. Royal Soc. B: Biol. Sci. 365(1554):3065-3081.
35. Rembold F, Hodges R, Bernard M, Knipschild H, Léo O (2011). The African Postharvest Losses Information System (APHLIS): An innovative framework to analyse and compute quantitative postharvest losses for cereals under different farming and environmental conditions in East and Southern Africa. JRC Scientific and Technical Reports. pp. 1-72.

36. Tefera T (2012). Post-harvest losses in African maize in the face of increasing food shortage. Food Secur. 4:267-277.
37. Tefera T, Kanampiu F, De Groote H, Hellin J, Mugo S, Kimenju S, Beyene Y, Boddupalli PM, Shiferaw B, Banziger M (2011). The metal silo: An effective grain storage technology for reducing post-harvest insect and pathogen losses in maize while improving smallholder farmers' food security in developing countries. Crop Prot. 30:240-245.
38. Victor K (2014). Post-harvest losses and strategies to reduce them. Technical paper on Post Harvest Losses. Action Contre la Faim (ACF), member of ACF International. pp. 2-25.
39. World Bank (2004). Ethiopia: A Country Status Report on Health and Poverty. Africa Region Human Development & Ministry of Health Ethiopia, Addis Ababa, Ethiopia.
40. World Bank (2004). Ethiopia: A Country Status Report on Health and Poverty. Africa Region Human Development & Ministry of Health Ethiopia, Addis Ababa, Ethiopia.
41. Wu CT (2010). An overview of postharvest biology and technology of fruits and vegetables. AARDO workshop: Technology on reducing postharvest losses and maintaining quality of fruits and vegetables 2-11, National Taiwan University, Taiwan.

Index